The Pow
Movement i

Charles Darwin, Sir Francis Darwin

Alpha Editions

This edition published in 2024

ISBN 9789361476464

Design and Setting By

Alpha Editions
www.alphaedis.com

Email - info@alphaedis.com

Contents

INTRODUCTION.

The chief object of the present work is to describe and connect together several large classes of movement, common to almost all plants. The most widely prevalent movement is essentially of the same nature as that of the stem of a climbing plant, which bends successively to all points of the compass, so that the tip revolves. This movement has been called by Sachs "revolving nutation;" but we have found it much more convenient to use the terms circumnutation and circumnutate. As we shall have to say much about this movement, it will be useful here briefly to describe its nature. If we observe a circumnutating stem, which happens at the time to be bent, we will say towards the north, it will be found gradually to bend more and more easterly, until it faces the east; and so onwards to the south, then to the west, and back again to the north. If the movement had been quite regular, the apex would have described a circle, or rather, as the stem is always growing upwards, a circular spiral. But it generally describes irregular elliptical or oval figures; for the apex, after pointing in any one direction, commonly moves back to the opposite side, not, however, returning along the same line. Afterwards other irregular ellipses or ovals are successively described, with their longer axes directed to different points of the compass. Whilst describing such figures, the apex often travels in a zigzag line, or makes small subordinate loops or triangles. In the case of leaves the ellipses are generally narrow.

Until recently the cause of all such bending movements was believed to be due to the increased growth of the side which becomes for a time convex; that this side does temporarily grow more quickly than the concave side has been well established; but De Vries has lately shown that such increased growth follows a previously increased state of turgescence on the convex side.[1] In the case of parts provided with a so-called joint, cushion or pulvinus, which consists of an aggregate of small cells that have ceased to increase in size from a very early age, we meet with similar movements; and here, as Pfeffer has shown[2] and as we shall see in the course of this work, the increased turgescence of the cells on opposite sides is not followed by increased growth. Wiesner denies in certain cases the accuracy of De Vries' conclusion about turgescence, and maintains[3] that the increased extensibility of the cell-walls is the more important element. That such extensibility must accompany increased turgescence in order that the part may bend is manifest, and this has been insisted on by several botanists; but in the case of unicellular plants it can hardly fail to be the more important element. On the whole we may at present conclude that increased growth, first on one side and then on another, is a secondary effect, and that the increased turgescence of the cells, together with the extensibility of their walls, is the primary cause of the movement of circumnutation.[4]

[1] Sachs first showed ('Lehrbuch,' etc., 4th edit. p. 452) the intimate connection between turgescence and growth. For De Vries' interesting essay, 'Wachsthumskrümmungen mehrzelliger Organe,' see 'Bot. Zeitung,' Dec. 19, 1879, p. 830.

[2] 'Die Periodischen Bewegungen der Blattorgane,' 1875.

[3] 'Untersuchungen über den Heliotropismus,' Sitzb. der K. Akad. der Wissenschaft. (Vienna), Jan. 1880.

[4] See Mr. Vines' excellent discussion ('Arbeiten des Bot. Instituts in Würzburg,' B. II. pp. 142, 143, 1878) on this intricate subject. Hofmeister's observations ('Jahreschrifte des Vereins für Vaterl. Naturkunde in Würtemberg,' 1874, p. 211) on the curious movements of Spirogyra, a plant consisting of a single row of cells, are valuable in relation to this subject.

In the course of the present volume it will be shown that apparently every growing part of every plant is continually circumnutating, though often on a small scale. Even the stems of seedlings before they have broken through the ground, as well as their buried radicles, circumnutate, as far as the pressure of the surrounding earth permits. In this universally present movement we have the basis or groundwork for the acquirement, according to the requirements of the plant, of the most diversified movements. Thus, the great sweeps made by the stems of twining plants, and by the tendrils of other climbers, result from a mere increase in the amplitude of the ordinary movement of circumnutation. The position which young leaves and other organs ultimately assume is acquired by the circumnutating movement being increased in some one direction. the leaves of various plants are said to sleep at night, and it will be seen that their blades then assume a vertical position through modified circumnutation, in order to protect their upper surfaces from being chilled through radiation. The movements of various organs to the light, which are so general throughout the vegetable kingdom, and occasionally from the light, or transversely with respect to it, are all modified forms of circumnutation; as again are the equally prevalent movements of stems, etc., towards the zenith, and of roots towards the centre of the earth. In accordance with these conclusions, a considerable difficulty in the way of evolution is in part removed, for it might have been asked, how did all these diversified movements for the most different purposes first arise? As the case stands, we know that there is always movement in progress, and its amplitude, or direction, or both, have only to be modified for the good of the plant in relation with internal or external stimuli.

Besides describing the several modified forms of circumnutation, some other subjects will be discussed. The two which have interested us most are, firstly, the fact that with some seedling plants the uppermost part alone is sensitive to light, and transmits an influence to the lower part, causing it to bend. If therefore the upper part be wholly protected from light, the lower part may be exposed for hours to it, and yet does not become in the least bent, although this would have occurred quickly if the upper part had been excited by light. Secondly, with the radicles of seedlings, the tip is sensitive to various stimuli, especially to very slight pressure, and when thus excited, transmits an influence to the upper part, causing it to bend from the pressed side. On the other hand, if the tip is subjected to the vapour of water proceeding from one side, the upper part of the radicle bends towards this side. Again it is the tip, as stated by Ciesielski, though denied by others, which is sensitive to the attraction of gravity, and by transmission causes the adjoining parts of the radicle to bend towards the centre of the earth. These several cases of the effects of contact, other irritants, vapour, light, and the attraction of gravity being transmitted from the excited part for some little distance along the organ in question, have an important bearing on the theory of all such movements.

Terminology.—A brief explanation of some terms which will be used, must here be given. With seedlings, the stem which supports the cotyledons (i.e. the organs which represent the first leaves) has been called by many botanists the hypocotyledonous stem, but for brevity sake we will speak of it merely as the hypocotyl: the stem immediately above the cotyledons will be called the epicotyl or plumule. The radicle can be distinguished from the hypocotyl only by the presence of root-hairs and the nature of its covering. The meaning of the word circumnutation has already been explained. Authors speak of positive and negative heliotropism,[5]—that is, the bending of an organ to or from the light; but it is much more convenient to confine the word heliotropism to bending towards the light, and to designate as apheliotropism bending from the light. There is another reason for this change, for writers, as we have observed, occasionally drop the adjectives positive and negative, and thus introduce confusion into their discussions. Diaheliotropism may express a position more or less transverse to the light and induced by it. In like manner positive geotropism, or bending towards the centre of the earth, will be called by us geotropism; apogeotropism will mean bending in opposition to gravity or from the centre of the earth; and diageotropism, a position more or less transverse to the radius of the earth. The words heliotropism and geotropism properly mean the act of moving in relation to the light or the earth; but in the same manner as gravitation, though defined as "the act of tending to the centre," is often used to express the cause of a body falling, so it will be found convenient occasionally to employ heliotropism and geotropism, etc., as the cause of the movements in question.

[5] The highly useful terms of Heliotropism and Geotropism were first used by Dr. A. B. Frank: see his remarkable 'Beiträge zur Pflanzenphysiologie,' 1868.

The term epinasty is now often used in Germany, and implies that the upper surface of an organ grows more quickly than the lower surface, and thus causes it to bend downwards. Hyponasty is the reverse, and implies increased growth along the lower surface, causing the part to bend upwards.[6]

[6] These terms are used in the sense given them by De Vries, 'Würzburg Arbeiten,' Heft ii 1872, p. 252.

Methods of Observation.—The movements, sometimes very small and sometimes considerable in extent, of the various organs observed by us, were traced in the manner which after many trials we found to be best, and which must be described. Plants growing in pots were protected wholly from the light, or had light admitted from above, or on one side as the case might require, and were covered above by a large horizontal sheet of glass, and with another vertical sheet on one side. A glass filament, not thicker than a horsehair, and from a quarter to three-quarters of an inch in length, was affixed to the part to be observed by means of shellac dissolved in alcohol. The solution was allowed to evaporate, until it became so thick that it set hard in two or three seconds, and it never injured the tissues, even the tips of tender radicles, to which it was applied. To the end of the glass filament an excessively minute bead of black sealing-wax was cemented, below or behind which a bit of card with a black dot was fixed to a stick driven into the ground. The weight of the filament was so slight that even small leaves were not perceptibly pressed down. another method of observation, when much magnification of the movement was not required, will presently be described. The bead and the dot on the card were viewed through the horizontal or vertical glass-plate (according to the position of the object), and when one exactly covered the other, a dot was made on the glass-plate with a sharply pointed stick dipped in thick Indian-ink. Other dots were made at short intervals of time and these were afterwards joined by straight lines. The figures thus traced were therefore angular; but if dots had been made every 1 or 2 minutes, the lines would have been more curvilinear, as occurred when radicles were allowed to trace their own courses on smoked glass-plates. To make the dots accurately was the sole difficulty, and required some practice. Nor could this be done quite accurately, when the movement was much magnified, such as 30 times and upwards; yet even in this case the general course may be trusted. To test the accuracy of the above method of observation, a filament was fixed to an inanimate object which was made to slide along a straight edge and dots were repeatedly made on a glass-plate; when these were joined, the result ought to have been a perfectly straight line, and the line was very nearly straight. It may be added that when the dot on the card was placed half-an-inch below or behind the bead of sealing-wax, and when the glass-plate (supposing it to have been properly curved) stood at a distance of 7 inches in front (a common distance), then the tracing represented the movement of the bead magnified 15 times.

Whenever a great increase of the movement was not required, another, and in some respects better, method of observation was followed. This consisted in fixing two minute triangles of thin paper, about 1/20 inch in height, to the two ends of the attached glass filament; and when their tips were brought into a line so that they covered one another, dots were made as before on the glass-plate. If we suppose the glass-plate to stand at a distance of seven inches from the end of the shoot bearing the filament, the dots when joined, will give nearly the same figure as if a filament seven inches long, dipped in ink, had been fixed to the moving shoot, and had inscribed its own course on the plate. The movement is thus considerably magnified; for instance, if a shoot one inch in length were bending, and the glass-plate stood at the distance of seven inches, the movement would be magnified eight times. It would, however, have been very difficult to have ascertained in each case how great a length of the shoot was bending; and this is indispensable for ascertaining the degree to which the movement is magnified.

After dots had been made on the glass-plates by either of the above methods, they were copied on tracing paper and joined by ruled lines, with arrows showing the direction of the movement. The nocturnal courses are represented by straight broken lines. the first dot is always made larger than the others, so as to catch the eye, as may be seen in the diagrams. The figures on the glass-plates were often drawn on too large a scale to be reproduced on the pages of this volume, and the proportion in which they have been reduced is always given.[7] Whenever it could be approximately told how much the movement had been magnified, this is stated. We have perhaps introduced a superfluous number of diagrams; but they take up less space than a full description of the movements. Almost all the sketches of plants asleep, etc., were carefully drawn for us by Mr. George Darwin.

[7] We are much indebted to Mr. Cooper for the care with which he has reduced and engraved our diagrams.

As shoots, leaves, etc., in circumnutating bend more and more, first in one direction and then in another, they were necessarily viewed at different times more or less obliquely; and as the dots were made on a flat surface, the apparent amount of movement is exaggerated according to the degree of obliquity of the point of view. It would, therefore, have been a much better plan to have used hemispherical glasses, if we had possessed them of all sizes, and if the bending part of the shoot had been distinctly hinged and could have been placed so as to have formed one of the radii of the sphere. But even in this case it would have been necessary afterwards to have projected the figures on paper; so that complete accuracy could not have been attained. From the distortion of our figures, owing to the above causes, they are of no use to any one who wishes to know the exact amount of movement, or the exact course pursued; but they serve excellently for ascertaining whether or not the part moved at all, as well as the general character of the movement.

In the following chapters, the movements of a considerable number of plants are described; and the species have been arranged according to the system adopted by Hooker in Le Maout and Decaisne's 'Descriptive Botany.' No one who is not investigating the present subject need read all the details, which, however, we have thought it advisable to give. To save the reader trouble, the conclusions and most of the more important parts have been printed in larger type than the other parts. He may, if he thinks fit, read the last chapter first, as it includes a summary of the whole volume; and he will thus see what points interest him, and on which he requires the full evidence.

Finally, we must have the pleasure of returning our sincere thanks to Sir Joseph Hooker and to Mr. W. Thiselton Dyer for their great kindness, in not only sending us plants from Kew, but in procuring others from several sources when they were required for our observations; also, for naming many species, and giving us information on various points.

CHAPTER I.
THE CIRCUMNUTATING MOVEMENTS OF SEEDLING PLANTS.

Brassica oleracea, circumnutation of the radicle, of the arched hypocotyl whilst still buried beneath the ground, whilst rising above the ground and straightening itself, and when erect—Circumnutation of the cotyledons—Rate of movement—Analogous observations on various organs in species of Githago, Gossypium, Oxalis, Tropaeolum, Citrus, Æsculus, of several Leguminous and Cucurbitaceous genera, Opuntia, Helianthus, Primula, Cyclamen, Stapelia, Cerinthe, Nolana, Solanum, Beta, Ricinus, Quercus, Corylus, Pinus, Cycas, Canna, Allium, Asparagus, Phalaris, Zea, Avena, Nephrodium, and Selaginella.

The following chapter is devoted to the circumnutating movements of the radicles, hypocotyls, and cotyledons of seedling plants; and, when the cotyledons do not rise above the ground, to the movements of the epicotyl. But in a future chapter we shall have to recur to the movements of certain cotyledons which sleep at night.

Brassica oleracea (Cruciferae)'.—Fuller details will be given with respect to the movements in this case than in any other, as space and time will thus ultimately be saved.

Radicle.—A seed with the radicle projecting .05 inch was fastened with shellac to a little plate of zinc, so that the radicle stood up vertically; and a fine glass filament was then fixed near its base, that is, close to the seed-coats. The seed was surrounded by little bits of wet sponge, and the movement of the bead at the end of the filament was traced (Fig. 1) during sixty hours. In this time the radicle increased in length from .05 to .11 inch. Had the filament been attached at first close to the apex of the radicle, and if it could have remained there all the time, the movement exhibited would have been much greater, for at the close of our observations the tip, instead of standing vertically upwards, had become bowed downwards through geotropism, so as almost to touch the zinc plate. As far as we could roughly ascertain by measurements made with compasses on other seeds, the tip alone, for a length of only 2/100 to 3/100 of an inch, is acted on by geotropism. But the tracing shows that the basal part of the radicle continued to circumnutate irregularly during the whole time. The actual extreme amount of movement of the bead at the end of the filament was nearly .05 inch, but to what extent the movement of the radicle was magnified by the filament, which was nearly 3/4 inch in length, it was impossible to estimate.

Fig. 1. Brassica oleracea: circumnutation of radicle, traced on horizontal glass, from 9 A.M. Jan. 31st to 9 P.M. Feb. 2nd. Movement of bead at end of filament magnified about 40 times.

Another seed was treated and observed in the same manner, but the radicle in this case protruded .1 inch, and was not fastened so as to project quite vertically upwards. The filament was affixed close to its base. The tracing (Fig. 2, reduced by half) shows the movement from 9 A.M. Jan. 31st to 7 A.M. Feb. 2nd; but it continued to move during the whole of the 2nd in the same general direction, and in a similar zigzag manner. From the radicle not being quite perpendicular when the filament was affixed geotropism came into play at once; but the irregular zigzag course shows that there was growth (probably preceded by turgescence), sometimes on one and sometimes on another side. Occasionally the bead remained stationary for about an hour, and then probably growth occurred on the side opposite to that which caused the geotropic curvature. In the case previously described the basal part of the very short radicle from being turned vertically upwards, was at first very little affected by geotropism. Filaments were affixed in two other instances to rather longer radicles protruding obliquely from seeds which had been turned upside down; and in these cases the lines traced on the horizontal glasses were only slightly zigzag, and the movement was always in the same general direction, through the action of geotropism. All these observations are liable to several causes of error, but we believe, from what will hereafter be shown with respect to the movements of the radicles of other plants, that they may be largely trusted.

Fig. 2. Brassica oleracea: circumnutating and geotropic movement of radicle, traced on horizontal glass during 46 hours.

Hypocotyl.—The hypocotyl protrudes through the seed-coats as a rectangular projection, which grows rapidly into an arch like the letter U turned upside down; the cotyledons being still enclosed within the seed. In whatever position the seed may be embedded in the earth or otherwise fixed, both legs of the arch bend upwards through apogeotropism, and thus rise vertically above the ground. As soon as this has taken place, or even earlier, the inner or concave surface of the arch grows more quickly than the upper or convex surface; and this tends to separate the two legs and aids in drawing the cotyledons out of the buried seed-coats. By the growth of the whole arch the cotyledons are ultimately dragged from beneath the ground, even from a considerable depth; and now the hypocotyl quickly straightens itself by the increased growth of the concave side.

Even whilst the arched or doubled hypocotyl is still beneath the ground, it circumnutates as much as the pressure of the surrounding soil will permit; but this was difficult to observe, because as soon as the arch is freed from lateral pressure the two legs begin to separate, even at a very early age, before the arch would naturally have reached the surface. Seeds were allowed to germinate on the surface of damp earth, and after they had fixed themselves by their radicles, and after the, as yet, only slightly arched hypocotyl had become nearly vertical, a glass filament was affixed on two occasions near to the base of the basal leg (i.e. the one in connection with the radicle), and its movements were traced in darkness on a horizontal glass. The result was that long lines were formed running in nearly the plane of the vertical arch, due to the early separation of the two legs now freed from pressure; but as the lines were zigzag, showing lateral movement, the arch must have been circumnutating, whilst it was straightening itself by growth along its inner or concave surface.

A somewhat different method of observation was next followed: as soon as the earth with seeds in a pot began to crack, the surface was removed in parts to the depth of .2 inch; and a filament was fixed to the basal leg of a buried and arched hypocotyl, just above the summit of the radicle. The cotyledons were still almost completely enclosed within the much-cracked seed-coats; and these were again covered up with damp adhesive soil pressed pretty firmly down. The movement of the filament was traced (Fig. 3) from 11 A.M. Feb. 5th till 8 A.M. Feb. 7th. By this latter period the cotyledons had been dragged from beneath the pressed-down earth, but the upper part of the hypocotyl still formed nearly a right angle with the lower part. The tracing shows that the arched hypocotyl tends at this early age to circumnutate irregularly. On the first day the greater movement (from right to left in the figure) was not in the plane of the vertical and arched hypocotyl, but at right angles to it, or in the plane of the two cotyledons, which were still in close contact. The basal leg of the arch at the time when the filament was affixed to it, was already bowed considerably backwards, or from the cotyledons; had the filament been affixed before this bowing occurred, the chief movement would have been at right angles to that shown in the figure. A filament was attached to another buried hypocotyl of the same age, and it moved in a similar general manner, but the line traced was not so complex. This hypocotyl became almost straight, and the cotyledons were dragged from beneath the ground on the evening of the second day.

Fig. 3. Brassica oleracea: circumnutating movement of buried and arched hypocotyl (dimly illuminated from above), traced on horizontal glass during

45 hours. Movement of bead of filament magnified about 25 times, and here reduced to one-half of original scale.

Fig. 4. Brassica oleracea: circumnutating movement of buried and arched hypocotyl, with the two legs of the arch tied together, traced on horizontal glass during 33½ hours. Movement of the bead of filament magnified about 26 times, and here reduced to one-half original scale.

Before the above observations were made, some arched hypocotyls buried at the depth of a quarter of an inch were uncovered; and in order to prevent the two legs of the arch from beginning to separate at once, they were tied together with fine silk. This was done partly because we wished to ascertain how long the hypocotyl, in its arched condition, would continue to move, and whether the movement when not masked and disturbed by the straightening process, indicated circumnutation. Firstly a filament was fixed to the basal leg of an arched hypocotyl close above the summit of the radicle. The cotyledons were still partially enclosed within the seed-coats. The movement was traced (Fig. 4) from 9.20 A.M. on Dec. 23rd to 6.45 A.M. on Dec. 25th. No doubt the natural movement was much disturbed by the two legs having been tied together; but we see that it was distinctly zigzag, first in one direction and then in an almost opposite one. After 3 P.M. on the 24th the arched hypocotyl sometimes remained stationary for a considerable time, and when moving, moved far slower than before. Therefore, on the morning of the 25th, the glass filament was removed from the base of the basal leg, and was fixed horizontally on the summit of the arch, which, from the legs having been tied, had grown broad and almost flat. The movement was now traced during 23 hours (Fig. 5), and we see that the course was still zigzag, which indicates a tendency to circumnutation. The base of the basal leg by this time had almost completely ceased to move.

Fig. 5. Brassica oleracea: circumnutating movement of the crown of a buried and arched hypocotyl, with the two legs tied together, traced on a horizontal glass during 23 hours. Movement of the bead of the filament magnified about 58 times, and here reduced to one-half original scale.

As soon as the cotyledons have been naturally dragged from beneath the ground, and the hypocotyl has straightened itself by growth along the inner or concave surface, there is nothing to interfere with the free movements of the parts; and the circumnutation now becomes much more regular and clearly displayed, as shown in the following cases:—A seedling was placed in front and near a north-east window with a line joining the two cotyledons parallel to the window. It was thus left the whole day so as to accommodate itself to the light. On the following morning a filament was fixed to the midrib of the larger and taller cotyledon (which enfolds the other and smaller one, whilst still within the seed), and a mark being placed close behind, the movement of the whole plant, that is, of the hypocotyl and cotyledon, was traced greatly magnified on a vertical glass. At first the plant bent so much towards the light that it was useless to attempt to trace the movement; but at 10 A.M. heliotropism almost wholly ceased and the first dot was made on the glass. The last was made at 8.45 P.M.; seventeen dots being altogether made in this interval of 10 h. 45 m. (see Fig. 6). It should be noticed that when I looked shortly after 4 P.M. the bead was pointing off the glass, but it came on again at 5.30 P.M., and the course during this interval of 1 h. 30 m. has been filled up by imagination, but cannot be far from correct. The bead moved seven times from side to side, and thus described 3½ ellipses in 10 3/4 h.; each being completed on an average in 3 h. 4 m.

Fig. 6. Brassica oleracea: conjoint circumnutation of the hypocotyl and cotyledons during 10 hours 45 minutes. Figure here reduced to one-half original scale.

On the previous day another seedling had been observed under similar conditions, excepting that the plant was so placed that a line joining the two cotyledons pointed towards the window; and the filament was attached to the smaller cotyledon on the side furthest from the window. Moreover the plant was now for the first time placed in this position. The cotyledons bowed themselves greatly towards the light from 8 to 10.50 A.M., when the first dot was made (Fig. 7). During the next 12 hours the bead swept obliquely up and down 8 times and described 4 figures representing ellipses; so that it travelled at nearly the same rate as in the previous case. during the night it moved upwards, owing to the sleep-movement of the cotyledons, and continued to move in the same direction till 9 A.M. on the following morning; but this latter movement would not have occurred with seedlings under their natural conditions fully exposed to the light.

Fig. 7. Brassica oleracea: conjoint circumnutation of the hypocotyl and cotyledons, from 10.50 A.M. to 8 A.M. on the following morning. Tracing made on a vertical glass.

By 9.25 A.M. on this second day the same cotyledon had begun to fall, and a dot was made on a fresh glass. The movement was traced until 5.30 P.M. as shown in (Fig. 8), which is given, because the course followed was much more irregular than on the two previous occasions. During these 8 hours the bead changed its course greatly 10 times. The upward movement of the cotyledon during the afternoon and early part of the night is here plainly shown.

Fig. 8. Brassica oleracea: conjoint circumnutation of the hypocotyl and cotyledons during 8 hours. Figure here reduced to one-third of the original scale, as traced on a vertical glass.

As the filaments were fixed in the three last cases to one of the cotyledons, and as the hypocotyl was left free, the tracings show the movement of both organs conjoined; and we now wished to ascertain whether both circumnutated. Filaments were therefore fixed horizontally to two hypocotyls close beneath the petioles of their cotyledons. These seedlings had stood for two days in the same position before a north-east window. In the morning, up to about 11 A.M., they moved in zigzag lines towards the light; and at night they again became almost upright through apogeotropism. After about 11 A.M. they moved a little back from the light, often crossing and recrossing their former path in zigzag lines. the sky on this day varied much in brightness, and these observations merely proved that the hypocotyls were continually moving in a manner resembling circumnutation. On a previous day which was uniformly cloudy, a hypocotyl was firmly secured to a little stick, and a filament was fixed to the larger of the two cotyledons, and its movement was traced on a vertical glass. It fell greatly from 8.52 A.M., when the first dot was made, till 10.55 A.M.; it then rose greatly until 12.17 P.M. Afterwards it fell a little and made a loop, but by 2.22 P.M. it had risen a little and continued rising till 9.23 P.M., when it made another loop, and at 10.30 P.M. was again rising. These observations show that the cotyledons move vertically up and down all day long, and as there was some slight lateral movement, they circumnutated.

Fig. 9. Brassica oleracea: circumnutation of hypocotyl, in darkness, traced on a horizontal glass, by means of a filament with a bead fixed across its summit, between 9.15 A.M. and 8.30 A.M. on the following morning. Figure here reduced to one-half of original scale.

The cabbage was one of the first plants, the seedlings of which were observed by us, and we did not then know how far the circumnutation of the different parts was affected by light. Young seedlings were therefore kept in complete darkness except for a minute or two during each observation, when they were illuminated by a small wax taper held almost vertically above them. During the first day the hypocotyl of one changed its course 13 times (see Fig. 9); and it deserves notice that the longer axes of the figures described often cross one another at right or nearly right angles. Another seedling was observed in the same manner, but it was much older, for it had formed a true leaf a quarter of an inch in length, and the hypocotyl was 1 3/8 inch in height. The figure traced was a very complex one, though the movement was not so great in extent as in the last case.

The hypocotyl of another seedling of the same age was secured to a little stick, and a filament having been fixed to the midrib of one of the cotyledons, the movement of the bead was traced during 14 h. 15 m. (see Fig. 10) in darkness. It should be noted that the chief movement of the cotyledons, namely, up and down, would be shown on a horizontal glass-plate only by the lines in the direction of the midrib (that is, up and down, as Fig. 10 here stands) being a little lengthened or shortened; whereas any lateral movement would be well exhibited. The present tracing shows that the cotyledon did thus move laterally (that is, from side to side in the tracing) 12 times in the 14 h. 15 m. of observation. Therefore the cotyledons certainly circumnutated, though the chief movement was up and down in a vertical plane.

Fig. 10. Brassica oleracea: circumnutation of a cotyledon, the hypocotyl having been secured to a stick, traced on a horizontal glass, in darkness, from 8.15 A.M. to 10.30 P.M. Movement of the bead of the filament magnified 13 times.

Rate of Movement.—The movements of the hypocotyls and cotyledons of seedling cabbages of different ages have now been sufficiently illustrated. With respect to the rate, seedlings were placed under the microscope with the stage removed, and with a micrometer eye-piece so adjusted that each division equalled 1/500 inch; the plants were illuminated by light passing through a solution of bichromate of potassium so as to eliminate heliotropism. Under these circumstances it was interesting to observe how rapidly the circumnutating apex of a cotyledon passed across the divisions of the micrometer. Whilst travelling in any direction the apex generally oscillated backwards and forwards to the extent of 1/500 and sometimes of nearly 1/250 of an inch. These oscillations were quite different from the trembling caused by any disturbance in the same room or by the shutting of a distant door. The first seedling observed was nearly two inches in height and had been etiolated by having been grown in darkness. The tip of the cotyledon passed across 10 divisions of the micrometer, that is, 1/50 of an inch, in 6 m. 40 s. Short glass filaments were then fixed vertically to the hypocotyls of several seedlings so as to project a little above the cotyledons, thus exaggerating the rate of movement; but only a few of the observations thus made are worth giving. The most remarkable fact was the oscillatory movement above described, and the difference of rate at which the point crossed the divisions of the micrometer, after short intervals of time. For instance, a tall not-etiolated seedling had been kept for 14 h. in darkness; it was exposed before a north-east window for only two or three minutes whilst a glass filament was fixed vertically to the hypocotyl; it was then again placed in darkness for half an hour and afterwards observed by light passing through bichromate of potassium.

The point, oscillating as usual, crossed five divisions of the micrometer (i.e. 1/100 inch) in 1 m. 30 s. The seedling was then left in darkness for an hour, and now it required 3 m. 6 s. to cross one division, that is, 15 m. 30 s. to have crossed five divisions. Another seedling, after being occasionally observed in the back part of a northern room with a very dull light, and left in complete darkness for intervals of half an hour, crossed five divisions in 5 m. in the direction of the window, so that we concluded that the movement was heliotropic. But this was probably not the case, for it was placed close to a north-east window and left there for 25 m., after which time, instead of moving still more quickly towards the light, as might have been expected, it travelled only at the rate of 12 m. 30 s. for five divisions. It was then again left in complete darkness for 1 h., and the point now travelled in the same direction as before, but at the rate of 3 m. 18 s. for five divisions.

We shall have to recur to the cotyledons of the cabbage in a future chapter, when we treat of their sleep-movements. The circumnutation, also, of the leaves of fully-developed plants will hereafter be described.

Fig. 11. Githago segetum: circumnutation of hypocotyl, traced on a horizontal glass, by means of a filament fixed transversely across its summit, from 8.15 A.M. to 12.15 P.M. on the following day. Movement of bead of filament magnified about 13 times, here reduced to one-half the original scale.

Githago segetum (Caryophylleae).—A young seedling was dimly illuminated from above, and the circumnutation of the hypocotyl was observed during 28 h., as shown in Fig. 11. It moved in all directions; the lines from right and to left in the figure being parallel to the blades of the cotyledons. The actual distance travelled from side to side by the summit of the hypocotyl was about .2 of an inch; but it was impossible to be accurate on this head, as the more obliquely the plant was viewed, after it had moved for some time, the more the distances were exaggerated.

We endeavoured to observe the circumnutation of the cotyledons, but as they close together unless kept exposed to a moderately bright light, and as the hypocotyl is extremely heliotropic, the necessary arrangements were too troublesome. We shall recur to the nocturnal or sleep-movements of the cotyledons in a future chapter.

Fig. 12. Gossypium: circumnutation of hypocotyl, traced on a horizontal glass, from 10.30 A.M. to 9.30 A.M. on following morning, by means of a filament fixed across its summit. Movement of bead of filament magnified about twice; seedling illuminated from above.

Gossypium (var. Nankin cotton) (Malvaceae).—The circumnutation of a hypocotyl was observed in the hot-house, but the movement was so much exaggerated that the bead twice passed for a time out of view. It was, however, manifest that two somewhat irregular ellipses were nearly completed in 9 h. Another seedling, 1½ in. in height, was then observed during 23 h.; but the observations were not made at sufficiently short intervals, as shown by the few dots in Fig. 12, and the tracing was not now sufficiently enlarged. Nevertheless there could be no doubt about the circumnutation of the hypocotyl, which described in 12 h. a figure representing three irregular ellipses of unequal sizes.

The cotyledons are in constant movement up and down during the whole day, and as they offer the unusual case of moving downwards late in the evening and in the early part of the night, many observations were made on them. A filament was fixed along the middle of one, and its movement traced on a vertical glass; but the tracing is not given, as the hypocotyl was not secured, so that it was impossible to distinguish clearly between its movement and that of the cotyledon. The cotyledons rose from 10.30 A.M. to about 3 P.M.; they then sank till 10 P.M., rising, however, greatly in the latter part of the night. The angles above the horizon at which the cotyledons of another seedling stood at different hours is recorded in the following short table:—

Oct. 20 2.50 P.M...25° above horizon. Oct. 20 4.20 P.M...22° above horizon. Oct. 20 5.20 P.M...15° above horizon. Oct. 20 10.40 P.M...8° above horizon. Oct. 21 8.40 A.M...28° above horizon. Oct. 21 11.15 A.M...35° above horizon. Oct. 21 9.11 P.M...10° below horizon.

The position of the two cotyledons was roughly sketched at various hours with the same general result.

In the following summer, the hypocotyl of a fourth seedling was secured to a little stick, and a glass filament with triangles of paper having been fixed to one of the cotyledons, its movements were traced on a vertical glass under a double skylight in the house. The first dot was made at 4.20 P.M. June 20th; and the cotyledon fell till 10.15 P.M. in a nearly straight line. Just past midnight it was found a little lower and somewhat to one side. By the early morning, at 3.45 A.M., it had risen greatly, but by 6.20 A.M. had fallen a little. During the whole of this day (21st) it fell in a slightly zigzag line, but its normal course was disturbed by the want of sufficient illumination, for during the night it rose only a little, and travelled irregularly during the whole of the following day and night of June 22nd. The ascending and descending lines traced during the three days did not coincide, so that the movement was one of circumnutation. This seedling was then taken back to the hot-house, and after five days was inspected at 10 P.M., when the cotyledons were found hanging so nearly vertically down, that they might justly be said to have been asleep. On the following morning they had resumed their usual horizontal position.

Oxalis rosea (Oxalideae).—The hypocotyl was secured to a little stick, and an extremely thin glass filament, with two triangles of paper, was attached to one of the cotyledons, which was .15 inch in length. In this and the following species the end of the petiole, where united to the blade, is developed into a pulvinus. The apex of the cotyledon stood only 5 inches from the vertical glass, so that its movement was not greatly exaggerated as long as it remained nearly horizontal; but in the course of the day it both rose considerably above and fell beneath a horizontal position, and then of course the movement was much exaggerated. In Fig. 13 its course is shown from 6.45 A.M. on June 17th, to 7.40 A.M. on the following morning; and we see that during the daytime, in the course of 11 h. 15 m., it travelled thrice down and twice up. After 5.45 P.M. it moved rapidly downwards, and in an hour or two depended vertically; it thus remained all night asleep. This position could not be represented on the vertical glass nor in the figure here given. By 6.40 A.M. on the following morning (18th) both cotyledons had risen greatly, and they continued to rise until 8 A.M., when they stood almost horizontally. Their movement was traced during the whole of this day and until the next morning; but a tracing is not given, as it was closely similar to Fig. 13, excepting that the lines were more zigzag. The cotyledons moved 7 times, either upwards or downwards; and at about 4 P.M. the great nocturnal sinking movement commenced.

Fig. 13. Oxalis rosea: circumnutation of cotyledons, the hypocotyl being secured to a stick; illuminated from above. Figure here given one-half of original scale.

Another seedling was observed in a similar manner during nearly 24 h., but with the difference that the hypocotyl was left free. The movement also was less magnified. Between 8.12 A.M. and 5 P.M. on the 18th, the apex of the cotyledon moved 7 times upwards or downwards (Fig. 14). The nocturnal sinking movement, which is merely a great increase of one of the diurnal oscillations, commenced about 4 P.M.

Oxalis Valdiviana.—This species is interesting, as the cotyledons rise perpendicularly upwards at night so as to come into close contact, instead of sinking vertically downwards, as in the case of O. rosea. A glass filament was fixed to a cotyledon, .17 of an inch in length, and the hypocotyl was left free. On the first day the seedling was placed too far from the vertical glass; so that the tracing was enormously exaggerated and the movement could not be traced when the cotyledon either rose or sank much; but it was clearly seen that the cotyledons rose thrice and fell twice between 8.15

A.M. and 4.15 P.M. Early on the following morning (June 19th) the apex of a cotyledon was placed only 1 7/8 inch from the vertical glass. At 6.40 A.M. it stood horizontally; it then fell till 8.35, and then rose. Altogether in the course of 12 h. it rose thrice and fell thrice, as may be seen in Fig. 15. The great nocturnal rise of the cotyledons usually commences about 4 or 5 P.M., and on the following morning they are expanded or stand horizontally at about 6.30 A.M. In the present instance, however, the great nocturnal rise did not commence till 7 P.M.; but this was due to the hypocotyl having from some unknown cause temporarily bent to the left side, as is shown in the tracing. To ascertain positively that the hypocotyl circumnutated, a mark was placed at 8.15 P.M. behind the two now closed and vertical cotyledons; and the movement of a glass filament fixed upright to the top of the hypocotyl was traced until 10.40 P.M. During this time it moved from side to side, as well as backwards and forwards, plainly showing circumnutation; but the movement was small in extent. Therefore Fig. 15 represents fairly well the movements of the cotyledons alone, with the exception of the one great afternoon curvature to the left.

Fig. 14. Oxalis rosea: conjoint circumnutation of the cotyledons and hypocotyl, traced from 8.12 A.M. on June 18th to 7.30 A.M. 19th. The apex of the cotyledon stood only 3 3/4 inches from the vertical glass. Figure here given one-half of original scale.

Fig. 15. Oxalis Valdiviana: conjoint circumnutation of a cotyledon and the hypocotyl, traced on vertical glass, during 24 hours. Figure here given one-half of original scale; seedling illuminated from above.

Oxalis corniculata (var. cuprea).—The cotyledons rise at night to a variable degree above the horizon, generally about 45°: those on some seedlings between 2 and 5 days old were found to be in continued movement all day long; but the movements were more simple than in the last two species. This may have partly resulted from their not being sufficiently illuminated whilst being observed, as was shown by their not beginning to rise until very late in the evening.

Oxalis (Biophytum) sensitiva.—The cotyledons are highly remarkable from the amplitude and rapidity of their movements during the day. The angles at which they stood above or beneath the horizon were measured at short intervals of time; and we regret that their course was not traced during the whole day. We will give only a few of the measurements, which were made whilst the seedlings were exposed to a temperature of 22½° to 24½° C. One cotyledon rose 70° in 11 m.; another, on a distinct seedling, fell 80° in 12 m. Immediately before this latter fall the same cotyledon had risen from a vertically downward to a vertically upward position in 1 h. 48 m., and had therefore passed through 180° in under 2 h. We have met with no other instance of a circumnutating movement of such great amplitude as 180°; nor of such rapidity of movement as the passage through 80° in 12 m. The cotyledons of this plant sleep at night by rising vertically and coming into close contact. This upward movement differs from one of the great diurnal oscillations above described only by the position being permanent during the night and by its periodicity, as it always commences late in the evening.

Tropaeolum minus (?) (var. Tom Thumb) (Tropaeoleae).—The cotyledons are hypogean, or never rise above the ground. By removing the soil a buried epicotyl or plumule was found, with its summit arched abruptly downwards, like the arched hypocotyl of the cabbage previously described. A glass filament with a bead at its end was affixed to the basal half or leg, just above the hypogean cotyledons, which were again almost surrounded by loose earth. The tracing (Fig. 16) shows the course of the bead during 11 h. After the last dot given in the figure, the bead moved to a great distance, and finally off the glass, in the direction indicated by the broken line. This great movement, due to increased growth along the concave surface of the arch, was caused by the basal leg bending backwards from the upper part, that is in a direction opposite to the dependent tip, in the same manner as occurred with the hypocotyl of the cabbage. Another buried and arched epicotyl was observed in the same manner, excepting that the two legs of the arch were tied together with fine silk for the sake of preventing the great movement just mentioned. It moved, however, in the evening in the same direction as before, but the line followed was not so straight. During the morning the tied arch moved in an irregularly circular, strongly zigzag course, and to a greater distance than in the previous case, as was shown in a tracing, magnified 18 times. The movements of a young plant bearing a few leaves and of a mature plant, will hereafter be described.

Fig. 16. Tropaeolum minus (?): circumnutation of buried and arched epicotyl, traced on a horizontal glass, from 9.20 A.M. to 8.15 P.M. Movement of bead of filament magnified 27 times.

Citrus aurantium (Orange) (Aurantiaceae).—The cotyledons are hypogean. The circumnutation of an epicotyl, which at the close of our observations was .59 of an inch (15 mm.) in height above the ground, is shown in the annexed figure (Fig. 17), as observed during a period of 44 h. 40 m.

Fig. 17. Citrus aurantium: circumnutation of epicotyl with a filament fixed transversely near its apex, traced on a horizontal glass, from 12.13 P.M. on Feb. 20th to 8.55 A.M. on 22nd. The movement of the bead of the filament was at first magnified 21 times, or 10½, in figure here given, and afterwards 36 times, or 18 as here given; seedling illuminated from above.

Æsculus hippocastanum (Hippocastaneae).—Germinating seeds were placed in a tin box, kept moist internally, with a sloping bank of damp argillaceous sand, on which four smoked glass-plates rested, inclined at angles of 70° and 65° with the horizon. The tips of the radicles were placed so as just to touch the upper end of the glass-plates, and, as they grew downwards they pressed lightly, owing to geotropism, on the smoked surfaces, and left tracks of their course. In the middle part of each track the glass was swept clean, but the margins were much blurred and irregular. Copies of two of these tracks (all four being nearly alike) were made on tracing paper placed over the glass-plates after they had been varnished; and they are as exact as possible considering the nature of the margins (Fig. 18). They suffice to show that there was some lateral, almost serpentine movement, and that the tips in their downward course pressed with unequal force on the plates, as the tracks varied in breadth. The more perfectly serpentine tracks made by the radicles of Phaseolus multiflorus and Vicia faba (presently to be described), render it almost certain that the radicles of the present plant circumnutated.

Fig. 18. Æsculus hippocastanum: outlines of tracks left on inclined glass-plates by tips of radicles. In A the plate was inclined at 70° with the horizon, and the radicle was 1.9 inch in length, and .23 inch in diameter at base. In B the plate was inclined 65° with the horizon, and the radicle was a trifle larger.

Phaseolus multiflorus (Leguminosae).—Four smoked glass-plates were arranged in the same manner as described under Æsculus, and the tracks left by the tips of four radicles of the present plant, whilst growing downwards, were photographed as transparent objects. Three of them are here exactly copied (Fig. 19). Their serpentine courses show that the tips moved regularly from side to side; they also pressed alternately with greater or less force on the plates, sometimes rising up and leaving them altogether for a very short distance; but this was better seen on the original plates than in the copies. These radicles therefore were continually moving in all directions—that is, they circumnutated. The distance between the extreme right and left positions of the radicle A, in its lateral movement, was 2 mm., as ascertained by measurement with an eye-piece micrometer.

Fig. 19. Phaseolus multiflorus: tracks left on inclined smoked glass-plates by tips of radicles in growing downwards. A and C, plates inclined at 60°, B inclined at 68° with the horizon.

Vicia faba (Common Bean) (Leguminosae).—Radicle.—Some beans were allowed to germinate on bare sand, and after one had protruded its radicle to a length of .2 of an inch, it was turned upside down, so that the radicle, which was kept in damp air, now stood upright. A filament, nearly an inch in length, was affixed obliquely near its tip; and the movement of the terminal bead was traced from 8.30 A.M. to 10.30 P.M., as shown in Fig. 18. The radicle at first changed its course twice abruptly, then made a small loop and then a larger zigzag curve. During the night and till 11 A.M. on the following morning, the bead moved to a great distance in a nearly straight line, in the direction indicated by the broken line in the figure. This resulted from the tip bending quickly downwards, as it had now become much declined, and had thus gained a position highly favourable for the action of geotropism.

Fig. 20. Vicia faba: circumnutation of a radicle, at first pointing vertically upwards, kept in darkness, traced on a horizontal glass, during 14 hours. Movement of bead of filament magnified 23 times, here reduced to one-half of original scale.

Fig. 21. Vicia faba: tracks left on inclined smoked glass-plates, by tips of radicles in growing downwards. Plate C was inclined at 63°, plates A and D at 71°, plate B at 75°, and plate E at a few degrees beneath the horizon.

We next experimented on nearly a score of radicles by allowing them to grow downwards over inclined plates of smoked glass, in exactly the same manner as with Æsculus and Phaseolus. Some of the plates were inclined only a few degrees beneath the horizon, but most of them between 60° and 75°. In the latter cases the radicles in growing downwards were deflected only a little from the direction which they had followed whilst germinating in sawdust, and they pressed lightly on the glass-plates (Fig. 21). Five of the most distinct tracks are here copied, and they are all slightly sinuous, showing circumnutation. Moreover, a close examination of almost every one of the tracks clearly showed that the tips in their downward course had alternately pressed with greater or less force on the plates, and had sometimes risen up so as nearly to leave them for short intervals. The distance between the extreme right and left positions of the radicle A was 0.7 mm., ascertained in the same manner as in the case of Phaseolus.

Epicotyl.—At the point where the radicle had protruded from a bean laid on its side, a flattened solid lump projected .1 of an inch, in the same horizontal plane with the bean. This protuberance consisted of the convex summit of the arched epicotyl; and as it became developed the two legs of the arch curved themselves laterally upwards, owing to apogeotropism, at such a rate that the arch stood highly inclined after 14 h., and vertically in 48 h. A filament was fixed to the crown of the protuberance before any arch was visible, but the basal half grew so quickly that on the second morning the end of the filament was bowed greatly downwards. It was therefore removed and fixed lower down. The line traced during these two days extended in the same general direction, and was in parts nearly straight, and in others plainly zigzag, thus giving some evidence of circumnutation.

As the arched epicotyl, in whatever position it may be placed, bends quickly upwards through apogeotropism, and as the two legs tend at a very early age to separate from one another, as soon as they are relieved from the pressure of the surrounding earth, it was difficult to ascertain positively whether the epicotyl, whilst remaining arched, circumnutated. Therefore some rather deeply buried beans were uncovered, and the two legs of the arches were tied together, as had been done with the epicotyl of Tropaeolum and the hypocotyl of the Cabbage. The movements of the tied arches were traced in the usual manner on two occasions during three days. But the tracings made under such unnatural conditions are not worth giving; and it need only be said that the lines were decidedly zigzag, and that small loops were occasionally formed. We may therefore conclude that the epicotyl circumnutates whilst still arched and before it has grown tall enough to break through the surface of the ground.

In order to observe the movements of the epicotyl at a somewhat more advanced age, a filament was fixed near the base of one which was no longer arched, for its upper half now formed a right angle with the lower half. This bean had germinated on bare damp sand, and the epicotyl began to straighten itself much sooner than would have occurred if it had been properly planted. The course pursued during 50 h. (from 9 A.M. Dec. 26th, to 11 A.M. 28th) is here shown (Fig. 22); and we see that the epicotyl circumnutated during the whole time. Its basal part grew so much during the 50 h. that the filament at the end of our observations was attached at the height of .4 inch above the upper surface of the bean, instead of close to it. If the bean had been properly planted, this part of the epicotyl would still have been beneath the soil.

Fig. 22. Vicia faba: circumnutation of young epicotyl, traced in darkness during 50 hours on a horizontal glass. Movement of bead of filament magnified 20 times, here reduced to one-half of original scale.

Late in the evening of the 28th, some hours after the above observations were completed, the epicotyl had grown much straighter, for the upper part now formed a widely open angle with the lower part. A filament was fixed to the upright basal part, higher up than before, close beneath the lowest scale-like process or homologue of a leaf; and its movement was traced during 38 h. (Fig. 23). We here again have plain evidence of continued circumnutation. Had the bean been properly planted, the part of the epicotyl to which the filament was attached, the movement of which is here shown, would probably have just risen above the surface of the ground.

Fig. 23. Vicia faba: circumnutation of the same epicotyl as in Fig. 22, a little more advanced in age, traced under similar conditions as before, from 8.40 A.M. Dec. 28th, to 10.50 A.M. 30th. Movement of bead here magnified 20 times.

Lathyrus nissolia (Leguminosae).—This plant was selected for observation from being an abnormal form with grass-like leaves. The cotyledons are hypogean, and the epicotyl breaks through the ground in an arched form. The movements of a stem, 1.2 inch in height, consisting of three internodes, the lower one almost wholly subterranean, and the upper one bearing a short, narrow leaf, is shown during 24 h., in Fig. 24. No glass filament was employed, but a mark was placed beneath the apex of the leaf. The actual length of the longer of the two ellipses described by the stem was about .14 of an inch. On the previous day the chief line of movement was nearly at right angles to that shown in the present figure, and it was more simple.

Fig. 24. Lathyrus nissolia: circumnutation of stem of young seedling, traced in darkness on a horizontal glass, from 6.45 A.M. Nov. 22nd, to 7 A.M. 23rd. Movement of end of leaf magnified about 12 times, here reduced to one-half of original scale.

Cassia tora[1] (Leguminosae).—A seedling was placed before a north-east window; it bent very little towards it, as the hypocotyl which was left free was rather old, and therefore not highly heliotropic. A filament had been fixed to the midrib of one of the cotyledons, and the movement of the whole seedling was traced during two days. The circumnutation of the hypocotyl is quite insignificant compared with that of the cotyledons. These rise up vertically at night and come into close contact; so that they may be said to sleep. This seedling was so old that a very small true leaf had been developed, which at night was completely hidden by the closed cotyledons. On Sept. 24th, between 8 A.M. and 5 P.M., the cotyledons moved five times up and five times down; they therefore described five irregular ellipses in the course of the 9 h. The great nocturnal rise commenced about 4.30 P.M.

[1] Seeds of this plant, which grew near the sea-side, were sent to us by Fritz Müller from S. Brazil. The seedlings did not flourish or flower well with us; they were sent to Kew, and were pronounced not to be distinguishable from C. tora.

Fig. 25. Cassia tora: conjoint circumnutation of cotyledons and hypocotyl, traced on vertical glass, from 7.10 A.M. Sept. 25th to 7.30 A.M. 26th. Figure here given reduced to one-half of original scale.

On the following morning (Sept. 25th) the movement of the same cotyledon was again traced in the same manner during 24 h.; and a copy of the tracing is here given (Fig. 25). The morning was cold, and the window had been accidentally left open for a short time, which must have chilled the plant; and this probably prevented it from moving quite as freely as on the previous day; for it rose only four and sank only four times during the day, one of the oscillations being very small. At 7.10 A.M., when the first dot was made, the cotyledons were not fully open or awake; they continued to open till about 9 A.M., by which time they had sunk a little beneath the horizon: by 9.30 A.M. they had risen, and then they oscillated up and down; but the upward and downward lines never quite coincided. At about 4.30 P.M. the great nocturnal rise commenced. At 7 A.M. on the following morning (Sept. 26th) they occupied nearly the same level as on the previous morning, as shown in the diagram: they then began to open or sink in the usual manner. The diagram leads to the belief that the great periodical daily rise and fall does not differ essentially, excepting in amplitude, from the oscillations during the middle of the day.

Lotus Jacoboeus (Leguminosae).—The cotyledons of this plant, after the few first days of their life, rise so as to stand almost, though rarely quite, vertically at night. They continue to act in this manner for a long time even after the development of some of the true leaves. With seedlings, 3 inches in height, and bearing five or six leaves, they rose at night about 45°. They continued to act thus for about an additional fortnight. Subsequently they remained horizontal at night, though still green and at last dropped off. Their rising at night so as to stand almost vertically appears to depend largely on temperature; for when the seedlings were kept in a cool house, though they still continued to grow, the cotyledons did not become vertical at night. It is remarkable that the cotyledons do not generally rise at night to any conspicuous extent during the first four or five days after germination; but the period was extremely variable with seedlings kept under the same conditions; and many were observed. Glass filaments with minute triangles of paper were fixed to the cotyledons (1½ mm. in breadth) of two seedlings, only 24 h. old, and the hypocotyl was secured to a stick; their movements greatly magnified were traced, and they certainly circumnutated the whole time on a small scale, but they did not exhibit any distinct nocturnal and diurnal movement. The hypocotyls, when left free, circumnutated over a large space.

Another and much older seedling, bearing a half-developed leaf, had its movements traced in a similar manner during the three first days and nights of June; but seedlings at this age appear to be very sensitive to a deficiency of light; they were observed under a rather dim skylight, at a temperature of between 16° to 17½° C.' and apparently, in consequence of these conditions, the great daily movement of the cotyledons ceased on the third day. During the first two days they began rising in the early afternoon in a nearly straight line, until between 6 and 7 P.M., when they stood vertically. During the latter part of the night, or more probably in the early morning, they began to fall or open, so that by 6.45 A.M. they stood fully expanded and horizontal. They continued to fall slowly for some time, and during the second day described a single small ellipse, between 9 A.M. and 2 P.M., in addition to the great diurnal movement. The course pursued during the whole 24 h. was far less complex than in the foregoing case of Cassia. On the third morning they fell very much, and then circumnutated on a small scale round the same spot; by 8.20 P.M. they showed no tendency to rise at night. Nor did the cotyledons of any of the many other seedlings in the same pot rise; and so it was on the following night of June 5th. The pot was then taken back into the hot-house, where it was exposed to the sun, and on the succeeding night all the cotyledons rose again to a high angle, but did not stand quite vertically. On each of the above days the line representing the great nocturnal rise did not coincide with that of the great diurnal fall, so that narrow ellipses were described, as is the usual rule with circumnutating organs. The cotyledons are provided with a pulvinus, and its development will hereafter be described.

Mimosa pudica (Leguminosae).—The cotyledons rise up vertically at night, so as to close together. Two seedlings were observed in the greenhouse (temp. 16° to 17° C. or 63° to 65° F.). Their hypocotyls were secured to sticks, and glass filaments bearing little triangles of paper were affixed to the cotyledons of both. Their movements were traced on a vertical glass during 24 h. on November 13th. The pot had stood for some time in the same position, and they were chiefly illuminated through the glass-roof. The cotyledons of one of these seedlings moved downward in the morning till 11.30 A.M., and then rose, moving rapidly in the evening until they stood vertically, so that in this case there was simply a single great daily fall and rise. The other seedling behaved rather differently, for it fell in the morning until 11.30 A.M., and then rose, but after 12.10 P.M. again fell; and the great evening rise did not begin until 1.22 P.M. On the following morning this cotyledon had fallen greatly from its vertical position by 8.15 A.M. Two other seedlings (one seven and the other eight days old) had been previously observed under unfavourable circumstances, for they had been brought into a room and placed before a north-east window, where the temperature was between only 56° and 57° F. They had, moreover, to be protected from lateral light, and perhaps were not sufficiently illuminated. Under these circumstances the cotyledons moved simply downwards from 7 A.M. till 2 P.M., after which hour and during a large part of the night they continued to rise. Between 7 and 8 A.M. on the following morning they fell again; but on this second and likewise on the third day the movements became irregular, and between 3 and 10.30 P.M. they circumnutated to a small extent about the same spot; but they did not rise at night. Nevertheless, on the following night they rose as usual.

Cytisus fragrans (Leguminosae).—Only a few observations were made on this plant. The hypocotyl circumnutated to a considerable extent, but in a simple manner—namely, for two hours in one direction, and then much more slowly back again in a zigzag course, almost parallel to the first line, and beyond the starting-point. It moved in the same direction all night, but next morning began to return. The cotyledons continually move both up and down and laterally; but they do not rise up at night in a conspicuous manner.

Lupinus luteus (Leguminosae).—Seedlings of this plant were observed because the cotyledons are so thick (about .08 of an inch) that it seemed unlikely that they would move. Our observations were not very successful, as the seedlings are strongly heliotropic, and their circumnutation could not be accurately observed near a north-east window, although they had been kept during the previous day in the same position. A seedling was then placed in darkness with the hypocotyl secured to a stick; both cotyledons rose a little at first, and then fell during the rest of the day; in the evening between 5 and 6 P.M. they moved very slowly; during the night one continued to fall and the other rose, though only a little. The tracing was not much magnified, and as the lines were plainly zigzag, the cotyledons must have moved a little laterally, that is, they must have circumnutated.

The hypocotyl is rather thick, about .12 of inch; nevertheless it circumnutated in a complex course, though to a small extent. The movement of an old seedling with two true leaves partially developed, was observed in the dark. As the movement was magnified about 100 times it is not trustworthy and is not given; but there could be no doubt that the hypocotyl moved in all directions during the day, changing its course 19 times. The extreme actual distance from side to side through which the upper part of the hypocotyl passed in the course of 14½ hours was only 1/60 of an inch; it sometimes travelled at the rate of 1/50 of an inch in an hour.

Cucurbita ovifera (Cucurbitaceæ).—Radicle: a seed which had germinated on damp sand was fixed so that the slightly curved radicle, which was only .07 inch in length, stood almost vertically upwards, in which position geotropism would act at first with little power. A filament was attached near to its base, and projected at about an angle of 45° above the horizon. The general course followed during the 11 hours of observation and during the following night is shown in the accompanying diagram (Fig. 26), and was plainly due to geotropism; but it was also clear that the radicle circumnutated. By the next morning the tip had curved so much downwards that the filament, instead of projecting at 45° above the horizon, was nearly horizontal. Another germinating seed was turned upside down and covered with damp sand; and a filament was fastened to the radicle so as to project at an angle of about 50° above the horizon; this radicle was .35 of an inch in length and a little curved. The course pursued was mainly governed, as in the last case, by geotropism, but the line traced during 12 hours and magnified as before was more strongly zigzag, again showing circumnutation.

Fig. 26. Cucurbita ovifera: course followed by a radicle in bending geotropically downwards, traced on a horizontal glass, between 11.25 A.M. and 10.25 P.M.; the direction during the night is indicated by the broken line. Movement of bead magnified 14 times.

Four radicles were allowed to grow downwards over plates of smoked glass, inclined at 70° to the horizon, under the same conditions as in the cases of Æsculus, Phaseolus, and Vicia. Facsimiles are here given (Fig. 27) of two of these tracks; and a third short one was almost as plainly serpentine as that at A. It was also manifest by a greater or less amount of soot having been swept off the glasses, that the tips had pressed alternately with greater and less force on them. There must, therefore, have been movement in at least two planes at right angles to one another. These radicles were so delicate that they rarely had the power to sweep the glasses quite clean. One of them had developed some lateral or secondary rootlets, which projected a few degrees beneath the horizon; and it is an important fact that three of them left distinctly serpentine tracks on the smoked surface, showing beyond doubt that they had circumnutated like the main or primary radicle. But the tracks were so slight that they could not be traced and copied after the smoked surface had been varnished.

Fig. 27. Cucurbita ovifera: tracks left by tips of radicles in growing downwards over smoked glass-plates, inclined at 70° to the horizon.

Fig. 28. Cucurbita ovifera: circumnutation of arched hypocotyl at a very early age, traced in darkness on a horizontal glass, from 8 A.M. to 10.20 A.M. on the following day. The movement of the bead magnified 20 times, here reduced to one-half of original scale.

Fig. 29. Cucurbita ovifera: circumnutation of straight and vertical hypocotyl, with filament fastened transversely across its upper end, traced in darkness on a horizontal glass, from 8.30 A.M. to 8.30 P.M. The movement of the terminal bead originally magnified about 18 times, here only 4½ times.

Hypocotyl.—A seed lying on damp sand was firmly fixed by two crossed wires and by its own growing radicle. The cotyledons were still enclosed within the seed-coats; and the short hypocotyl, between the summit of the radicle and the cotyledons, was as yet only slightly arched. A filament (.85 of inch in length) was attached at an angle of 35° above the horizon to the side of the arch adjoining the cotyledons. This part would ultimately form the upper end of the hypocotyl, after it had grown straight and vertical. Had the seed been properly planted, the hypocotyl at this stage of growth would have been deeply buried beneath the surface. The course followed by the bead of the filament is shown in Fig. 28. The chief lines of movement from left to right in the figure were parallel to the plane of the two united cotyledons and of the flattened seed; and this movement would aid in dragging them out of the seed-coats, which are held down by a special structure hereafter to be described. The movement at right angles to the above lines was due to the arched hypocotyl becoming more arched as it increased in height. The foregoing observations apply to the leg of the arch next to the cotyledons, but the other leg adjoining the radicle likewise circumnutated at an equally early age.

The movement of the same hypocotyl after it had become straight and vertical, but with the cotyledons only partially expanded, is shown in Fig. 29. The course pursued during 12 h. apparently represents four and a half ellipses or ovals, with the longer axis of the first at nearly right angles to that of the others. The longer axes of all were oblique to a line joining the opposite cotyledons. The actual extreme distance from side to side over which the summit of the tall hypocotyl passed in the course of 12 h. was .28 of an inch. The original figure was traced on a large scale, and from the obliquity of the line of view the outer parts of the diagram are much exaggerated.

Cotyledons.—On two occasions the movements of the cotyledons were traced on a vertical glass, and as the ascending and descending lines did not quite coincide, very narrow ellipses were formed; they therefore circumnutated. Whilst young they rise vertically up at night, but their tips always remain reflexed; on the following morning they sink down again. With a seedling kept in complete darkness they moved in the same manner, for they sank from 8.45 A.M. to 4.30 P.M.; they then began to rise and remained close together until 10 P.M., when they were last observed. At 7 A.M. on the following morning they were as much expanded as at any hour on the previous day. The cotyledons of another young seedling, exposed to the light, were fully open for the first time on a certain day, but were found completely closed at 7 A.M. on the following morning. They soon began to expand again, and continued doing so till about 5 P.M.; they then began to rise, and by 10.30 P.M. stood vertically and were almost closed. At 7 A.M.

on the third morning they were nearly vertical, and again expanded during the day; on the fourth morning they were not closed, yet they opened a little in the course of the day and rose a little on the following night. By this time a minute true leaf had become developed. Another seedling, still older, bearing a well-developed leaf, had a sharp rigid filament affixed to one of its cotyledons (85 mm. in length), which recorded its own movements on a revolving drum with smoked paper. The observations were made in the hot-house, where the plant had lived, so that there was no change in temperature or light. The record commenced at 11 A.M. on February 18th; and from this hour till 3 P.M. the cotyledon fell; it then rose rapidly till 9 P.M., then very gradually till 3 A.M. February 19th, after which hour it sank gradually till 4.30 P.M.; but the downward movement was interrupted by one slight rise or oscillation about 1.30 P.M. After 4.30 P.M. (19th) the cotyledon rose till 1 A.M. (in the night of February 20th) and then sank very gradually till 9.30 A.M., when our observations ceased. The amount of movement was greater on the 18th than on the 19th or on the morning of the 20th.

Cucurbita aurantia.—An arched hypocotyl was found buried a little beneath the surface of the soil; and in order to prevent it straightening itself quickly, when relieved from the surrounding pressure of the soil, the two legs of the arch were tied together. The seed was then lightly covered with loose damp earth. A filament with a bead at the end was affixed to the basal leg, the movements of which were observed during two days in the usual manner. On the first day the arch moved in a zigzag line towards the side of the basal leg. On the next day, by which time the dependent cotyledons had been dragged above the surface of the soil, the tied arch changed its course greatly nine times in the course of 14½ h. It swept a large, extremely irregular, circular figure, returning at night to nearly the same spot whence it had started early in the morning. The line was so strongly zigzag that it apparently represented five ellipses, with their longer axes pointing in various directions. With respect to the periodical movements of the cotyledons, those of several young seedlings formed together at 4 P.M. an angle of about 60°, and at 10 P.M. their lower parts stood vertically and were in contact; their tips, however, as is usual in the genus, were permanently reflexed. These cotyledons, at 7 A.M. on the following morning, were again well expanded.

Lagenaria vulgaris (var. miniature Bottle-gourd) (Cucurbitaceæ).—A seedling opened its cotyledons, the movements of which were alone observed, slightly on June 27th and closed them at night: next day, at noon (28th), they included an angle of 53°, and at 10 P.M. they were in close contact, so that each had risen 26½°. At noon, on the 29th, they included an angle of 118°, and at 10 P.M. an angle of 54°, so each had risen 32°. On the following day they were still more open, and the nocturnal rise was greater, but the angles were not measured. Two other seedlings were observed, and behaved during three days in a closely similar manner. The cotyledons, therefore, open more and more on each succeeding day, and rise each night about 30°; consequently during the first two nights of their life they stand vertically and come into contact.

Fig. 30. Lagenaria vulgaris: circumnutation of a cotyledon, 1½ inch in length, apex only 4 3/4 inches from the vertical glass, on which its movements were traced from 7.35 A.M. July 11th to 9.5 A.M. on the 14th. Figure here given reduced to one-third of original scale.

In order to ascertain more accurately the nature of these movements, the hypocotyl of a seedling, with its cotyledons well expanded, was secured to a little stick, and a filament with triangles of paper was affixed to one of the cotyledons. The observations were made under a rather dim skylight, and the temperature during the whole time was between 17½° to 18° C. (63° to 65° F.). Had the temperature been higher and the light brighter, the movements would probably have been greater. On July 11th (see Fig. 30), the cotyledon fell from 7.35 A.M. till 10 A.M.; it then rose (rapidly after 4 P.M.) till it stood quite vertically at 8.40 P.M. During the early morning of the next day (12th) it fell, and continued to fall till 8 A.M., after which hour it rose, then fell, and again rose, so that by 10.35 P.M. it stood much higher than it did in the morning, but was not vertical as on the preceding night. During the following early morning and whole day (13th) it fell and circumnutated, but had not risen when observed late in the evening; and this was probably due to the deficiency of heat or light, or of both. We thus see that the cotyledons became more widely open at noon on each succeeding day; and that they rose considerably each night, though not acquiring a vertical position, except during the first two nights.

Cucumis dudaim (Cucurbitaceæ).—Two seedlings had opened their cotyledons for the first time during the day,—one to the extent of 90° and the other rather more; they remained in nearly the same position until 10.40 P.M.; but by 7 A.M. on the following morning the one which had been previously open to the extent of 90° had its cotyledons vertical and completely shut; the other seedling had them nearly shut. Later in the morning they opened in the ordinary manner. It appears therefore that the cotyledons of this plant close and open at somewhat different periods from those of the foregoing species of the allied genera of Cucurbita and Lagenaria.

Fig. 31. Opuntia basilaris: conjoint circumnutation of hypocotyl and cotyledon; filament fixed longitudinally to cotyledon, and movement traced during 66 h. on horizontal glass. Movement of the terminal bead magnified about 30 times, here reduced to one-third scale. Seedling kept in hot-house, feebly illuminated from above.

Opuntia basilaris (Cacteæ).—A seedling was carefully observed, because, considering its appearance and the nature of the mature plant, it seemed very unlikely that either the hypocotyl or cotyledons would circumnutate to an appreciable extent. The cotyledons were well developed, being .9 of an inch in length, .22 in breadth, and .15 in thickness. The almost cylindrical hypocotyl, now bearing a minute spinous bud on its summit, was only .45 of an inch in height, and .19 in diameter. The tracing (Fig. 31) shows the combined movement of the hypocotyl and of one of the cotyledons, from 4.45 P.M. on May 28th to 11 A.M. on the 31st. On the 29th a nearly perfect ellipse was completed. On the 30th the hypocotyl moved, from some unknown cause, in the same general direction in a zigzag line; but between 4.30 and 10 P.M. almost completed a second small ellipse. The cotyledons move only a little up and down: thus at 10.15 P.M. they stood only 10° higher than at noon. The chief seat of movement therefore, at least when the cotyledons are rather old as in the present case, lies in the hypocotyl. The ellipse described on the 29th had its longer axis directed at nearly right angles to a line joining the two cotyledons. The actual amount of movement of the bead at the end of the filament was, as far as could be ascertained, about .14 of an inch.

Fig. 32. Helianthus annuus: circumnutation of hypocotyl, with filament fixed across its summit, traced on a horizontal glass in darkness, from 8.45 A.M. to 10.45 P.M., and for an hour on following morning. Movement of bead magnified 21 times, here reduced to one-half of original scale.

Helianthus annuus (Compositæ).—The upper part of the hypocotyl moved during the day-time in the course shown in the annexed figure (Fig. 32). As the line runs in various directions, crossing itself several times, the movement may be considered as one of circumnutation. The extreme actual distance travelled was at least .1 of an inch. The movements of the cotyledons of two seedlings were observed; one facing a north-east window, and the other so feebly illuminated from above us as to be almost in darkness. They continued to sink till about noon, when they began to rise; but between 5 and 7 or 8 P.M. they either sank a little, or moved laterally, and then again began to rise. At 7 A.M. on the following morning those on the plant before the north-east window had opened so little that they stood at an angle of 73° above the horizon, and were not observed any longer. Those on the seedling which had been kept in almost complete darkness, sank during the whole day, without rising about mid-day, but rose during the night. On the third and fourth days they continued sinking without any alternate ascending movement; and this, no doubt, was due to the absence of light.

Primula Sinensis (Primulaceae).—A seedling was placed with the two cotyledons parallel to a north-east window on a day when the light was nearly uniform, and a filament was affixed to one of them. From observations subsequently made on another seedling with the stem secured to a stick, the greater part of the movement shown in the annexed figure (Fig. 33), must have been that of the hypocotyl, though the cotyledons certainly move up and down to a certain extent both during the day and night. The movements of the same seedling were traced on the following day with nearly the same result; and there can be no doubt about the circumnutation of the hypocotyl.

Fig. 33. Primula Sinensis: conjoint circumnutation of hypocotyl and cotyledon, traced on vertical glass, from 8.40 A.M. to 10.45 P.M. Movements of bead magnified about 26 times.

Cyclamen Persicum (Primulaceae).—This plant is generally supposed to produce only a single cotyledon, but Dr. H. Gressner[2] has shown that a second one is developed after a long interval of time. The hypocotyl is converted into a globular corm, even before the first cotyledon has broken through thc ground with its blade closely enfolded and with its petiole in the form of an arch, like the arched hypocotyl or epicotyl of any ordinary dicotyledonous plant. A glass filament was affixed to a cotyledon, .55 of an inch in height, the petiole of which had straightened itself and stood nearly vertical, but with the blade not as yet fully expanded. Its movements were traced during 24½ h. on a horizontal glass, magnified 50 times; and in this interval it described two irregular small circles; it therefore circumnutates, though on an extremely small scale.

[2] 'Bot. Zeitung,' 1874, p. 837.

Fig. 34. Stapelia sarpedon: circumnutation of hypocotyl, illuminated from above, traced on horizontal glass, from 6.45 A.M. June 26th to 8.45 A.M. 28th. Temp. 23–24° C. Movement of bead magnified 21 times.

Stapelia sarpedon (Asclepiadeae).—This plant, when mature, resembles a cactus. The flattened hypocotyl is fleshy, enlarged in the upper part, and bears two rudimentary cotyledons. It breaks through the ground in an arched form, with the rudimentary cotyledons closed or in contact. A filament was affixed almost vertically to the hypocotyl of a seedling half an inch high; and its movements were traced during 50 h. on a horizontal glass (Fig. 34). From some unknown cause it bowed itself to one side, and as this was effected by a zigzag course, it probably circumnutated; but with hardly any other seedling observed by us was this movement so obscurely shown.

Ipomœa caerulea vel Pharbitis nil (Convolvulaceae).—Seedlings of this plant were observed because it is a twiner, the upper internodes of which circumnutate conspicuously; but like other twining plants, the first few internodes which rise above the ground are stiff enough to support themselves, and therefore do not circumnutate in any plainly recognisable manner.[3] In this particular instance the fifth internode (including the hypocotyl) was the first which plainly circumnutated and twined round a stick. We therefore wished to learn whether circumnutation could be observed in the hypocotyl if carefully observed in our usual manner. Two seedlings were kept in the dark with filaments fixed to the upper part of their hypocotyls; but from circumstances not worth explaining their movements were traced for only a short time. One moved thrice forwards and twice backwards in nearly opposite directions, in the course of 3 h. 15 m.; and the other twice forwards and twice backwards in 2 h. 22 m. The hypocotyl therefore circumnutated at a remarkably rapid rate. It may here be added that a filament was affixed transversely to the summit of the second internode above the cotyledons of a little plant 3½ inches in height; and its movements were traced on a horizontal glass. It circumnutated, and the actual distance travelled from side to side was a quarter of an inch, which was too small an amount to be perceived without the aid of marks.

[3] 'Movements and Habits of Climbing Plants,' p. 33, 1875.

The movements of the cotyledons are interesting from their complexity and rapidity, and in some other respects. The hypocotyl (2 inches high) of a vigorous seedling was secured to a stick, and a filament with triangles of paper was affixed to one of the cotyledons. The plant was kept all day in the hot-house, and at 4.20 P.M. (June 20th) was placed under a skylight in the house, and observed occasionally during the evening and night. It fell in a slightly zigzag line to a moderate extent from 4.20 P.M. till 10.15 P.M. When looked at shortly after midnight (12.30 P.M.) it had risen a very little, and considerably by 3.45 A.M. When again looked at, at 6.10 A.M. (21st), it had fallen largely. A new tracing was now begun (see Fig. 35), and soon afterwards, at 6.42 A.M., the cotyledon had risen a little. During the forenoon it was observed about every hour; but between 12.30 and 6 P.M. every half-hour. If the observations had been made at these short intervals during the whole day, the figure would have been too intricate to have been copied. As it was, the cotyledon moved up and down in the course of 16 h. 20 m. (i.e. between 6.10 A.M. and 10.30 P.M.) thirteen times.

Fig. 35. Ipomœa caerulea: circumnutation of cotyledon, traced on vertical glass, from 6.10 A.M. June 21st to 6.45 A.M. 22nd. Cotyledon with petiole 1.6 inch in length, apex of blade 4.1 inch from the vertical glass; so movement not greatly magnified; temp. 20° C.

The cotyledons of this seedling sank downwards during both evenings and the early part of the night, but rose during the latter part. As this is an unusual movement, the cotyledons of twelve other seedlings were observed; they stood almost or quite horizontally at mid-day, and at 10 P.M. were all declined at various angles. The most usual angle was between 30° and 35°; but three stood at about 50° and one at even 70° beneath the horizon. The blades of all these cotyledons had attained almost their full size, viz. from 1 to 1½ inches in length, measured along their midribs. It is a remarkable fact that whilst young—that is, when less than half an inch in length, measured in the same manner—they do not sink downwards in the evening. Therefore their weight, which is considerable when almost fully developed, probably came into play in originally determining the downward movement. The periodicity of this movement is much influenced by the degree of light to which the seedlings have been exposed during the day; for three kept in an obscure place began to sink about noon, instead of late in the evening; and those of another seedling were almost paralysed by having been similarly kept during two whole days. The cotyledons of several other species of Ipomœa likewise sink downwards late in the evening.

Cerinthe major (Boragineae).—The circumnutation of the hypocotyl of a young seedling with the cotyledons hardly expanded, is shown in the annexed figure (Fig. 36), which apparently represents four or five irregular ellipses, described in the course of a little over 12 hours. Two older seedlings were similarly observed, excepting that one of them was kept in the dark; their hypocotyls also circumnutated, but in a more simple manner. The cotyledons on a seedling exposed to the light fell from the early morning until a little after noon, and then continued to rise until 10.30 P.M. or later. The cotyledons of this same seedling acted in the same general manner during the two following days. It had previously been tried in the dark, and after being thus kept for only 1 h. 40 m. the cotyledons began at 4.30 P.M. to sink, instead of continuing to rise till late at night.

Fig. 36. Cerinthe major: circumnutation of hypocotyl, with filament fixed across its summit, illuminated from above, traced on horizontal glass, from 9.26 A.M. to 9.53 P.M. on Oct. 25th. Movement of the bead magnified 30 times, here reduced to one-third of original scale.

Nolana prostrata (Nolaneae).—The movements were not traced, but a pot with seedlings, which had been kept in the dark for an hour, was placed under the microscope, with the micrometer eye-piece so adjusted that each division equalled 1/500th of an inch. The apex of one of the cotyledons crossed rather obliquely four divisions in 13 minutes; it was also sinking, as shown by getting out of focus. The seedlings were again placed in darkness for another hour, and the apex now crossed two divisions in 6 m. 18 s.; that is, at very nearly the same rate as before. After another interval of an hour in darkness, it crossed two divisions in 4 m. 15 s., therefore at a quicker rate. In the afternoon, after a longer interval in the dark, the apex was motionless, but after a time it recommenced moving, though slowly; perhaps the room was too cold. Judging from previous cases, there can hardly be a doubt that this seedling was circumnutating.

Solanum lycopersicum (Solaneae).—The movements of the hypocotyls of two seedling tomatoes were observed during seven hours, and there could be no doubt that both circumnutated. They were illuminated from above, but by an accident a little light entered on one side, and in the accompanying figure (Fig. 37) it may be seen that the hypocotyl moved to this side (the upper one in the figure), making small loops and zigzagging in its course. The movements of the cotyledons were also traced both on vertical and horizontal glasses; their angles with the horizon were likewise measured at various hours. They fell from 8.30 A.M. (October 17th) to about noon; then moved laterally in a zigzag line, and at about 4 P.M. began to rise; they continued to do so until 10.30 P.M., by which hour they stood vertically and were asleep. At what hour of the night or early morning they began to fall was not ascertained. Owing to the lateral movement shortly after mid-day, the descending and ascending lines did not coincide, and irregular ellipses were described during each 24 h. The regular periodicity of these movements is destroyed, as we shall hereafter see, if the seedlings are kept in the dark.

Fig. 37. Solanum lycopersicum: circumnutation of hypocotyl, with filament fixed across its summit, traced on horizontal glass, from 10 A.M. to 5 P.M. Oct. 24th. Illuminated obliquely from above. Movement of bead magnified about 35 times, here reduced to one-third of original scale.

Solanum palinacanthum.—Several arched hypocotyls rising nearly .2 of an inch above the ground, but with the cotyledons still buried beneath the surface, were observed, and the tracings showed that they circumnutated. Moreover, in several cases little open circular spaces or cracks in the argillaceous sand which surrounded the arched hypocotyls were visible, and these appeared to have been made by the hypocotyls having bent first to one and then to another side whilst growing upwards. In two instances the vertical arches were observed to move to a considerable distance backwards from the point where the cotyledons lay buried; this movement, which has been noticed in some other cases, and which seems to aid in extracting the cotyledons from the buried seed-coats, is due to the commencement of the straightening of the hypocotyl. In order to prevent this latter movement, the two legs of an arch, the summit of which was on a level with the surface of the soil, were tied together; the earth having been previously removed to a little depth all round. The movement of the arch during 47 hours under these unnatural circumstances is exhibited in the annexed figure.

Fig. 38. Solanum palinacanthum: circumnutation of an arched hypocotyl, just emerging from the ground, with the two legs tied together, traced in darkness on a horizontal glass, from 9.20 A.M. Dec. 17th to 8.30 A.M. 19th. Movement of bead magnified 13 times; but the filament, which was affixed obliquely to the crown of the arch, was of unusual length.

The cotyledons of some seedlings in the hot-house were horizontal about noon on December 13th; and at 10 P.M. had risen to an angle of 27° above the horizon; at 7 A.M. on the following morning, before it was light, they had risen to 59° above the horizon; in the afternoon of the same day they were found again horizontal.

Beta vulgaris (Chenopodeae).—The seedlings are excessively sensitive to light, so that although on the first day they were uncovered only during two or three minutes at each observation, they all moved steadily towards the side of the room whence the light proceeded, and the tracings consisted only of slightly zigzag lines directed towards the light. On the next day the plants were placed in a completely darkened room, and at each observation were illuminated as much as possible from vertically above by a small wax taper. The annexed figure (Fig. 39) shows the movement of the hypocotyl during 9 h. under these circumstances. A second seedling was similarly observed at the same time, and the tracing had the same peculiar character, due to the hypocotyl often moving and returning in nearly parallel lines. The movement of a third hypocotyl differed greatly.

Fig. 39. Beta vulgaris: circumnutation of hypocotyl, with filament fixed obliquely across its summit, traced in darkness on horizontal glass, from 8.25 A.M. to 5.30 P.M. Nov. 4th. Movement of bead magnified 23 times, here reduced to one-third of original scale.

We endeavoured to trace the movements of the cotyledons, and for this purpose some seedlings were kept in the dark, but they moved in an abnormal manner; they continued rising from 8.45 A.M. to 2 P.M., then moved laterally, and from 3 to 6 P.M. descended; whereas cotyledons which have been exposed all the day to the light rise in the evening so as to stand vertically at night; but this statement applies only to young seedlings. For instance, six seedlings in the greenhouse had their cotyledons partially open for the first time on the morning of November 15th, and at 8.45 P.M. all were completely closed, so that they might properly be said to be asleep. Again, on the morning of November 27th, the cotyledons of four other seedlings, which were surrounded by a collar of brown paper so that they received light only from above, were open to the extent of 39°; at 10 P.M. they were completely closed; next morning (November 28th) at 6.45 A.M. whilst it was still dark, two of them were partially open and all opened in the course of the morning; but at 10.20 P.M. all four (not to mention nine others which had been open in the morning and six others on another occasion) were again completely closed. On the morning of the 29th they were open, but at night only one of the four was closed, and this only partially; the three others had their cotyledons much more raised than during the day. On the night of the 30th the cotyledons of the four were only slightly raised.

Ricinus Borboniensis (Euphorbiaceae).—Seeds were purchased under the above name—probably a variety of the common castor-oil plant. As soon as an arched hypocotyl had risen clear above the ground, a filament was attached to the upper leg bearing the cotyledons which were still buried beneath the surface, and the movement of the bead was traced on a horizontal glass during a period of 34 h. The lines traced were strongly zigzag, and as the bead twice returned nearly parallel to its former course in two different directions, there could be no doubt that the arched hypocotyl circumnutated. At the close of the 34 h. the upper part began to rise and straighten itself, dragging the cotyledons out of the ground, so that the movements of the bead could no longer be traced on the glass.

Quercus (American sp.) (Cupuliferae).—Acorns of an American oak which had germinated at Kew were planted in a pot in the greenhouse. This transplantation checked their growth; but after a time one grew to a height of five inches, measured to the tips of the small partially unfolded leaves on the summit, and now looked vigorous. It consisted of six very thin internodes of unequal lengths. Considering these circumstances and the nature of the plant, we hardly expected that it would circumnutate; but the annexed figure (Fig. 40) shows that it did so in a conspicuous manner, changing its course many times and travelling in all directions during the 48 h. of observation. The figure seems to represent 5 or 6 irregular ovals or ellipses. The actual amount of movement from side to side (excluding one great bend to the left) was about .2 of an inch; but this was difficult to estimate, as owing to the rapid growth of the stem, the attached filament was much further from the mark beneath at the close than at the commencement of the observations. It deserves notice that the pot was placed in a north-east room within a deep box, the top of which was not at first covered up, so that the inside facing the windows was a little more illuminated than the opposite side; and during the first morning the stem travelled to a greater distance in this direction (to the left in the figure) than it did afterwards when the box was completely protected from light.

Fig. 40. Quercus (American sp.): circumnutation of young stem, traced on horizontal glass, from 12.50 P.M. Feb. 22nd to 12.50 P.M. 24th. Movement of bead greatly magnified at first, but slightly towards the close of the observations—about 10 times on an average.

Quercus robur.—Observations were made only on the movements of the radicles from germinating acorns, which were allowed to grow downwards in the manner previously described, over plates of smoked glass, inclined at angles between 65° and 69° to the horizon. In four cases the tracks left were almost straight, but the tips had pressed sometimes with more and sometimes with less force on the glass, as shown by the varying thickness of the tracks and by little bridges of soot left across them. In the fifth case the track was slightly serpentine, that is, the tip had moved a little from side to side. In the sixth case (Fig. 41, A) it was plainly serpentine, and the tip had pressed almost equably on the glass in its whole course. In the seventh case (B) the tip had moved both laterally and had pressed alternately with unequal force on the glass; so that it had moved a little in two planes at right angles to one another. In the eighth and last case (C) it had moved very little laterally, but had alternately left the glass and come into contact with it again. There can be no doubt that in the last four cases the radicle of the oak circumnutated whilst growing downwards.

Fig. 41. Quercus robur: tracks left on inclined smoked glass-plates by tips of radicles in growing downwards. Plates A and C inclined at 65° and plate B at 68° to the horizon.

Corylus avellana (Corylaceae).—The epicotyl breaks through the ground in an arched form; but in the specimen which was first examined, the apex had become decayed, and the epicotyl grew to some distance through the soil, in a tortuous, almost horizontal direction, like a root. In consequence of this injury it had emitted near the hypogean cotyledons two secondary shoots, and it was remarkable that both of these were arched, like the normal epicotyl in ordinary cases. The soil was removed from around one of these arched secondary shoots, and a glass filament was affixed to the basal leg. The whole was kept damp beneath a metal-box with a glass lid, and was thus illuminated only from above. Owing apparently to the lateral pressure of the earth being removed, the terminal and bowed-down part of the shoot began at once to move upwards, so that after 24 h. it formed a right angle with the lower part. This lower part, to which the filament was attached, also straightened itself, and moved a little backwards from the upper part. Consequently a long line was traced on the horizontal glass; and this was in parts straight and in parts decidedly zigzag, indicating circumnutation.

On the following day the other secondary shoot was observed; it was a little more advanced in age, for the upper part, instead of depending vertically downwards, stood at an angle of 45° above the horizon. The tip of the shoot projected obliquely .4 of an inch above the ground, but by the close of our observations, which lasted 47 h., it had grown, chiefly towards its base, to a height of .85 of an inch. The filament was fixed transversely to the basal and almost upright half of the shoot, close beneath the lowest scale-like appendage. The circumnutating course pursued is shown in the accompanying figure (Fig. 42). The actual distance traversed from side to side was about .04 of an inch.

Fig. 42. Corylus avellana: circumnutation of a young shoot emitted from the epicotyl, the apex of which had been injured, traced on a horizontal glass, from 9 A.M. Feb. 2nd to 8 A.M. 4th. Movement of bead magnified about 27 times.

Pinus pinaster (Coniferæ).—A young hypocotyl, with the tips of the cotyledons still enclosed within the seed-coats, was at first only .35 of an inch in height; but the upper part grew so rapidly that at the end of our observations it was .6 in height, and by this time the filament was attached some way down the little stem. From some unknown cause, the hypocotyl moved far towards the left, but there could be no doubt (Fig. 43) that it circumnutated. Another hypocotyl was similarly observed, and it likewise moved in a strongly zigzag line to the same side. This lateral movement was not caused by the attachment of the glass filaments, nor by the action of light; for no light was allowed to enter when each observation was made, except from vertically above.

Fig. 43. Pinus pinaster: circumnutation of hypocotyl, with filament fixed across its summit, traced on horizontal glass, from 10 A.M. March 21st to 9 A.M. 23rd. Seedling kept in darkness. Movement of bead magnified about 35 times.

The hypocotyl of a seedling was secured to a little stick; it bore nine in appearance distinct cotyledons, arranged in a circle. The movements of two nearly opposite ones were observed. The tip of one was painted white, with a mark placed below, and the figure described (Fig. 44, A) shows that it made an irregular circle in the course of about 8 h. during the night it travelled to a considerable distance in the direction indicated by the broken line. A glass filament was attached longitudinally to the other cotyledon, and this nearly completed (Fig, 44, B) an irregular circular figure in about 12 hours. During the night it also moved to a considerable distance, in the direction indicated by the broken line. The cotyledons therefore circumnutate independently of the movement of the hypocotyl. Although they moved much during the night, they did not approach each other so as to stand more vertically than during the day.

Fig. 44. Pinus pinaster: circumnutation of two opposite cotyledons, traced on horizontal glass in darkness, from 8.45 A.M. to 8.35 P.M. Nov. 25th. Movement of tip in A magnified about 22 times, here reduced to one-half of original scale.

Cycas pectinata (Cycadeæ).—The large seeds of this plant in germinating first protrude a single leaf, which breaks through the ground with the petiole bowed into an arch and with the leaflets involuted. A leaf in this condition, which at the close of our observations was 2½ inches in height, had its movements traced in a warm greenhouse by means of a glass filament bearing paper triangles attached across its tip. The tracing (Fig. 45) shows how large, complex, and rapid were the circumnutating movements. The extreme distance from side to side which it passed over amounted to between .6 and .7 of an inch.

Fig. 45. Cycas pectinata: circumnutation of young leaf whilst emerging from the ground, feebly illuminated from above, traced on vertical glass, from 5 P.M. May 28th to 11 A.M. 31st. Movement magnified 7 times, here reduced to two-thirds of original scale.

Canna Warscewiczii (Cannaceae).—A seedling with the plumule projecting one inch above the ground was observed, but not under fair conditions, as it was brought out of the hot-house and kept in a room not sufficiently warm. Nevertheless the tracing (Fig. 46) shows that it made two or three incomplete irregular circles or ellipses in the course of 48 hours. The plumule is straight; and this was the first instance observed by us of the part that first breaks through the ground not being arched.

Fig. 46. Canna Warscewiczii: circumnutation of plumule with filament affixed obliquely to outer sheath-like leaf, traced in darkness on horizontal glass from 8.45 A.M. Nov. 9th to 8.10 A.M. 11th. Movement of bead magnified 6 times.

Allium cepa (Liliaceae).—The narrow green leaf, which protrudes from the seed of the common onion as a cotyledon,[4] breaks through the ground in the form of an arch, in the same manner as the hypocotyl or epicotyl of a dicotyledonous plant. Long after the arch has risen above the surface the apex remains within the seed-coats, evidently absorbing the still abundant contents. The summit or crown of the arch, when it first protrudes from the seed and is still buried beneath the ground, is simply rounded; but before it reaches the surface it is developed into a conical protuberance of a white colour (owing to the absence of chlorophyll), whilst the adjoining parts are green, with the epidermis apparently rather thicker and tougher than elsewhere. We may therefore conclude that this conical protuberance is a special adaptation for breaking through the ground,[5] and answers the same end as the knife-like white crest on the summit of the straight cotyledon of the Gramineæ. After a time the apex is drawn out of the empty seed-coats, and rises up, forming a right angle, or more commonly a still larger angle with the lower part, and occasionally the whole becomes nearly straight. The conical protuberance, which originally formed the crown of the arch, is now seated on one side, and appears like a joint or knee, which from acquiring chlorophyll becomes green, and increases in size. In rarely or never becoming perfectly straight, these cotyledons differ remarkably from the ultimate condition of the arched hypocotyls or epicotyls of dicotyledons. It is, also, a singular circumstance that the attenuated extremity of the upper bent portion invariably withers and dies.

[4] This is the expression used by Sachs in his 'Text-book of Botany.'

[5] Haberlandt has briefly described ('Die Schutzeinrichtungen...Keimpflanze,' 1877, p. 77) this curious structure and the purpose which it subserves. He states that good figures of the cotyledon of the onion have been given by Tittmann and by Sachs in his 'Experimental Physiologie,' p. 93.

A filament, 1.7 inch in length, was affixed nearly upright beneath the knee to the basal and vertical portion of a cotyledon; and its movements were traced during 14 h. in the usual manner. The tracing here given (Fig. 47) indicates circumnutation. The movement of the upper part above the knee of the same cotyledon, which projected at about an angle of 45° above the horizon, was observed at the same time. A filament was not affixed to it, but a mark was placed beneath the apex, which was almost white from beginning to wither, and its movements were thus traced. The figure described resembled pretty closely that above given; and this shows that the chief seat of movement is in the lower or basal part of the cotyledon.

Fig. 47. Allium cepa: circumnutation of basal half of arched cotyledon, traced in darkness on horizontal glass, from 8.15 A.M. to 10 P.M. Oct. 31st. Movement of bead magnified about 17 times.

Asparagus officinalis (Asparageae).—The tip of a straight plumule or cotyledon (for we do not know which it should be called) was found at a depth of .1 inch beneath the surface, and the earth was then removed all round to the dept of .3 inch. a glass filament was affixed obliquely to it, and the movement of the bead, magnified 17 times, was traced in darkness. During the first 1 h. 15 m. the plumule moved to the right, and during the next two hours it returned in a roughly parallel but strongly zigzag course. From some unknown cause it had grown up through the soil in an inclined direction, and now through apogeotropism it moved during nearly 24 h. in the same general direction, but in a slightly zigzag manner, until it became upright. On the following morning it changed its course completely. There can therefore hardly be a doubt that the plumule circumnutates, whilst buried beneath the ground, as much as the pressure of the surrounding earth will permit. The surface of the soil in the pot was now covered with a thin layer of very fine argillaceous sand, which was kept damp; and after the tapering seedlings had grown a few tenths of an inch in height, each was found surrounded by a little open space or circular crack; and this could be accounted for only by their having circumnutated and thus pushed away the sand on all sides; for there was no vestige of a crack in any other part.

In order to prove that there was circumnutation, the movements of five seedlings, varying in height from .3 inch to 2 inches, were traced. They were placed within a box and illuminated from above; but in all five cases the longer axes of the figures described were directed to nearly the same point; so that more light seemed to have come through the glass roof of the greenhouse on one side than on any other. All five tracings resembled each other to a certain extent, and it will suffice to give two of them. In A (Fig. 48) the seedling was only .45 of an inch in height, and consisted of a single internode bearing a bud on its summit. The apex described between 8.30 A.M. and 10.20 P.M. (i.e. during nearly 14 hours) a figure which would probably have consisted of 3½ ellipses, had not the stem been drawn to one side until 1 P.M., after which hour it moved backwards. On the following morning it was not far distant from the point whence it had first started. The actual amount of movement of the apex from side to side was very small, viz. about 1/18th of an inch. The seedling of which the movements are shown in Fig. 48, B, was 1 3/4 inch in height, and consisted of three internodes besides the bud on the summit. The figure, which was described during 10 h., apparently represents two irregular and unequal ellipses or circles. The actual amount of movement of the apex, in the line not influenced by the light, was .11 of an inch, and in that thus influenced .37 of an inch. With a seedling 2 inches in height it was obvious, even without the aid of any tracing, that the uppermost part of the stem bent successively to all points of the compass, like the stem of a twining plant. A little increase in the power of circumnutating and in the flexibility of the stem, would convert the common asparagus into a twining plant, as has occurred with one species in this genus, namely, A. scandens.

Fig. 48. Asparagus officinalis: circumnutation of plumules with tips whitened and marks placed beneath, traced on a horizontal glass. A, young plumule; movement traced from 8.30 A.M. Nov. 30th to 7.15 A.M. next morning; magnified about 35 times. B, older plumule; movement traced from 10.15 A.M. to 8.10 P.M. Nov. 29th; magnified 9 times, but here reduced to one-half of original scale.

Phalaris Canariensis (Gramineæ).—With the Gramineæ the part which first rises above the ground has been called by some authors the pileole; and various views have been expressed on its homological nature. It is considered by some great authorities to be a cotyledon, which term we will use without venturing to express any opinion on the subject.[6] It consists in the present case of a slightly flattened reddish sheath, terminating upwards in a sharp white edge; it encloses a true green leaf, which protrudes from the sheath through a slit-like orifice, close beneath and at right angles to the sharp edge on the summit. The sheath is not arched when it breaks through the ground.

[6] We are indebted to the Rev. G. Henslow for an abstract of the views which have been held on this subject, together with references.

The movements of three rather old seedlings, about 1½ inch in height, shortly before the protrusion of the leaves, were first traced. They were illuminated exclusively from above; for, as will hereafter be shown, they are excessively sensitive to the action of light; and if any enters even temporarily on one side, they merely bend to this side in slightly zigzag lines. Of the three tracings one alone (Fig. 49) is here given. Had the observations been more frequent during the 12 h. two oval figures would have been described with their longer axes at right angles to one another. The actual amount of movement of the apex from side to side was about .3 of an inch. The figures described by the other two seedlings resembled to a certain extent the one here given.

Fig. 49. Phalaris Canariensis: circumnutation of a cotyledon, with a mark placed below the apex, traced on a horizontal glass, from 8.35 A.M. Nov. 26th to 8.45 A.M. 27th. Movement of apex magnified 7 times, here reduced to one-half scale.

A seedling which had just broken through the ground and projected only 1/20th of an inch above the surface, was next observed in the same manner as before. It was necessary to clear away the earth all round the seedling to a little depth in order to place a mark beneath the apex. The figure (Fig. 50) shows that the apex moved to one side, but changed its course ten times in the course of the ten hours of observation; so that there can be no doubt about its circumnutation. The cause of the general movement in one direction could hardly be attributed to the entrance of lateral light, as this was carefully guarded against; and we suppose it was in some manner connected with the removal of the earth round the little seedling.

Fig. 50. Phalaris Canariensis: circumnutation of a very young cotyledon, with a mark placed below the apex, traced on a horizontal glass, from 11.37 A.M. to 9.30 P.M. Dec. 13th. Movement of apex greatly magnified, here reduced to one-fourth of original scale.

Lastly, the soil in the same pot was searched with the aid of a lens, and the white knife-like apex of a seedling was found on an exact level with that of the surrounding surface. The soil was removed all round the apex to the depth of a quarter of an inch, the seed itself remaining covered. The pot, protected from lateral light, was placed under the microscope with a micrometer eye-piece, so arranged that each division equalled 1/500th of an inch. After an interval of 30 m. the apex was observed, and it was seen to cross a little obliquely two divisions of the micrometer in 9 m. 15 s.; and after a few minutes it crossed the same space in 8 m. 50s. The seedling was again observed after an interval of three-quarters of an hour, and now the apex crossed rather obliquely two divisions in 10 m. We may therefore conclude that it was travelling at about the rate of 1/50th of an inch in 45 minutes. We may also conclude from these and the previous observations, that the seedlings of Phalaris in breaking through the surface of the soil circumnutate as much as the surrounding pressure will permit. This fact accounts (as in the case before given of the asparagus) for a circular, narrow, open space or crack being distinctly visible round several seedlings which had risen through very fine argillaceous sand, kept uniformly damp.

Fig. 51. Zea mays: circumnutation of cotyledon, traced on horizontal glass, from 8.30 A.M. Feb. 4th to 8 A.M. 6th. Movement of bead magnified on an average about 25 times.

Zea mays (Gramineæ).—A glass filament was fixed obliquely to the summit of a cotyledon, rising .2 of an inch above the ground; but by the third morning it had grown to exactly thrice this height, so that the distance of the bead from the mark below was greatly increased, consequently the tracing (Fig. 51) was much more magnified on the first than on the second day. The upper part of the cotyledon changed its course by at least as much as a rectangle six times on each of the two days. The plant was illuminated by an obscure light from vertically above. This was a necessary precaution, as on the previous day we had traced the movements of cotyledons placed in a deep box, the inner side of which was feebly illuminated on one side from a distant north-east window, and at each observation by a wax taper held for a minute or two on the same side; and the result was that the cotyledons travelled all day long to this side, though making in their course some conspicuous flexures, from which fact alone we might have concluded that they were circumnutating; but we thought it advisable to make the tracing above given.

Radicles.—Glass filaments were fixed to two short radicles, placed so as to stand almost upright, and whilst bending downwards through geotropism their courses were strongly zigzag; from this latter circumstance circumnutation might have been inferred, had not their tips become slightly withered after the first 24 h., though they were watered and the air kept very damp. Nine radicles were next arranged in the manner formerly described, so that in growing downwards they left tracks on smoked glass-plates, inclined at various angles between 45° and 80° beneath the horizon. Almost every one of these tracks offered evidence in their greater or less breadth in different parts, or in little bridges of soot being left, that the apex had come alternately into more and less close contact with the glass. In the accompanying figure (Fig. 52) we have an accurate copy of one such track. In two instances alone (and in these the plates were highly inclined) there was some evidence of slight lateral movement. We presume therefore that the friction of the apex on the smoked surface, little as this could have been, sufficed to check the movement from side to side of these delicate radicles.

Fig. 52. Zea mays: track left on inclined smoked glass-plate by tip of radicle in growing downwards.

Avena sativa (Gramineæ).—A cotyledon, 1½ inch in height, was placed in front of a north-east window, and the movement of the apex was traced on a horizontal glass during two days. It moved towards the light in a slightly zigzag line from 9 to 11.30 A.M. on October 15th; it then moved a little backwards and zigzagged much until 5 P.M., after which hour, and curing the night, it continued to move towards the window. On the following morning the same movement was continued in a nearly straight line until 12.40 P.M., when the sky remained until 2.35 extraordinarily dark from thunder-clouds. During this interval of 1 h. 55 m., whilst the light was obscure, it was interesting to observe how circumnutation overcame heliotropism, for the apex, instead of continuing to move towards the window in a slightly zigzag line, reversed its course four times, making two small narrow ellipses. A diagram of this case will be given in the chapter on Heliotropism.

A filament was next fixed to a cotyledon only 1/4 of an inch in height, which was illuminated exclusively from above, and as it was kept in a warm greenhouse, it grew rapidly; and now there could be no doubt about its circumnutation, for it described a figure of 8 as well as two small ellipses in 5½ hours.

Nephrodium molle (Filices).—A seedling fern of this species came up by chance in a flowerpot near its parent. The frond, as yet only slightly lobed, was only .16 of an inch in length and .2 in breadth, and was supported on a rachis as fine as a hair and .23 of an inch in height. A very thin glass filament, which projected for a length of .36 of an inch, was fixed to the end of the frond. The movement was so highly magnified that the figure (Fig. 53) cannot be fully trusted; but the frond was constantly moving in a complex manner, and the bead greatly changed its course eighteen times in the 12 hours of observation. Within half an hour it often returned in a line almost parallel to its former course. The greatest amount of movement occurred between 4 and 6 P.M. The circumnutation of this plant is interesting, because the species in the genus Lygodium are well known to circumnutate conspicuously and to twine round any neighbouring object.

Fig. 53. Nephrodium molle: circumnutation of very young frond, traced in darkness on horizontal glass, from 9 A.M. to 9 P.M. Oct. 30th. Movement of bead magnified 48 times.

Selaginella Kraussii (?) (Lycopodiaceæ).—A very young plant, only .4 of an inch in height, had sprung up in a pot in the hot-house. An extremely fine glass filament was fixed to the end of the frond-like stem, and the movement of the bead traced on a horizontal glass. It changed its course several times, as shown in Fig. 54, whilst observed during 13 h. 15 m., and returned at night to a point not far distant from that whence it had started in the morning. There can be no doubt that this little plant circumnutated.

Fig. 54. Selaginella Kraussii (?): circumnutation of young plant, kept in darkness, traced from 8.45 A.M. to 10 P.M. Oct. 31st.

CHAPTER II.
GENERAL CONSIDERATIONS ON THE MOVEMENTS AND GROWTH OF SEEDLING PLANTS.

Generality of the circumnutating movement—Radicles, their circumnutation of service—Manner in which they penetrate the ground—Manner in which hypocotyls and other organs break through the ground by being arched—Singular manner of germination in Megarrhiza, etc.—Abortion of cotyledons—Circumnutation of hypocotyls and epicotyls whilst still buried and arched—Their power of straightening themselves—Bursting of the seed-coats—Inherited effect of the arching process in hypogean hypocotyls—Circumnutation of hypocotyls and epicotyls when erect—Circumnutation of cotyledons—Pulvini or joints of cotyledons, duration of their activity, rudimentary in Oxalis corniculata, their development—Sensitiveness of cotyledons to light and consequent disturbance of their periodic movements—Sensitiveness of cotyledons to contact.

The circumnutating movements of the several parts or organs of a considerable number of seedling plants have been described in the last chapter. A list is here appended of the Families, Cohorts, Sub-classes, etc., to which they belong, arranged and numbered according to the classification adopted by Hooker.[1] Any one who will consider this list will see that the young plants selected for observation, fairly represent the whole vegetable series excepting the lowest cryptogams, and the movements of some of the latter when mature will hereafter be described. As all the seedlings which were observed, including Conifers, Cycads and Ferns, which belong to the most ancient types amongst plants, were continually circumnutating, we may infer that this kind of movement is common to every seedling species.

[1] As given in the 'General System of Botany,' by Le Maout and Decaisne, 1873.

SUB-KINGDOM I.—Phaenogamous Plants.

Class I.—DICOTYLEDONS.

Sub-class I.—Angiosperms. Family. Cohort. 14. Cruciferae. II. PARIETALES. 26. Caryophylleae. IV. CARYOPHYLLALES. 36. Malvaceae. VI MALVALES. 41. Oxalideae. VII. GERANIALES. 49. Tropaeoleae. DITTO 52. Aurantiaceae. DITTO 70. Hippocastaneae. X. SAPINDALES. 75. Leguminosae. XI. ROSALES. 106. Cucurbitaceæ. XII. PASSIFLORALES. 109. Cacteæ. XIV. FICOIDALES. 122. Compositæ. XVII. ASTRALES. 135. Primulaceae. XX. PRIMULALES. 145. Asclepiadeae. XXII. GENTIANALES. 151. Convolvulaceae. XXIII. POLEMONIALES. 154. Boragineae. DITTO 156. Nolaneae. DITTO 157. Solaneae. XXIV. SOLANALES. 181. Chenopodieae. XXVII. CHENOPODIALES. 202. Euphorbiaceae. XXXII. EUPHORBIALES. 211. Cupuliferae. XXXVI. QUERNALES. 212. Corylaceae. DITTO

Sub-class II.—Gymnosperms. 223. Coniferæ. 224. Cycadeæ.

Class II.—MONOCOTYLEDONS. 2. Cannaceae. II. AMOMALES. 34. Liliaceae. XI. LILIALES. 41. Asparageae. DITTO 55. Gramineæ. XV. GLUMALES.

SUB-KINGDOM II.—Cryptogamic Plants.

1. Filices. I. FILICALES. 6. Lycopodiaceæ. DITTO

Radicles.—In all the germinating seeds observed by us, the first change is the protrusion of the radicle, which immediately bends downwards and endeavours to penetrate the ground. In order to effect this, it is almost necessary that the seed should be pressed down so as to offer some resistance, unless indeed the soil is extremely loose; for otherwise the seed is lifted up, instead of the radicle penetrating the surface. But seeds often get covered by earth thrown up by burrowing quadrupeds or scratching birds, by the castings of earth-worms, by heaps of excrement, the decaying branches of trees, etc., and will thus be pressed down; and they must often fall into cracks when the ground is dry, or into holes. Even with seeds lying on the bare surface, the first developed root-hairs, by becoming attached to stones or other objects on the surface, are able to hold down the upper part of the radicle, whilst the tip penetrates the ground. Sachs has shown[2] how well and closely root-hairs adapt themselves by growth to the most irregular particles in the soil, and become firmly attached to them. This attachment seems to be effected by the softening or liquefaction of the outer surface of the wall of the hair and its subsequent consolidation, as will be on some future occasion more fully described. This intimate union plays an important part, according to Sachs, in the absorption of water and of the inorganic matter dissolved in it. The mechanical aid afforded by the root-hairs in penetrating the ground is probably only a secondary service.

[2] 'Physiologie Végétale,' 1868, pp. 199, 205.

The tip of the radicle, as soon as it protrudes from the seed-coats, begins to circumnutate, and the whole growing part continues to do so, probably for as long as growth continues. This movement of the radicle has been described in Brassica, Æsculus, Phaseolus, Vicia, Cucurbita, Quercus and Zea. The probability of its occurrence was inferred by Sachs,[3] from radicles placed vertically upwards being acted on by geotropism (which we likewise found to be the case), for if they had remained absolutely perpendicular, the attraction of gravity could not have caused them to bend to any one side. Circumnutation was observed in the above specified cases, either by means of extremely fine filaments of glass affixed to the radicles in the manner previously described, or by their being allowed to grow downwards over inclined smoked glass-plates, on which they left their tracks. In the latter cases the serpentine course (see Figs. 19, 21, 27, 41) showed unequivocally that the apex had continually moved from side to side. This lateral movement was small in extent, being in the case of Phaseolus at most about 1 mm. from a medial line to both sides. But there was also movement in a vertical plane at right angles to the inclined glass-plates. This was shown by the tracks often being alternately a little broader and narrower, due to the radicles having alternately pressed with greater and less force on the plates. Occasionally little bridges of soot were left across the tracks, showing that the apex had at these spots been lifted up. This latter fact was especially apt to occur xwhen the radicle instead of travelling straight down the glass made a semicircular bend; but Fig. 52 shows that this may occur when the track is rectilinear. The apex by thus rising, was in one instance able to surmount a bristle cemented across an inclined glass-plate; but slips of wood only 1/40 of an inch in thickness always caused the radicles to bend rectangularly to one side, so that the apex did not rise to this small height in opposition to geotropism.

[3] 'Ueber das Wachsthum der Wurzeln: Arbeiten des bot. Instituts in Würzburg,' Heft iii. 1873, p. 460. This memoir, besides its intrinsic and great interest, deserves to be studied as a model of careful investigation, and we shall have occasion to refer to it repeatedly. Dr. Frank had previously remarked ('Beiträge zur Pflanzenphysiologie, 1868, p. 81) on the fact of radicles placed vertically upwards being acted on by geotropism, and he explained it by the supposition that their growth was not equal on all sides.

In those cases in which radicles with attached filaments were placed so as to stand up almost vertically, they curved downwards through the action of geotropism, circumnutating at the same time, and their courses were consequently zigzag. Sometimes, however, they made great circular sweeps, the lines being likewise zigzag.

Radicles closely surrounded by earth, even when this is thoroughly soaked and softened, may perhaps be quite prevented from circumnutating. Yet we should remember that the circumnutating sheath-like cotyledons of Phalaris, the hypocotyls of Solanum, and the epicotyls of Asparagus formed round themselves little circular cracks or furrows in a superficial layer of damp argillaceous sand. They were also able, as well as the hypocotyls of Brassica, to form straight furrows in damp sand, whilst circumnutating and bending towards a lateral light. In a future chapter it will be shown that the rocking or circumnutating movement of the flower-heads of Trifolium subterraneum aids them in burying themselves. It is therefore probable that the circumnutation of the tip of the radicle aids it slightly in penetrating the ground; and it may be observed in several of the previously given diagrams, that the movement is more strongly pronounced in radicles when they first protrude from the seed than at a rather later period; but whether this is an accidental or an adaptive coincidence we do not pretend to decide. Nevertheless, when young radicles of *Phaseolus multiflorus* were fixed vertically close over damp sand, in the expectation that as soon as they reached it they would form circular furrows, this did not occur,—a fact which may be accounted for, as we believe, by the furrow being filled up as soon as formed by the rapid increase of thickness in the apex of the radicle. Whether or not a radicle, when surrounded by softened earth, is aided in forming a passage for itself by circumnutating, this movement can hardly fail to be of high importance, by guiding the radicle along a line of least resistance, as will be seen in the next chapter when we treat of the sensibility of the tip to contact. If, however, a radicle in its downward growth breaks obliquely into any crevice, or a hole left by a decayed root, or one made by the larva of an insect, and more especially by worms, the circumnutating movement of the tip will materially aid it in following such open passage; and we have observed that roots commonly run down the old burrows of worms.[4]

[4] See, also, Prof. Hensen's statements ('Zeitschrift für Wissen, Zool.,' B. xxviii. p. 354, 1877) to the same effect. He goes so far as to believe that roots are able to penetrate the ground to a great depth only by means of the burrows made by worms.

When a radicle is placed in a horizontal or inclined position, the terminal growing part, as is well known, bends down towards the centre of the earth; and Sachs[5] has shown that whilst thus bending, the growth of the lower surface is greatly retarded, whilst that of the upper surface continues at the normal rate, or may be even somewhat increased. He has further shown by attaching a thread, running over a pulley, to a horizontal radicle of large size, namely that of the common bean, that it was able to pull up a weight of only one gramme, or 15.4 grains. We may therefore conclude that geotropism does not give a radicle force sufficient to penetrate the ground, but merely tells it (if such an expression may be used) which course to pursue. Before we knew of Sachs' more precise observations we covered a flat surface of damp sand with the thinnest tin-foil which we could procure (.02 to .03 mm., or .00012 to .00079 of an inch in thickness), and placed a radicle close above, in such a position that it grew almost perpendicularly downwards. When the apex came into contact with the polished level surface it turned at right angles and glided over it without leaving any impression; yet the tin-foil was so flexible, that a little stick of soft wood, pointed to the same degree as the end of the radicle and gently loaded with a weight of only a quarter of an ounce (120 grains) plainly indented the tin-foil.

[5] 'Arbeiten des bot. Inst. Würzburg,' vol. i. 1873, p. 461. See also p. 397 for the length of the growing part, and p. 451 on the force of geotropism.

Radicles are able to penetrate the ground by the force due to their longitudinal and transverse growth; the seeds themselves being held down by the weight of the superincumbent soil. In the case of the bean the apex, protected by the root-cap, is sharp, and the growing part, from 8 to 10 mm. in length, is much more rigid, as Sachs has proved, than the part immediately above, which has ceased to increase in length. We endeavoured to ascertain the downward pressure of the growing part, by placing germinating beans between two small metal plates, the upper one of which was loaded with a known weight; and the radicle was then allowed to grow into a narrow hole in wood, 2 or 3 tenths of an inch in depth, and closed at the bottom. The wood was so cut that the short space of radicle between the mouth of the hole and the bean could not bend laterally on three sides; but it was impossible to protect the fourth side, close to the bean. Consequently, as long as the radicle continued to increase in length and remained straight, the weighted bean would be lifted up after the tip had reached the bottom of the shallow hole. Beans thus arranged, surrounded by damp sand, lifted up a quarter of a pound in 24 h. after the tip of the radicle had entered the hole. With a greater weight the radicles themselves always became bent on the one unguarded side; but this probably would not have occurred if they had been closely surrounded on all sides by compact earth. There was, however, a possible, but not probable, source of error in these trials, for it was not ascertained whether the beans themselves go on swelling for several days after they have germinated, and after having been treated in the manner in which ours had been; namely, being first left for 24 h. in water, then allowed to germinate in very damp air, afterwards placed over the hole and almost surrounded by damp sand in a closed box.

Fig. 55. Outline of piece of stick (reduced to one-half natural size) with a hole through which the radicle of a bean grew. Thickness of stick at narrow end .08 inch, at broad end .16; depth of hole .1 inch.

We succeeded better in ascertaining the force exerted transversely by these radicles. Two were so placed as to penetrate small holes made in little sticks, one of which was cut into the shape here exactly copied (Fig. 55). The short end of the stick beyond the hole was purposely split, but not the opposite end. As the wood was highly elastic, the split or fissure closed immediately after being made. After six days the stick and bean were dug out of the damp sand, and the radicle was found to be much enlarged above and beneath the hole. The fissure which was at first quite closed, was now open to a width of 4 mm.; as soon as the radicle was extracted, it immediately closed to a width of 2 mm. The stick was then suspended horizontally by a fine wire passing through the hole lately filled by the radicle, and a little saucer was suspended beneath to receive the weights; and it required 8 lbs. 8 ozs. to open the fissure to the width of 4 mm.—that is, the width before the root was extracted. But the part of the radicle (only .1 of an inch in length) which was embedded in the hole, probably exerted a greater transverse strain even than 8 lbs. 8 ozs., for it had split the solid wood for a length of rather more than a quarter of an inch (exactly .275 inch), and this fissure is shown in Fig. 55. A second stick was tried in the same manner with almost exactly the same result.

Fig. 56. Wooden pincers, kept closed by a spiral brass spring, with a hole (.14 inch in diameter and .6 inch in depth) bored through the narrow closed part, through which a radicle of a bean was allowed to grow. Temp. 50°–60° F.

We then followed a better plan. Holes were bored near the narrow end of two wooden clips or pincers (Fig. 56), kept closed by brass spiral springs. Two radicles in damp sand were allowed to grow through these holes. The pincers rested on glass-plates to lessen the friction from the sand. The holes were a little larger (viz..14 inch) and considerably deeper (viz..6 inch) than in the trials with the sticks; so that a greater length of a rather thicker radicle exerted a transverse strain. After 13 days they were taken up. The distance of two dots (see the figure) on the longer ends of the pincers was now carefully measured; the radicles were then extracted from the holes, and the pincers of course closed. They were then suspended horizontally in the same manner as were the bits of sticks, and a weight of 1500 grams (or 3 pounds 4 ounces) was necessary with one of the pincers to open them to the same extent as had been effected by the transverse growth of the radicle. As soon as this radicle had slightly opened the pincers, it had grown into a flattened form and had escaped a little beyond the hole; its diameter in one direction being 4.2 mm., and at rightangles 3.5 mm. If this escape and flattening could have been prevented, the radicle would probably have exerted a greater strain than the 3 pounds 4 ounces. With the other pincers the radicle escaped still further out of the hole; and the weight required to open them to the same extent as had been effected by the radicle, was only 600 grams.

With these facts before us, there seems little difficulty in understanding how a radicle penetrates the ground. The apex is pointed and is protected by the root-cap; the terminal growing part is rigid, and increases in length with a force equal, as far as our observations can be trusted, to the pressure of at least a quarter of a pound, probably with a much greater force when prevented from bending to any side by the surrounding earth. Whilst thus increasing in length it increases in thickness, pushing away the damp earth on all sides, with a force of above 8 pounds in one case, of 3 pounds in another case. It was impossible to decide whether the actual apex exerts, relatively to its diameter, the same transverse strain as the parts a little higher up; but there seems no reason to doubt that this would be the case. The growing part therefore does not act like a nail when hammered into a board, but more like a wedge of wood, which whilst slowly driven into a crevice continually expands at the same time by the absorption of water; and a wedge thus acting will split even a mass of rock.

Manner in which Hypocotyls, Epicotyls, etc., rise up and break through the ground.—After the radicle has penetrated the ground and fixed the seed, the hypocotyls of all the dicotyledonous seedlings observed by us, which lift their cotyledons above the surface, break through the ground in the form of an arch. When the cotyledons are hypogean, that is, remain buried in the soil, the hypocotyl is hardly developed, and the epicotyl or plumule rises in like manner as an arch through the ground. In all, or at least in most of such cases, the downwardly bent apex remains for a time enclosed within the seed-coats. With Corylus avellena the cotyledons are hypogean, and the epicotyl is arched; but in the particular case described in the last chapter its apex had been injured, and it grew laterally through the soil like a root; and in consequence of this it had emitted two secondary shoots, which likewise broke through the ground as arches.

Cyclamen does not produce any distinct stem, and only a single cotyledon appears at first;[6] its petiole breaks through the ground as an arch (Fig. 57). Abronia has only a single fully developed cotyledon, but in this case it is the hypocotyl which first emerges and is arched. Abronia umbellata, however, presents this peculiarity, that the enfolded blade of the one developed cotyledon (with the enclosed endosperm) whilst still beneath the surface has its apex upturned and parallel to the descending leg of the arched hypocotyl; but it is dragged out of the ground by the continued growth of the hypocotyl, with the apex pointing downward. With Cycas pectinata the cotyledons are hypogean, and a true leaf first breaks through the ground with its petiole forming an arch.

[6] This is the conclusion arrived at by Dr. H. Gressner ('Bot. Zeitung,' 1874, p. 837), who maintains that what has been considered by other botanists as the first true leaf is really the second cotyledon, which is greatly delayed in its development.

Fig. 57. Cyclamen Persicum: seedling, figure enlarged: c, blade of cotyledon, not yet expanded, with arched petiole beginning to straighten itself; h, hypocotyl developed into a corm; r, secondary radicles.

Fig. 58. Acanthus mollis: seedling with the hypogean cotyledon on the near side removed and the radicles cut off; a, blade of first leaf beginning to expand, with petiole still partially arched; b, second and opposite leaf, as yet very imperfectly developed; c, hypogean cotyledon on the opposite side.

In the genus Acanthus the cotyledons are likewise hypogean. In A. mollis, a single leaf first breaks through the ground with its petiole arched, and with the opposite leaf much less developed, short, straight, of a yellowish colour, and with the petiole at first not half as thick as that of the other. The undeveloped leaf is protected by standing beneath its arched fellow; and it is an instrucive fact that it is not arched, as it has not to force for itself a passage through the ground. In the accompanying sketch (Fig. 58) the petiole of the first leaf has already partially straightened itself, and the blade is beginning to unfold. The small second leaf ultimately grows to an equal size with the first, but this process is effected at very different rates in different individuals: in one instance the second leaf did not appear fully above the ground until six weeks after the first leaf. As the leaves in the whole family of the Acanthaceae stand either opposite one another or in whorls, and as these are of equal size, the great inequality between the first two leaves is a singular fact. We can see how this inequality of development and the arching of the petiole could have been gradually acquired, if they were beneficial to the seedlings by favouring their emergence; for with A. candelabrum, spinosus, and latifolius there was a great variability in the inequality between the two first leaves and in the arching of their petioles. In one seedling of A. candelabrum the first leaf was arched and nine times as long as the second, which latter consisted of a mere little, yellowish-white, straight, hairy style. In other seedlings the difference in length between the two leaves was as 3 to 2, or as 4 to 3, or as only .76 to .62 inch. In these latter cases the first and taller leaf was not properly arched. Lastly, in another seedling there was not the least difference in size between the two first leaves, and both of them had their petioles straight; their laminae were enfolded and pressed against each other, forming a lance or wedge, by which means they had broken through the ground. Therefore in different individuals of this same species of Acanthus the first pair of leaves breaks through the ground by two widely different methods; and if either had proved decidedly advantageous or disadvantageous, one of them no doubt would soon have prevailed.

Asa Gray has described[7] the peculiar manner of germination of three widely different plants, in which the hypocotyl is hardly at all developed. These were therefore observed by us in relation to our present subject.

[7] 'Botanical Text-Book,' 1879, p. 22.

Delphinium nudicaule.—The elongated petioles of the two cotyledons are confluent (as are sometimes their blades at the base), and they break through the ground as an arch. They thus resemble in a most deceptive manner a hypocotyl. At first they are solid, but after a time become tubular; and the basal part beneath the ground is enlarged into a hollow chamber, within which the young leaves are developed without any prominent plumule. Externally root-hairs are formed on the confluent petioles, either a little above, or on a level with, the plumule. The first leaf at an early period of its growth and whilst within the chamber is quite straight, but the petiole soon becomes arched; and the swelling of this part (and probably of the blade) splits open one side of the chamber, and the leaf then emerges. The slit was found in one case to be 3.2 mm. in length, and it is seated on the line of confluence of the two petioles. The leaf when it first escapes from the chamber is buried beneath the ground, and now an upper part of the petiole near the blade becomes arched in the usual manner. The second leaf comes out of the slit either straight or somewhat arched, but afterwards the upper part of the petiole,—certainly in some, and we believe in all cases,—arches itself whilst forcing a passage through the soil.

Megarrhiza Californica.—The cotyledons of this Gourd never free themselves from the seed-coats and are hypogean. Their petioles are completely confluent, forming a tube which terminates downwards in a little solid point, consisting of a minute radicle and hypocotyl, with the likewise minute plumule enclosed within the base of the tube. This structure was well exhibited in an abnormal specimen, in which one of the two cotyledons failed to produce a petiole, whilst the other produced one consisting of an open semicylinder ending in a sharp point, formed of the parts just described. As soon as the confluent petioles protrude from the seed they bend down, as they are strongly geotropic, and penetrate the ground. The seed itself retains its original position, either on the surface or buried at some depth, as the case may be. If, however, the point of the confluent petioles meets with some obstacle in the soil, as appears to have occurred with the seedlings described and figured by Asa Gray,[8] the cotyledons are lifted up above the ground. The petioles are clothed with root-hairs like those on a true radicle, and they likewise resemble radicles in becoming brown when immersed in a solution of permanganate of potassium. Our seeds were subjected to a high temperature, and in the course of three or four days the petioles penetrated the soil perpendicularly to a depth of from 2 to 2½ inches; and not until then did the true radicle begin to grow. In one specimen which was closely observed, the petioles in 7 days after their first protrusion attained a length of 2½ inches, and the radicle by this time had also become well developed. The plumule, still enclosed within the tube, was now .3 inch in length, and was quite straight; but from having increased in thickness it had just begun to split open the lower part of the petioles on one side, along the line of their confluence. By the following morning the upper part of the plumule had arched itself into a right angle, and the convex side or elbow had thus been forced out through the slit. Here then the arching of the plumule plays the same part as in the case of the petioles of the Delphinium. As the plumule continued to grow, the tip became more arched, and in the course of six days it emerged through the 2½ inches of superincumbent soil, still retaining its arched form. After reaching the surface it straightened itself in the usual manner. In the accompanying figure (Fig. 58, A) we have a sketch of a seedling in this advanced state of development; the surface of the ground being represented by the line G...........G.

[8] 'American Journal of Science,' vol. xiv. 1877, p. 21.

Fig. 58, A. Megarrhiza Californica: sketch of seedling, copied from Asa Gray, reduced to one-half scale: c, cotyledons within seed-coats; p, the two confluent petioles; h and r, hypocotyl and radicle; p1, plumule; G..........G, surface of soil.

The germination of the seeds in their native Californian home proceeds in a rather different manner, as we infer from an interesting letter from Mr. Rattan, sent to us by Prof. Asa Gray. The petioles protrude from the seeds soon after the autumnal rains, and penetrate the ground, generally in a vertical direction, to a depth of from 4 to even 6 inches. they were found in this state by Mr. Rattan during the Christmas vacation, with the plumules still enclosed within the tubes; and he remarks that if the plumules had been at once developed and had reached the surface (as occurred with our seeds which were exposed to a high temperature), they would surely have been killed by the frost. As it is, they lie dormant at some depth beneath the surface, and are thus protected from the cold; and the root-hairs on the petioles would supply them with sufficient moisture. We shall hereafter see that many seedlings are protected from frost, but by a widely different process, namely, by being drawn beneath the surface by the contraction of their radicles. We may, however, believe that the extraordinary manner of germination of Megarrhiza has another and secondary advantage. The radicle begins in a few weeks to enlarge into a little tuber, which then abounds with starch and is only slightly bitter. It would therefore be very liable to be devoured by animals, were it not protected by being buried whilst young and tender, at a depth of some inches beneath the surface. Ultimately it grows to a huge size.

Ipomœa leptophylla.—In most of the species of this genus the hypocotyl is well developed, and breaks through the ground as an arch. But the seeds of the present species in germinating behave like those of Megarrhiza, excepting that the elongated petioles of the cotyledons are not confluent. After they have protruded from the seed, they are united at their lower ends with the undeveloped hypocotyl and undeveloped radicle, which together form a point only about .1 inch in length. They are at first highly geotropic, and penetrate the ground to a depth of rather above half an inch. The radicle then begins to grow. On four occasions after the petioles had grown for a short distance vertically downwards, they were placed in a horizontal position in damp air in the dark, and in the course of 4 hours they again became curved vertically downwards, having passed through 90° in this time. But their sensitiveness to geotropism lasts for only 2 or 3 days; and the terminal part alone, for a length of between .2 and .4 inch, is thus sensitive. Although the petioles of our specimens did not penetrate the ground to a greater depth than about ½ inch, yet they continued for some time to grow rapidly, and finally attained the great length of about 3 inches. The upper part is apogeotropic, and therefore grows vertically upwards, excepting a short portion close to the blades, which at an early period bends downwards and becomes arched, and thus breaks through the ground. Afterwards this portion straightens itself, and the cotyledons then free themselves from the seed-coats. Thus we here have in different parts of the same organ widely different kinds of movement and of sensitiveness; for the basal part is geotropic, the upper part apogeotropic, and a portion near the blades temporarily and spontaneously arches itself. The plumule is not developed for some little time; and as it rises between the bases of the parallel and closely approximate petioles of the cotyledons, which in breaking through the ground have formed an almost open passage, it does not require to be arched and is consequently always straight. Whether the plumule remains buried and dormant for a time in its native country, and is thus protected from the cold of winter, we do not know. The radicle, like that of the Megarrhiza, grows into a tuber-like mass, which ultimately attains a great size. So it is with Ipomœa pandurata, the germination of which, as Asa Gray informs us, resembles that of I. leptophylla.

The following case is interesting in connection with the root-like nature of the petioles. The radicle of a seedling was cut off, as it was completely decayed, and the two now separated cotyledons were planted. They emitted roots from their bases, and continued green and healthy for two months. The blades of both then withered, and on removing the earth the bases of the petioles (instead of the radicle) were found enlarged into little tubers. Whether these would have had the power of producing two independent plants in the following summer, we do not know.

In Quercus virens, according to Dr. Engelmann,[9] both the cotyledons and their petioles are confluent. The latter grow to a length "of an inch or even more;" and, if we understand rightly, penetrate the ground, so that they must be geotropic. The nutriment within the cotyledons is then quickly transferred to the hypocotyl or radicle, which thus becomes developed into a fusiform tuber. The fact of tubers being formed by the foregoing three widely distinct plants, makes us believe that their protection from animals at an early age and whilst tender, is one at least of the advantages gained by the remarkable elongation of the petioles of the cotyledons, together with their power of penetrating the ground like roots under the guidance of geotropism.

[9] 'Transact. St. Louis Acad. Science,' vol. iv. p. 190.

The following cases may be here given, as they bear on our present subject, though not relating to seedlings. The flower-stem of the parasitic Lathraea squamaria, which is destitute of true leaves, breaks through the ground as an arch;[10] so does the flower-stem of the parasitic and leafless Monotropa hypopitys. With Helleborus niger, the flower-stems, which rise up independently of the leaves, likewise break through the ground as arches. This is also the case with the greatly elongated flower-stems, as well as with the petioles of Epimedium pinnatum. So it is with the petioles of Ranunculus ficaria, when they have to break through the ground, but when they arise from the summit of the bulb above ground, they are from the first quite straight; and this is a fact which deserves notice. The rachis of the bracken fern (Pteris aquilina), and of some, probably many, other ferns, likewise rises above ground under the form of an arch. No doubt other analogous instances could be found by careful search. In all ordinary cases of bulbs, rhizomes, root-stocks, etc., buried beneath the ground, the surface is broken by a cone formed by the young imbricated leaves, the combined growth of which gives them force sufficient for the purpose.

[10] The passage of the flower-stem of the Lathraea through the ground cannot fail to be greatly facilitated by the extraordinary quantity of water secreted at this period of the year by the subterranean scale-like leaves; not that there is any reason to suppose that the secretion is a special adaptation for this purpose: it probably follows from the great quantity of sap absorbed in the early spring by the parasitic roots. After a long period without any rain, the earth had become light-coloured and very dry, but it

was dark-coloured and damp, even in parts quite wet, for a distance of at least six inches all round each flower-stem. The water is secreted by glands (described by Cohn, 'Bericht. Bot. Sect. der Schlesischen Gesell.,' 1876, p. 113) which line the longitudinal channels running through each scale-like leaf. A large plant was dug up, washed so as to remove the earth, left for some time to drain, and then placed in the evening on a dry glass-plate, covered with a bell-glass, and by next morning it had secreted a large pool of water. The plate was wiped dry, and in the course of the succeeding 7 or 8 hours another little pool was secreted, and after 16 additional hours several large drops. A smaller plant was washed and placed in a large jar, which was left inclined for an hour, by which time no more water drained off. The jar was then placed upright and closed: after 23 hours two drachms of water were collected from the bottom, and a little more after 25 additional hours. The flower-stems were now cut off, for they do not secrete, and the subterranean part of the plant was found to weigh 106.8 grams (1611 grains), and the water secreted during the 48 hours weighed 11.9 grams (183 grains),—that is, one-ninth of the whole weight of the plant, excluding the flower-stems. We should remember that plants in a state of nature would probably secrete in 48 hours much more than the above large amount, for their roots would continue all the time absorbing sap from the plant on which they were parasitic.

With germinating monocotyledonous seeds, of which, however, we did not observe a large number, the plumules, for instance, those of Asparagus and Canna, are straight whilst breaking through the ground. With the Gramineæ, the sheath-like cotyledons are likewise straight; they, however, terminate in a sharp crest, which is white and somewhat indurated; and this structure obviously facilitates their emergence from the soil: the first true leaves escape from the sheath through a slit beneath the chisel-like apex and at right angles to it. In the case of the onion (Allium cepa) we again meet with an arch; the leaf-like cotyledon being abruptly bowed, when it breaks through the ground, with the apex still enclosed within the seed-coats. The crown of the arch, as previously described, is developed into a white conical protuberance, which we may safely believe to be a special adaptation for this office.

The fact of so many organs of different kinds—hypocotyls and epicotyls, the petioles of some cotyledons and of some first leaves, the cotyledons of the onion, the rachis of some ferns, and some flower-stems—being all arched whilst they break through the ground, shows how just are Dr. Haberlandt's[11] remarks on the importance of the arch to seedling plants. He attributes its chief importance to the upper, young, and more tender parts of the hypocotyl or epicotyl, being thus saved from abrasion and pressure whilst breaking through the ground. But we think that some importance may be attributed to the increased force gained by the hypocotyl, epicotyl, or other organ by being at first arched; for both legs of the arch increase in length, and both have points of resistance as long as the tip remains enclosed within the seed-coats; and thus the crown of the arch is pushed up through the earth with twice as much force as that which a straight hypocotyl, etc., could exert. As soon, however, as the upper end has freed itself, all the work has to be done by the basal leg. In the case of the epicotyl of the common bean, the basal leg (the apex having freed itself from the seed-coats) grew upwards with a force sufficient to lift a thin plate of zinc, loaded with 12 ounces. Two more ounces were added, and the 14 ounces were lifted up to a very little height, and then the epicotyl yielded and bent to one side.

[11] 'Die Schutzeinrichtungen in der Entwickelung der Keimpflanze,' 1877. We have learned much from this interesting essay, though our observations lead us to differ on some points from the author.

With respect to the primary cause of the arching process, we long thought in the case of many seedlings that this might be attributed to the manner in which the hypocotyl or epicotyl was packed and curved within the seed-coats; and that the arched shape thus acquired was merely retained until the parts in question reached the surface of the ground. But it is doubtful whether this is the whole of the truth in any case. For instance, with the common bean, the epicotyl or plumule is bowed into an arch whilst breaking through the seed-coats, as shown in Fig. 59 (p. 92). The plumule first protrudes as a solid knob (e in A), which after twenty-four hours' growth is seen (e in B) to be the crown of an arch. Nevertheless, with several beans which germinated in damp air, and had otherwise been treated in an unnatural manner, little plumules were developed in the axils of the petioles of both cotyledons, and these were as perfectly arched as the normal plumule; yet they had not been subjected to any confinement or pressure, for the seed-coats were completely ruptured, and they grew in the open air. This proves that the plumule has an innate or spontaneous tendency to arch itself.

In some other cases the hypocotyl or epicotyl protrudes from the seed at first only slightly bowed; but the bowing afterwards increases independently of any constraint. The arch is thus made narrow, with the two legs, which are sometimes much elongated, parallel and close together, and thus it becomes well fitted for breaking through the ground.

With many kinds of plants, the radicle, whilst still enclosed within the seed and likewise after its first protrusion, lies in a straight line with the future hypocotyl and with the longitudinal axis of the cotyledons. This is the case with Cucurbita ovifera: nevertheless, in whatever position the seeds were buried, the hypocotyl always came up arched in one particular direction. Seeds were planted in friable peat at a depth of about an inch in a vertical position, with the end from which the radicle protrudes downwards. Therefore all the parts occupied the same relative positions which they would ultimately hold after the seedlings had risen clear above the surface. Notwithstanding this fact, the hypocotyl arched itself; and as the arch grew upwards through the peat, the buried seeds were turned either upside down, or were laid horizontally, being afterwards dragged above the ground. Ultimately the hypocotyl straightened itself in the usual manner; and now after all these movements the several parts occupied the same position relatively to one another and to the centre of the earth, which they had done when the seeds were first buried. But it may be argued in this and other such cases that, as the hypocotyl grows up through the soil, the seed will almost certainly be tilted to one side; and then from the resistance which it must offer during its further elevation, the upper part of the hypocotyl will be doubled down and thus become arched. This view seems the more probable, because with Ranunculus ficaria only the petioles of the leaves which forced a passage through the earth were arched; and not those which arose from the summits of the bulbs above the ground. Nevertheless, this explanation does not apply to the Cucurbita, for when germinating seeds were suspended in damp air in various positions by pins passing through the cotyledons, fixed to the inside of the lids of jars, in which case the hypocotyls were not subjected to any friction or constraint, yet the upper part became spontaneously arched. This fact, moreover, proves that it is not the weight of the cotyledons which causes the arching. Seeds of Helianthus annuus and of two species of Ipomœa (those of 'I. bona nox' being for the genus large and heavy) were pinned in the same manner, and the hypocotyls became spontaneously arched; the radicles, which had been vertically dependent, assumed in consequence a horizontal position. In the case of Ipomœa leptophylla it is the petioles of the cotyledons which become arched whilst rising through the ground; and this occurred spontaneously when the seeds were fixed to the lids of jars.

It may, however, be suggested with some degree of probability that the arching was aboriginally caused by mechanical compulsion, owing to the confinement of the parts in question within the seed-coats, or to friction whilst they were being dragged upwards. But if this is so, we must admit from the cases just given, that a tendency in the upper part of the several specified organs to bend downwards and thus to become arched, has now become with many plants firmly inherited. The arching, to whatever cause it may be due, is the result of modified circumnutation, through increased growth along the convex side of the part; such growth being only temporary, for the part always straightens itself subsequently by increased growth along the concave side, as will hereafter be described.

It is a curious fact that the hypocotyls of some plants, which are but little developed and which never raise their cotyledons above the ground, nevertheless inherit a slight tendency to arch themselves, although this movement is not of the least use to them. We refer to a movement observed by Sachs in the hypocotyls of the bean and some other Leguminosae, and which is shown in the accompanying figure (Fig. 59), copied from his Essay.[12] The hypocotyl and radicle at first grow perpendicularly downwards, as at A, and then bend, often in the course of 24 hours, into the position shown at B. As we shall hereafter often have to recur to this movement, we will, for brevity sake, call it "Sachs' curvature." At first sight it might be thought that the altered position of the radicle in B was wholly due to the outgrowth of the epicotyl (e), the petiole (p) serving as a hinge; and it is probable that this is partly the cause; but the hypocotyl and upper part of the radicle themselves become slightly curved.

[12] 'Arbeiten des bot. Instit. Würzburg,' vol. i. 1873, p. 403.

The above movement in the bean was repeatedly seen by us; but our observations were made chiefly on Phaseolus multiflorus, the cotyledons of which are likewise hypogean. Some seedlings with well-developed radicles were first immersed in a solution of permanganate of potassium; and, judging from the changes of colour (though these were not very clearly defined), the hypocotyl is about .3 inch in length. Straight, thin, black lines of this length were now drawn from the bases of the short petioles along the hypocotyls of 23 germinating seeds, which were pinned to the lids of jars, generally with the hilum downwards, and with their radicles pointing to the centre of the earth. After an interval of from 24 to 48 hours the black lines on the hypocotyls of 16 out of the 23 seedlings became distinctly curved, but in very various degrees (namely, with radii between 20 and 80 mm. on Sachs' cyclometer) in the same relative direction as shown at B in Fig. 59. As geotropism will obviously tend to check this curvature, seven seeds were allowed to germinate with proper precautions for their growth in a klinostat,[13] by which means geotropism was eliminated. The position of the hypocotyls was observed during four successive days, and they continued to bend towards the hilum and lower surface of the seed. On the fourth day they were deflected by an average angle of 63° from a line perpendicular to the lower surface, and were therefore considerably more curved than the hypocotyl and radicle in the bean at B (Fig. 59), though in the same relative direction.

[13] An instrument devised by Sachs, consisting essentially of a slowly revolving horizontal axis, on which the plant under observation is supported: see 'Würzburg Arbeiten,' 1879, p. 209.

Fig. 59. Vicia faba: germinating seeds, suspended in damp air: A, with radicle growing perpendicularly downwards; B, the same bean after 24 hours and after the radicle has curved itself; r. radicle; h, short hypocotyl; e, epicotyl appearing as a knob in A and as an arch in B; p, petiole of the cotyledon, the latter enclosed within the seed-coats.

It will, we presume, be admitted that all leguminous plants with hypogean cotyledons are descended from forms which once raised their cotyledons above the ground in the ordinary manner; and in doing so, it is certain that their hypocotyls would have been abruptly arched, as in the case of every other dicotyledonous plant. This is especially clear in the case of Phaseolus, for out of five species, the seedlings of which we observed, namely, P. multiflorus, caracalla, vulgaris, Hernandesii and Roxburghii (inhabitants of the Old and New Worlds), the three last-named species have well-developed hypocotyls which break through the ground as arches. Now, if we imagine a seedling of the common bean or of P. multiflorus, to behave as its progenitors once did, the hypocotyl (h, Fig. 59), in whatever position the seed may have been buried, would become so much arched that the upper part would be doubled down parallel to the lower part; and this is exactly the kind of curvature which actually occurs in these two plants, though to a much less degree. Therefore we can hardly doubt that their short hypocotyls have retained by inheritance a tendency to curve themselves in the same manner as they did at a former period, when this movement was highly important to them for breaking through the ground, though now rendered useless by the cotyledons being hypogean. Rudimentary structures are in most cases highly variable, and we might expect that rudimentary or obsolete actions would be equally so; and Sachs' curvature varies extremely in amount, and sometimes altogether fails. This is the sole instance known to us of the inheritance, though in a feeble degree, of movements which have become superfluous from changes which the species has undergone.

Rudimentary Cotyledons.—A few remarks on this subject may be here interpolated. It is well known that some dicotyledonous plants produce only a single cotyledon; for instance, certain species of Ranunculus, Corydalis, Chaerophyllum; and we will here endeavour to show that the loss of one or both cotyledons is apparently due to a store of nutriment being laid up in some other part, as in the hypocotyl or one of the two cotyledons, or one of the secondary radicles. With the orange (Citrus aurantium) the cotyledons are hypogean, and one is larger than the other, as may be seen in A (Fig. 60). In B the inequality is rather greater, and the stem has grown between the points of insertion of the two petioles, so that they do not stand opposite to one another; in another case the separation amounted to one-fifth of an inch. The smaller cotyledon of one seedling was extremely thin, and not half the length of the larger one, so that it was clearly becoming rudimentary[14] In all these seedlings the hypocotyl was enlarged or swollen.

Fig. 60. Citrus aurantium: two young seedlings: c, larger cotyledon; c', smaller cotyledon; h, thickened hypocotyl; r, radicle. In A the epicotyl is still arched, in B it has become erect.

Fig. 61. Abronia umbellata: seedling twice natural size: c cotyledon; c', rudimentary cotyledon; h, enlarged hypocotyl, with a heel or projection (h') at the lower end; r, radicle.

[14] In Pachira aquatica, as described by Mr. R. I. Lynch ('Journal Linn. Soc. Bot.' vol. xvii. 1878, p. 147), one of the hypogean cotyledons is of immense size; the other is small and soon falls off; the pair do not always stand opposite. In another and very different water-plant, 'Trapa natans', one of the cotyledons, filled with farinaceous matter, is much larger than the other, which is scarcely visible, as is stated by Aug. de Candolle, 'Physiologie Veg.' tom. ii. p. 834, 1832.

With Abronia umbellata one of the cotyledons is quite rudimentary, as may be seen (c') in Fig. 61. In this specimen it consisted of a little green flap, 1/84th inch in length, destitute of a petiole and covered with glands like those on the fully developed cotyledon (c). At first it stood opposite to the larger cotyledon; but as the petiole of the latter increased in length and grew in the same line with the hypocotyl (h), the rudiment appeared in older seedlings as if seated some way down the hypocotyl. With Abronia arenaria there is a similar rudiment, which in one specimen was only 1/100th and in another 1/60th inch in length; it ultimately appeared as if seated halfway down the hypocotyl. In both these species the hypocotyl is so much enlarged, especially at a very early age, that it might almost be called a corm. The lower end forms a heel or projection, the use of which will hereafter be described.

In *Cyclamen Persicum* the hypocotyl, even whilst still within the seed, is enlarged into a regular corm,[15] and only a single cotyledon is at first developed (see former Fig. 57). With *Ranunculus ficaria* two cotyledons are never produced, and here one of the secondary radicles is developed at an early age into a so-called bulb.[16] Again, certain species of Chaerophyllum and Corydalis produce only a single cotyledon;[17] in the former the hypocotyl, and in the latter the radicle is enlarged, according to Irmisch, into a bulb.

[15] Dr. H. Gressner, 'Bot. Zeitung,' 1874, p. 824.

[16] Irmisch, 'Beiträge zur Morphologie der Pflanzen,' 1854, pp. 11, 12; 'Bot. Zeitung,' 1874, p. 805.

[17] Delpino, 'Rivista Botanica,' 1877, p. 21. It is evident from Vaucher's account ('Hist. Phys. des Plantes d'Europe,' tom. i. 1841, p. 149) of the germination of the seeds of several species of Corydalis, that the bulb or tubercule begins to be formed at an extremely early age.

In the several foregoing cases one of the cotyledons is delayed in its development, or reduced in size, or rendered rudimentary, or quite aborted; but in other cases both cotyledons are represented by mere rudiments. With Opuntia basilaris this is not the case, for both cotyledons are thick and large, and the hypocotyl shows at first no signs of enlargement; but afterwards, when the cotyledons have withered and disarticulated themselves, it becomes thickened, and from its tapering form, together with its smooth, tough, brown skin, appears, when ultimately drawn down to some depth into the soil, like a root. On the other hand, with several other Cacteæ, the hypocotyl is from the first much enlarged, and both cotyledons are almost or quite rudimentary. Thus with Cereus Landbeckii two little triangular projections, representing the cotyledons, are narrower than the hypocotyl, which is pear-shaped, with the point downwards. In Rhipsalis cassytha the cotyledons are represented by mere points on the enlarged hypocotyl. In Echinocactus viridescens the hypocotyl is globular, with two little prominences on its summit. In Pilocereus Houlletii the hypocotyl, much swollen in the upper part, is merely notched on the summit; and each side of the notch evidently represents a cotyledon. Stapelia sarpedon, a member of the very distinct family of the Asclepiadeae, is fleshy like a cactus; and here again the upper part of the flattened hypocotyl is much thickened and bears two minute cotyledons, which, measured internally, were only .15 inch in length, and in breadth not equal to one-fourth of the diameter of the hypocotyl in its narrow axis; yet these minute cotyledons are probably not quite useless, for when the hypocotyl breaks through the ground in the form of an arch, they are closed or pressed against one another, and thus protect the plumule. They afterwards open.

From the several cases now given, which refer to widely distinct plants, we may infer that there is some close connection between the reduced size of one or both cotyledons and the formation, by the enlargement of the hypocotyl or of the radicle, of a so-called bulb. But it may be asked, did the cotyledons first tend to abort, or did a bulb first begin to be formed? As all dicotyledons naturally produce two well-developed cotyledons, whilst the thickness of the hypocotyl and of the radicle differs much in different plants, it seems probable that these latter organs first became from some cause thickened—in several instances apparently in correlation with the fleshy nature of the mature plant—so as to contain a store of nutriment sufficient for the seedling, and then that one or both cotyledons, from being superfluous, decreased in size. It is not surprising that one cotyledon alone should sometimes have been thus affected, for with certain plants, for instance the cabbage, the cotyledons are at first of unequal size, owing apparently to the manner in which they are packed within the seed. It does not, however, follow from the above connection, that whenever a bulb is formed at an early age, one or both cotyledons will necessarily become superfluous, and consequently more or less rudimentary. Finally, these cases offer a good illustration of the principle of compensation or balancement of growth, or, as Goethe expresses it, "in order to spend on one side, Nature is forced to economise on the other side."

Circumnutation and other movements of Hypocotyls and Epicotyls, whilst still arched and buried beneath the ground, and whilst breaking through it.—According to the position in which a seed may chance to have been buried, the arched hypocotyl or epicotyl will begin to protrude in a horizontal, a more or less inclined, or in a vertical plane. Except when already standing vertically upwards, both legs of the arch are acted on from the earliest period by apogeotropism. Consequently they both bend upwards until the arch becomes vertical. During the whole of this process, even before the arch has broken through the ground, it is continually trying to circumnutate to a slight extent; as it likewise does if it happens at first to stand vertically up,—all which cases have been observed and described, more or less fully, in the last chapter. After the arch has grown to some height upwards the basal part ceases to circumnutate, whilst the upper part continues to do so.

That an arched hypocotyl or epicotyl, with the two legs fixed in the ground, should be able to circumnutate, seemed to us, until we had read Prof. Wiesner's observations, an inexplicable fact. He has shown[18] in the case of certain seedlings, whose tips are bent downwards (or which nutate), that whilst the posterior side of the upper or dependent portion grows quickest, the anterior and opposite side of the basal portion of the same internode grows quickest; these two portions being separated by an indifferent zone, where the growth is equal on all sides. There may be even more than one indifferent zone in the same internode; and the opposite sides of the parts above and below each such zone grow quickest. This peculiar manner of growth is called by Wiesner "undulatory nutation." Circumnutation depends on one side of an organ growing quickest (probably preceded by increased turgescence), and then another side, generally almost the opposite one, growing quickest. Now if we look at an arch like this [upside down U] and suppose the whole of one side—we will say the whole convex side of both legs—to increase in length, this would not cause the arch to bend to either side. But if the outer side or surface of the left leg were to increase in length the arch would be pushed over to the right, and this would be aided by the inner side of the right leg increasing in length. If afterwards the process were reversed, the arch would be pushed over to the opposite or left side, and so on alternately,—that is, it would circumnutate. As an arched hypocotyl, with the two legs fixed in the ground, certainly circumnutates, and as it consists of a single internode, we may conclude that it grows in the manner described by Wiesner. It may be added, that the crown of the arch does not grow, or grows very slowly, for it does not increase much in breadth, whilst the arch itself increases greatly in height.

[18] 'Die undulirende Nutation der Internodien,' Akad. der Wissench. (Vienna), Jan. 17th, 1878. Also published separately, see p. 32.

The circumnutating movements of arched hypocotyls and epicotyls can hardly fail to aid them in breaking through the ground, if this be damp and soft; though no doubt their emergence depends mainly on the force exerted by their longitudinal growth. Although the arch circumnutates only to a slight extent and probably with little force, yet it is able to move the soil near the surface, though it may not be able to do so at a moderate depth. A pot with seeds of Solanum palinacanthum, the tall arched hypocotyls of which had emerged and were growing rather slowly, was covered with fine argillaceous sand kept damp, and this at first closely surrounded the bases of the arches; but soon a narrow open crack was formed round each of them, which could be accounted for only by their having pushed away the sand on all sides; for no such cracks surrounded some little sticks and pins which had been driven into the sand. It has already been stated that the cotyledons of Phalaris and Avena, the plumules of Asparagus and the hypocotyls of Brassica, were likewise able to displace the same kind of sand, either whilst simply circumnutating or whilst bending towards a lateral light.

As long as an arched hypocotyl or epicotyl remains buried beneath the ground, the two legs cannot separate from one another, except to a slight extent from the yielding of the soil; but as soon as the arch rises above the ground, or at an earlier period if the pressure of the surrounding earth be artificially removed, the arch immediately begins to straighten itself. This no doubt is due to growth along the whole inner surface of both legs of the arch; such growth being checked or prevented, as long as the two legs of the arch are firmly pressed together. When the earth is removed all round an arch and the two legs are tied together at their bases, the growth on the under side of the crown causes it after a time to become much flatter and broader than naturally occurs. The straightening process consists of a modified form of circumnutation, for the lines described during this process (as with the hypocotyl of Brassica, and the epicotyls of Vicia and Corylus) were often plainly zigzag and sometimes looped. After hypocotyls or epicotyls have emerged from the ground, they quickly become perfectly straight. No trace is left of their former abrupt curvature, excepting in the case of Allium cepa, in which the cotyledon rarely becomes quite straight, owing to the protuberance developed on the crown of the arch.

The increased growth along the inner surface of the arch which renders it straight, apparently begins in the basal leg or that which is united to the radicle; for this leg, as we often observed, is first bowed backwards from the other leg. This movement facilitates the withdrawal of the tip of the epicotyl or of the cotyledons, as the case may be, from within the seed-coats and from the ground. But the cotyledons often emerge from the ground still tightly enclosed within the seed-coats, which apparently serve to protect them. The seed-coats are afterwards ruptured and cast off by the swelling of the closely conjoined cotyledons, and not by any movement or their separation from one another.

Nevertheless, in some few cases, especially with the Cucurbitaceæ, the seed-coats are ruptured by a curious contrivance, described by M. Flahault.[19] A heel or peg is developed on one side of the summit of the radicle or base of the hypocotyl; and this holds down the lower half of the seed-coats (the radicle being fixed into the ground) whilst the continued growth of the arched hypocotyl forced upwards the upper half, and tears asunder the seed-coats at one end, and the cotyledons are then easily withdrawn.

[19] 'Bull. Soc. Bot. de France,' tom. xxiv. 1877, p. 201.

The accompanying figure (Fig. 62) will render this description intelligible. Forty-one seeds of Cucurbita ovifera were laid on friable peat and were covered by a layer about an inch in thickness, not much pressed down, so that the cotyledons in being dragged up were subjected to very little friction, yet forty of them came up naked, the seed-coats being left buried in the peat. This was certainly due to the action of the peg, for when it was prevented from acting, the cotyledons, as we shall presently see, were lifted up still enclosed in their seed-coats. They were, however, cast off in the course of two or three days by the swelling of the cotyledons. Until this occurs light is excluded, and the cotyledons cannot decompose carbonic acid; but no one probably would have thought that the advantage thus gained by a little earlier casting off of the seed-coats would be sufficient to account for the development of the peg. Yet according to M. Flahault, seedlings which have been prevented from casting their seed-coats whilst beneath the ground, are inferior to those which have emerged with their cotyledons naked and ready to act.

Fig. 62. Cucurbita ovifera: germinating seed, showing the heel or peg projecting on one side from summit of radicle and holding down lower tip of seed-coats, which have been partially ruptured by the growth of the arched hypocotyl.

The peg is developed with extraordinary rapidity; for it could only just be distinguished in two seedlings, having radicles .35 inch in length, but after an interval of only 24 hours was well developed in both. It is formed, according to Flahault, by the enlargement of the layers of the cortical parenchyma at the base of the hypocotyl. If, however, we judge by the effects of a solution of permanganate of potassium, it is developed on the exact line of junction between the hypocotyl and radicle; for the flat lower surface, as well as the edges, were coloured brown like the radicle; whilst the upper slightly inclined surface was left uncoloured like the hypocotyl, excepting indeed in one out of 33 immersed seedlings in which a large part of the upper surface was coloured brown. Secondary roots sometimes spring from the lower surface of the peg, which thus seems in all respects to partake of the nature of the radicle. The peg is always developed on the side which becomes concave by the arching of the hypocotyl; and it would be of no service if it were formed on any other side. It is also always developed with the flat lower side, which, as just stated, forms a part of the radicle, at right angles to it, and in a horizontal plane. This fact was clearly shown by burying some of the thin flat seeds in the same position as in Fig. 62, excepting that they were not laid on their flat broad sides, but with one edge downwards. Nine seeds were thus planted, and the peg was developed in the same position, relatively to the radicle, as in the figure; consequently it did not rest on the flat tip of the lower half of the seed-coats, but was inserted like a wedge between the two tips. As the arched hypocotyl grew upwards it tended to draw up the whole seed, and the peg necessarily rubbed against both tips, but did not hold either down. The result was, that the cotyledons of five out of the nine seeds thus placed were raised above the ground still enclosed within their seed-coats. Four seeds were buried with the end from which the radicle protrudes pointing vertically downwards, and owing to the peg being always developed in the same position, its apex alone came into contact with, and rubbed against the tip on one side; the result was, that the cotyledons of all four emerged still within their seed-coats. These cases show us how the peg acts in co-ordination with the position which the flat, thin, broad seeds would almost always occupy when naturally sown. When the tip of the lower half of the seed-coats was cut off, Flahault found (as we did likewise) that the peg could not act, since it had nothing to press on, and the cotyledons were raised above the ground with their seed-coats not cast off. Lastly, nature shows us the use of the peg; for in the one Cucurbitaceous genus known to us, in which the cotyledons are hypogean and do not cast their seed-coats, namely, Megarrhiza, there is no vestige of a peg. This structure seems to be present in most of the other genera in the family, judging from Flahault's statements' we found it well-developed and properly acting in Trichosanthes anguina, in which we hardly expected to find it, as the cotyledons are somewhat thick and fleshy. Few cases can be advanced of a structure better adapted for a special purpose than the present one.

With Mimosa pudica the radicle protrudes from a small hole in the sharp edge of the seed; and on its summit, where united with the hypocotyl, a transverse ridge is developed at an early age, which clearly aids in splitting the tough seed-coats; but it does not aid in casting them off, as this is subsequently effected by the swelling of the cotyledons after they have been raised above the ground. The ridge or heel therefore acts rather differently from that of Cucurbita. Its lower surface and the edges were coloured brown by the permanganate of potassium, but not the upper surface. It is a singular fact that after the ridge has done its work and has escaped from the seed-coats, it is developed into a frill all round the summit of the radicle.[20]

[20] Our attention was called to this case by a brief statement by Nobbe in his 'Handbuch der Samenkunde,' 1876, p. 215, where a figure is also given of a seedling of Martynia with a heel or ridge at the junction of the radicle and hypocotyl. This seed possesses a very hard and tough coat, and would be likely to require aid in bursting and freeing the cotyledons.

At the base of the enlarged hypocotyl of Abronia umbellata, where it blends into the radicle, there is a projection or heel which varies in shape, but its outline is too angular in our former figure (Fig. 61). The radicle first protrudes from a small hole at one end of the tough, leathery, winged fruit. At this period the upper part of the radicle is packed within the fruit parallel to the hypocotyl, and the single cotyledon is doubled back parallel to the latter. The swelling of these three parts, and especially the rapid development of the thick heel between the hypocotyl and radicle at the point where they are doubled, ruptures the tough fruit at the upper end and allows the arched hypocotyl to emerge; and this seems to be the function of the heel. A seed was cut out of the fruit and allowed to germinate in damp air, and now a thin flat disc was developed all round the base of the hypocotyl and grew to an extraordinary breadth, like the frill described under Mimosa, but somewhat broader. Flahault says that with Mirabilis, a member of the same family with Abronia, a heel or collar is developed all round the base of the hypocotyl, but more on one side than on the other; and that it frees the cotyledons from their seed-coats. We observed only old seeds, and these were ruptured by the absorption of moisture, independently of any aid from the heel and before the protrusion of the radicle; but it does not follow from our experience that fresh and tough fruits would behave in a like manner.

In concluding this section of the present chapter it may be convenient to summarise, under the form of an illustration, the usual movements of the hypocotyls and epicotyls of seedlings, whilst breaking through the ground and immediately afterwards. We may suppose a man to be thrown down on his hands and knees, and at the same time to one side, by a load of hay falling on him. He would first endeavour to get his arched back upright, wriggling at the same time in all directions to free himself a little from the surrounding pressure; and this may represent the combined effects of apogeotropism and circumnutation, when a seed is so buried that the arched hypocotyl or epicotyl protrudes at first in a horizontal or inclined plane. The man, still wriggling, would then raise his arched back as high as he could; and this may represent the growth and continued circumnutation of an arched hypocotyl or epicotyl, before it has reached the surface of the ground. As soon as the man felt himself at all free, he would raise the upper part of his body, whilst still on his knees and still wriggling; and this may represent the bowing backwards of the basal leg of the arch, which in most cases aids in the withdrawal of the cotyledons from the buried and ruptured seed-coats, and the subsequent straightening of the whole hypocotyl or epicotyl—circumnutation still continuing.

Circumnutation of Hypocotyls and Epicotyls, when erect.—The hypocotyls, epicotyls, and first shoots of the many seedlings observed by us, after they had become straight and erect, circumnutated continuously. The diversified figures described by them, often during two successive days, have been shown in the woodcuts in the last chapter. It should be recollected that the dots were joined by straight lines, so that the figures are angular; but if the observations had been made every few minutes the lines would have been more or less curvilinear, and irregular ellipses or ovals, or perhaps occasionally circles, would have been formed. The direction of the longer axes of the ellipses made during the same day or on successive days generally changed completely, so as to stand at right angles to one another. The number of irregular ellipses or circles made within a given time differs much with different species. Thus with Brassica oleracea, Cerinthe major, and Cucurbita ovifera about four such figures were completed in 12 h.; whereas with Solanum palinacanthum and Opuntia basilaris, scarcely more than one. The figures likewise differ greatly in size; thus they were very small and in some degree doubtful in Stapelia, and large in Brassica, etc. The ellipses described by Lathyrus nissolia and Brassica were narrow, whilst those made by the Oak were broad. The figures are often complicated by small loops and zigzag lines.

As most seedling plants before the development of true leaves are of low, sometimes very low stature, the extreme amount of movement from side to side of their circumnutating stems was small; that of the hypocotyl of

Githago segetum was about .2 of an inch, and that of Cucurbita ovifera about .28. A very young shoot of Lathyrus nissolia moved about .14, that of an American oak .2, that of the common nut only .04, and a rather tall shoot of the Asparagus .11 of an inch. The extreme amount of movement of the sheath-like cotyledon of Phalaris Canariensis was .3 of an inch; but it did not move very quickly, the tip crossing on one occasion five divisions of the micrometer, that is, 1/100th of an inch, in 22 m. 5 s. A seedling Nolana prostrata travelled the same distance in 10 m. 38 s. Seedling cabbages circumnutate much more quickly, for the tip of a cotyledon crossed 1/100th of an inch on the micrometer in 3 m. 20 s.; and this rapid movement, accompanied by incessant oscillations, was a wonderful spectacle when beheld under the microscope.

The absence of light, for at least a day, does not interfere in the least with the circumnutation of the hypocotyls, epicotyls, or young shoots of the various dicotyledonous seedlings observed by us; nor with that of the young shoots of some monocotyledons. The circumnutation was indeed much plainer in darkness than in light, for if the light was at all lateral the stem bent towards it in a more or less zigzag course.

Finally, the hypocotyls of many seedlings are drawn during the winter into the ground, or even beneath it so that they disappear. This remarkable process, which apparently serves for their protection, has been fully described by De Vries.[21] He shows that it is effected by the contraction of the parenchyma-cells of the root. But the hypocotyl itself in some cases contracts greatly, and although at first smooth becomes covered with zigzag ridges, as we observed with Githago segetum. How much of the drawing down and burying of the hypocotyl of Opuntia basilaris was due to the contraction of this part and how much to that of the radicle, we did not observe.

[21] 'Bot. Zeitung,' 1879, p. 649. See also Winkler in 'Verhandl. des Bot. Vereins der P. Brandenburg,' Jahrg. xvi. p. 16, as quoted by Haberlandt, 'Schutzeinrichungen der Keimpflanze,' 1877, p. 52.

Circumnutation of Cotyledons.—With all the dicotyledonous seedlings described in the last chapter, the cotyledons were in constant movement, chiefly in a vertical plane, and commonly once up and once down in the course of the 24 hours. But there were many exceptions to such simplicity of movement; thus the cotyledons of Ipomœa caerulea moved 13 times either upwards or downwards in the course of 16 h.. 18 m. Those of Oxalis rosea moved in the same manner 7 times in the course of 24 h.; and those of Cassia tora described 5 irregular ellipses in 9 h. The cotyledons of some individuals of Mimosa pudica and of Lotus Jacobæus moved only once up and down in 24 h., whilst those of others performed within the same period an additional small oscillation. Thus with different species, and with different individuals of the same species, there were many gradations from a single diurnal movement to oscillations as complex as those of the Ipomœa and Cassia. The opposite cotyledons on the same seedling move to a certain extent independently of one another. This was conspicuous with those of Oxalis sensitiva, in which one cotyledon might be seen during the daytime rising up until it stood vertically, whilst the opposite one was sinking down.

Although the movements of cotyledons were generally in nearly the same vertical plane, yet their upward and downward courses never exactly coincided; so that ellipses, more or less narrow, were described, and the cotyledons may safely be said to have circumnutated. Nor could this fact be accounted for by the mere increase in length of the cotyledons through growth, for this by itself would not induce any lateral movement. That there was lateral movement in some instances, as with the cotyledons of the cabbage, was evident; for these, besides moving up and down, changed their course from right to left 12 times in 14 h. 15 m. With Solanum lycopersicum the cotyledons, after falling in the forenoon, zigzagged from side to side between 12 and 4 P.M., and then commenced rising. The cotyledons of Lupinus luteus are so thick (about .08 of an inch) and fleshy,[22] that they seemed little likely to move, and were therefore observed with especial interest; they certainly moved largely up and down, and as the line traced was zigzag there was some lateral movement. The nine cotyledons of a seedling Pinus pinaster plainly circumnutated; and the figures described approached more nearly to irregular circles than to irregular ovals or ellipses. The sheath-like cotyledons of the Gramineæ circumnutate, that is, move to all sides, as plainly as do the hypocotyls or epicotyls of any dicotyledonous plants. Lastly, the very young fronds of a Fern and of a Selaginella circumnutated.

[22] The cotyledons, though bright green, resemble to a certain extent hypogean ones; see the interesting discussion by Haberlandt ('Die Schutzeinrichtungen,' etc., 1877, p. 95), on the gradations in the Leguminosae between subaërial and subterranean cotyledons.

In a large majority of the cases which were carefully observed, the cotyledons sink a little downwards in the forenoon, and rise a little in the afternoon or evening. They thus stand rather more highly inclined during the night than during the mid-day, at which time they are expanded almost horizontally. The circumnutating movement is thus at least partially periodic, no doubt in connection, as we shall hereafter see, with the daily alternations of light and darkness. The cotyledons of several plants move up so much at night as to stand nearly or quite vertically; and in this latter case they come into close contact with one another. On the other hand, the cotyledons of a few plants sink almost or quite vertically down at night; and in this latter case they clasp the upper part of the hypocotyl. In the same genus Oxalis the cotyledons of certain species stand vertically up, and those of other species vertically down, at night. In all such cases the cotyledons may be said to sleep, for they act in the same manner as do the leaves of many sleeping plants. This is a movement for a special purpose, and will therefore be considered in a future chapter devoted to this subject.

In order to gain some rude notion of the proportional number of cases in which the cotyledons of dicotyledonous plants (hypogean ones being of course excluded) changed their position in a conspicuous manner at night, one or more species in several genera were cursorily observed, besides those described in the last chapter. Altogether 153 genera, included in as many families as could be procured, were thus observed by us. The cotyledons were looked at in the middle of the day and again at night; and those were noted as sleeping which stood either vertically or at an angle of at least 60° above or beneath the horizon. Of such genera there were 26; and in 21 of them the cotyledons of some of the species rose, and in only 6 sank at night; and some of these latter cases are rather doubtful from causes to be explained in the chapter on the sleep of cotyledons. When cotyledons which at noon were nearly horizontal, stood at night at more than 20° and less than 60° above the horizon, they were recorded as "plainly raised;" and of such genera there were 38. We did not meet with any distinct instances of cotyledons periodically sinking only a few degrees at night, although no doubt such occur. We have now accounted for 64 genera out of the 153, and there remain 89 in which the cotyledons did not change their position at night by as much as 20°—that is, in a conspicuous manner which could easily be detected by the unaided eye and by memory; but it must not be inferred from this statement that these cotyledons did not move at all, for in several cases a rise of a few degrees was recorded, when they were carefully observed. The number 89 might have been a little increased, for the cotyledons remained almost horizontal at night in some species in a few genera, for instance, Trifolium and Geranium, which are included amongst the sleepers, such genera might therefore have been added to the 89. Again, one species of Oxalis generally raised its cotyledons at night more than 20° and less than 60° above the horizon; so that this genus might have been included under two heads. But as several species in the same genus were not often observed, such double entries have been avoided.

In a future chapter it will be shown that the leaves of many plants which do not sleep, rise a few degrees in the evening and during the early part of the night; and it will be convenient to defer until then the consideration of the periodicity of the movements of cotyledons.

On the Pulvini or Joints of Cotyledons.—With several of the seedlings described in this and the last chapter, the summit of the petiole is developed into a pulvinus, cushion, or joint (as this organ has been variously called), like that with which many leaves are provided. It consists of a mass of small cells usually of a pale colour from the absence of chlorophyll, and with its outline more or less convex, as shown in the annexed figure. In the case of Oxalis sensitiva two-thirds of the petiole, and in that of Mimosa pudica, apparently the whole of the short sub-petioles of the leaflets have been converted into pulvini. With pulvinated leaves (i.e. those provided with a pulvinus) their periodical movements depend, according to Pfeffer,[23] on the cells of the pulvinus alternately expanding more quickly on one side than on the other; whereas the similar movements of leaves not provided with pulvini, depend on their growth being alternately more rapid on one side than on the other.[24] As long as a leaf provided with a pulvinus is young and continues to grow, its movement depends on both these causes combined;[25] and if the view now held by many botanists be sound, namely, that growth is always preceded by the expansion of the growing cells, then the difference between the movements induced by the aid of pulvini and without such aid, is reduced to the expansion of the cells not being followed by growth in the first case, and being so followed in the second case.

[23] 'Die Periodische Bewegungen der Blattorgane,' 1875.

[24] Batalin, 'Flora,' Oct. 1st, 1873

[25] Pfeffer, ibid. p. 5.

Fig. 63. Oxalis rosea: longitudinal section of a pulvinus on the summit of the petiole of a cotyledon, drawn with the camera lucida, magnified 75 times: p, p, petiole; f, fibro-vascular bundle: b, b, commencement of blade of cotyledon.

Dots were made with Indian ink along the midrib of both pulvinated cotyledons of a rather old seedling of Oxalis Valdiviana; their distances were repeatedly measured with an eye-piece micrometer during 8 3/4 days, and they did not exhibit the least trace of increase. It is therefore almost certain that the pulvinus itself was not then growing. Nevertheless, during this whole time and for ten days afterwards, these cotyledons rose vertically every night. In the case of some seedlings raised from seeds purchased under the name of Oxalis floribunda, the cotyledons continued for a long time to move vertically down at night, and the movement apparently depended exclusively on the pulvini, for their petioles were of nearly the same length in young, and in old seedlings which had produced true leaves. With some species of Cassia, on the other hand, it was obvious without any measurement that the pulvinated cotyledons continued to increase greatly in length during some weeks; so that here the expansion of the cells of the pulvini and the growth of the petiole were probably combined in causing their prolonged periodic movements. It was equally evident that the cotyledons of many plants, not provided with pulvini, increased rapidly in length; and their periodic movements no doubt were exclusively due to growth.

In accordance with the view that the periodic movements of all cotyledons depend primarily on the expansion of the cells, whether or not followed by growth, we can understand the fact that there is but little difference in the kind or form of movement in the two sets of cases. This may be seen by comparing the diagrams given in the last chapter. Thus the movements of the cotyledons of Brassica oleracea and of Ipomœa caerulea, which are not provided with pulvini, are as complex as those of Oxalis and Cassia which are thus provided. The pulvinated cotyledons of some individuals of Mimosa pudica and Lotus Jacobæus made only a single oscillation, whilst those of other individuals moved twice up and down in the course of 24 hours; so it was occasionally with the cotyledons of Cucurbita ovifera, which are destitute of a pulvinus. The movements of pulvinated cotyledons are generally larger in extent than those without a pulvinus; nevertheless some of the latter moved through an angle of 90°. There is, however, one important difference in the two sets of cases; the nocturnal movements of cotyledons without pulvini, for instance, those in the Cruciferae, Cucurbitaceæ, Githago, and Beta, never last even for a week, to any conspicuous degree. Pulvinated cotyledons, on the other hand, continue to rise at night for a much longer period, even for more than a month, as we shall now show. But the period no doubt depends largely on the temperature to which the seedlings are exposed and their consequent rate of development.

Oxalis Valdiviana.—Some cotyledons which had lately opened and were horizontal on March 6th at noon, stood at night vertically up; on the 13th the first true leaf was formed, and was embraced at night by the cotyledons; on April 9th, after an interval of 35 days, six leaves were developed, and yet the cotyledons rose almost vertically at night. The cotyledons of another seedling, which when first observed had already produced a leaf, stood vertically at night and continued to do so for 11 additional days. After 16 days from the first observation two leaves were developed, and the cotyledons were still greatly raised at night. After 21 days the cotyledons during the day were deflected beneath the horizon, but at night were raised 45° above it. After 24 days from the first observation (begun after a true leaf had been developed) the cotyledons ceased to rise at night.

Oxalis (Biophytum) sensitiva.—The cotyledons of several seedlings, 45 days after their first expansion, stood nearly vertical at night, and closely embraced either one or two true leaves which by this time had been formed. These seedlings had been kept in a very warm house, and their development had been rapid.

Oxalis corniculata.—The cotyledons do not stand vertical at night, but generally rise to an angle of about 45° above the horizon. They continued thus to act for 23 days after their first expansion, by which time two leaves had been formed; even after 29 days they still rose moderately above their horizontal or downwardly deflected diurnal position.

Mimosa pudica.—The cotyledons were expanded for the first time on Nov. 2nd, and stood vertical at night. On the 15th the first leaf was formed, and at night the cotyledons were vertical. On the 28th they behaved in the same manner. On Dec. 15th, that is after 44 days, the cotyledons were still considerably raised at night; but those of another seedling, only one day older, were raised very little.

Mimosa albida.—A seedling was observed during only 12 days, by which time a leaf had been formed, and the cotyledons were then quite vertical at night.

Trifolium subterraneum.—A seedling, 8 days old, had its cotyledons horizontal at 10.30 A.M. and vertical at 9.15 P.M. After an interval of two months, by which time the first and second true leaves had been developed, the cotyledons still performed the same movement. They had now increased greatly in size, and had become oval; and their petioles were actually .8 of an inch in length!

Trifolium strictum.—After 17 days the cotyledons still rose at night, but were not afterwards observed.

Lotus Jacoboeus.—The cotyledons of some seedlings having well-developed leaves rose to an angle of about 45° at night; and even after 3 or 4 whorls of leaves had been formed, the cotyledons rose at night considerably above their diurnal horizontal position.

Cassia mimosoides.—The cotyledons of this Indian species, 14 days after their first expansion, and when a leaf had been formed, stood during the day horizontal, and at night vertical.

Cassia sp? (a large S. Brazilian tree raised from seeds sent us by F. Müller).—The cotyledons, after 16 days from their first expansion, had increased greatly in size with two leaves just formed. They stood horizontally during the day and vertically at night, but were not afterwards observed.

Cassia neglecta (likewise a S. Brazilian species).—A seedling, 34 days after the first expansion of its cotyledons, was between 3 and 4 inches in height, with 3 well-developed leaves; and the cotyledons, which during the day were nearly horizontal, at night stood vertical, closely embracing the young stem. The cotyledons of another seedling of the same age, 5 inches in height, with 4 well-developed leaves, behaved at night in exactly the same manner.

It is known[26] that there is no difference in structure between the upper and lower halves of the pulvini of leaves, sufficient to account for their upward or downward movements. In this respect cotyledons offer an unusually good opportunity for comparing the structure of the two halves; for the cotyledons of Oxalis Valdiviana rise vertically at night, whilst those of O. rosea sink vertically; yet when sections of their pulvini were made, no clear difference could be detected between the corresponding halves of this organ in the two species which move so differently. With O. rosea, however, there were rather more cells in the lower than in the upper half, but this was likewise the case in one specimen of O. Valdiviana. the cotyledons of both species (3½ mm. in length) were examined in the morning whilst extended horizontally, and the upper surface of the pulvinus of O. rosea was then wrinkled transversely, showing that it was in a state of compression, and this might have been expected, as the cotyledons sink at night; with O. Valdiviana it was the lower surface which was wrinkled, and its cotyledons rise at night.

[26] Pfeffer, 'Die Period. Bewegungen,' 1875, p. 157.

Trifolium is a natural genus, and the leaves of all the species seen by us are pulvinated; so it is with the cotyledons of T. subterraneum and strictum, which stand vertically at night; whereas those of T. resupinatum exhibit not a trace of a pulvinus, nor of any nocturnal movement. This was ascertained by measuring the distance between the tips of the cotyledons of four seedlings at mid-day and at night. In this species, however, as in the others, the first-formed leaf, which is simple or not trifoliate, rises up and sleeps like the terminal leaflet on a mature plant.

In another natural genus, Oxalis, the cotyledons of O. Valdiviana, rosea, floribunda, articulata, and sensitiva are pulvinated, and all move at night into an upward or downward vertical position. In these several species the pulvinus is seated close to the blade of the cotyledon, as is the usual rule with most plants. Oxalis corniculata (var. Atro-purpurea) differs in several respects; the cotyledons rise at night to a very variable amount, rarely more than 45°; and in one lot of seedlings (purchased under the name of O. tropaeoloides, but certainly belonging to the above variety) they rose only from 5° to 15° above the horizon. The pulvinus is developed imperfectly and to an extremely variable degree, so that apparently it is tending towards abortion. No such case has hitherto, we believe, been described. It is coloured green from its cells containing chlorophyll; and it is seated nearly in the middle of the petiole, instead of at the upper end as in all the other species. The nocturnal movement is effected partly by its aid, and partly by the growth of the upper part of the petiole as in the case of plants destitute of a pulvinus. From these several reasons and from our having partially traced the development of the pulvinus from an early age, the case seems worth describing in some detail.

When the cotyledons of O. corniculata were dissected out of a seed from which they would soon have naturally emerged, no trace of a pulvinus could be detected; and all the cells forming the short petiole, 7 in number in a longitudinal row, were of nearly equal size. In seedlings one or two days old, the pulvinus was so indistinct that we thought at first that it did not exist; but in the middle of the petiole an ill-defined transverse zone of cells could be seen, which were much shorter than those both above and below, although of the same breadth with them. They presented the appearance of having been just formed by the transverse division of longer cells; and there can be little doubt that this had occurred, for the cells in the petiole which had been dissected out of the seed averaged in length 7 divisions of the micrometer (each division equalling .003 mm.), and were a little longer than those forming a well-developed pulvinus, which varied between 4 and 6 of these same divisions. After a few additional days the ill-defined zone of cells becomes distinct, and although it does not extend across the whole width of the petiole, and although the cells are of a green colour from containing chlorophyll, yet they certainly constitute a pulvinus, which as we shall presently see, acts as one. These small cells were arranged in longitudinal rows, and varied from 4 to 7 in number; and the cells themselves varied in length in different parts of the same pulvinus and in different individuals. In the accompanying figures, A and B (Fig. 64), we have views of the epidermis[27] in the middle part of the petioles of two seedlings, in which the pulvinus was for this species well developed. They offer a striking contrast with the pulvinus of O. rosea (see former Fig. 63), or of O. Valdiviana. With the seedlings, falsely called O. tropaeoloides, the cotyledons of which rise very little at night, the small cells were still fewer in number and in parts formed a single transverse row, and in other parts short longitudinal rows of only two or three. Nevertheless they sufficed to attract the eye, when the whole petiole was viewed as a transparent object beneath the microscope. In these seedlings there could hardly be a doubt that the pulvinus was becoming rudimentary and tending to disappear; and this accounts for its great variability in structure and function.

[27] Longitudinal sections show that the forms of the epidermic cells may be taken as a fair representation of those constituting the pulvinus.

Fig. 64. Oxalis corniculata: A and B the almost rudimentary pulvini of the cotyledons of two rather old seedlings, viewed as transparent objects. Magnified 50 times.

In the following Table some measurements of the cells in fairly well-developed pulvini of O. corniculata are given:—

Seedling 1 day old, with cotyledon 2.3 mm. in length. Divisions of Micrometer.[28] Average length of cells of pulvinus...6 to 7 Length of longest cell below the pulvinus................................... 13 Length of longest cell above the pulvinus.................................... 20

Seedling 5 days old, cotyledon 3.1 mm. in length, with the pulvinus quite distinct. Average length of cells of pulvinus.. 6 Length of longest cell below the pulvinus................................... 22 Length of longest cell above the pulvinus.................................... 40

Seedling 8 days old, cotyledon 5 mm. in length, with a true leaf formed but not yet expanded. Average length of cells of pulvinus.. 9 Length of longest cell below the pulvinus................................... 44 Length of longest cell above the pulvinus.................................... 70

Seedling 13 days old, cotyledon 4.5 mm. in length, with a small true leaf fully developed. Average length of cells of pulvinus.. 7 Length of longest cell below the pulvinus................................... 30 Length of longest cell above the pulvinus.................................... 60

---

[28] Each division equalled .003 mm.

We here see that the cells of the pulvinus increase but little in length with advancing age, in comparison with those of the petiole both above and below it; but they continue to grow in width, and keep equal in this respect with the other cells of the petiole. The rate of growth, however, varies in all parts of the cotyledons, as may be observed in the measurements of the 8-days' old seedling.

The cotyledons of seedlings only a day old rise at night considerably, sometimes as much as afterwards; but there was much variation in this respect. As the pulvinus is so indistinct at first, the movement probably does not then depend on the expansion of its cells, but on periodically unequal growth in the petiole. By the comparison of seedlings of different known ages, it was evident that the chief seat of growth of the petiole was in the upper part between the pulvinus and the blade; and this agrees with the fact (shown in the measurements above given) that the cells grow to a greater length in the upper than in the lower part. With a seedling 11 days old, the nocturnal rise was found to depend largely on the action of the pulvinus, for the petiole at night was curved upwards at this point; and during the day, whilst the petiole was horizontal, the lower surface of the pulvinus was wrinkled with the upper surface tense. Although the cotyledons at an advanced age do not rise at night to a higher inclination than whilst young, yet they have to pass through a larger angle (in one instance amounting to 63°) to gain their nocturnal position, as they are generally deflected beneath the horizon during the day. Even with the 11-days' old seedling the movement did not depend exclusively on the pulvinus, for the blade where joined to the petiole was curved upwards, and this must be attributed to unequal growth. Therefore the periodic movements of the cotyledons of 'O. corniculata' depend on two distinct but conjoint actions, namely, the expansion of the cells of the pulvinus and on the growth of the upper part of the petiole, including the base of the blade.

Lotus Jacoboeus.—The seedlings of this plant present a case parallel to that of Oxalis corniculata in some respects, and in others unique, as far as we have seen. The cotyledons during the first 4 or 5 days of their life do not exhibit any plain nocturnal movement; but afterwards they stand vertically or almost vertically up at night. There is, however, some degree of variability in this respect, apparently dependent on the season and on the degree to which they have been illuminated during the day. With older seedlings, having cotyledons 4 mm. in length, which rise considerably at night, there is a well-developed pulvinus close to the blade, colourless, and rather narrower than the rest of the petiole, from which it is abruptly separated. It is formed of a mass of small cells of an average length of .021 mm.; whereas the cells in the lower part of the petiole are about .06 mm., and those in the blade from .034 to .04 mm. in length. The epidermic cells in the lower part of the petiole project conically, and thus differ in shape from those over the pulvinus.

Turning now to very young seedlings, the cotyledons of which do not rise at night and are only from 2 to 2½ mm. in length, their petioles do not exhibit any defined zone of small cells, destitute of chlorophyll and differing in shape exteriorly from the lower ones. Nevertheless, the cells at the place where a pulvinus will afterwards be developed are smaller (being on an average .015 mm. in length) than those in the lower parts of the same petiole, which gradually become larger in proceeding downwards, the largest being .030 mm. in length. At this early age the cells of the blade are about .027 mm. in length. We thus see that the pulvinus is formed by the cells in the uppermost part of the petiole, continuing for only a short time to increase in length, then being arrested in their growth, accompanied by the loss of their chlorophyll grains; whilst the cells in the lower part of the petiole continue for a long time to increase in length, those of the epidermis becoming more conical. The singular fact of the cotyledons of this plant not sleeping at first is therefore due to the pulvinus not being developed at an early age.

We learn from these two cases of Lotus and Oxalis, that the development of a pulvinus follows from the growth of the cells over a small defined space of the petiole being almost arrested at an early age. With Lotus Jacobæus the cells at first increase a little in length; in Oxalis corniculata they decrease a little, owing to self-division. A mass of such small cells forming a pulvinus, might therefore be either acquired or lost without any special difficulty, by different species in the same natural genus: and we know that with seedlings of Trifolium, Lotus, and Oxalis some of the species have a well-developed pulvinus, and others have none, or one in a rudimentary condition. As the movements caused by the alternate turgescence of the cells in the two halves of a pulvinus, must be largely determined by the extensibility and subsequent contraction of their walls, we can perhaps understand why a large number of small cells will be more efficient than a small number of large cells occupying the same space. As a pulvinus is formed by the arrestment of the growth of its cells, movements dependent on their action may be long-continued without any increase in length of the part thus provided; and such long-continued movements seem to be one chief end gained by the development of a pulvinus. Long-continued movement would be impossible in any part, without an inordinate increase in its length, if the turgescence of the cells was always followed by growth.

Disturbance of the Periodic Movements of Cotyledons by Light.—The hypocotyls and cotyledons of most seedling plants are, as is well known, extremely heliotropic; but cotyledons, besides being heliotropic, are affected paratonically (to use Sachs' expression) by light; that is, their daily periodic movements are greatly and quickly disturbed by changes in its intensity or by its absence. It is not that they cease to circumnutate in darkness, for in all the many cases observed by us they continued to do so; but the normal order of their movements in relation to the alternations of day and night is much disturbed or quite annulled. This holds good with species the cotyledons of which rise or sink so much at night that they may be said to sleep, as well as with others which rise only a little. But different species are affected in very different degrees by changes in the light.

For instance, the cotyledons of Beta vulgaris, Solanum lycopersicum, Cerinthe major, and Lupinus luteus, when placed in darkness, moved down during the afternoon and early night, instead of rising as they would have done if they had been exposed to the light. All the individuals of the Solanum did not behave in the same manner, for the cotyledons of one circumnutated about the same spot between 2.30 and 10 P.M. The cotyledons of a seedling of Oxalis corniculata, which was feebly illuminated from above, moved downwards during the first morning in the normal manner, but on the second morning it moved upwards. The cotyledons of Lotus Jacoboeus were not affected by 4 h. of complete darkness, but when placed under a double skylight and thus feebly illuminated, they quite lost their periodical movements on the third morning. On the other hand, the cotyledons of Cucurbita ovifera moved in the normal manner during a whole day in darkness.

Seedlings of Githago segetum were feebly illuminated from above in the morning before their cotyledons had expanded, and they remained closed for the next 40 h. Other seedlings were placed in the dark after their cotyledons had opened in the morning and these did not begin to close until about 4 h. had elapsed. The cotyledons of Oxalis rosea sank vertically downwards after being left for 1 h. 20 m. in darkness; but those of some other species of Oxalis were not affected by several hours of darkness. The cotyledons of several species of Cassia are eminently susceptible to changes in the degree of light to which they are exposed: thus seedlings of an unnamed S. Brazilian species (a large and beautiful tree) were brought out of the hot-house and placed on a table in the middle of a room with two north-east and one north-west window, so that they were fairly well illuminated, though of course less so than in the hot-house, the day being moderately bright; and after 36 m. the cotyledons which had been horizontal rose up vertically and closed together as when asleep; after thus remaining on the table for 1 h. 13 m. they began to open. The cotyledons of young seedlings of another Brazilian species and of C. neglecta, treated in the same manner, behaved similarly, excepting that they did not rise up quite so much: they again became horizontal after about an hour.

Here is a more interesting case: seedlings of Cassia tora in two pots, which had stood for some time on the table in the room just described, had their cotyledons horizontal. One pot was now exposed for 2 h. to dull sunshine, and the cotyledons remained horizontal; it was then brought back to the table, and after 50 m. the cotyledons had risen 68° above the horizon. The other pot was placed during the same 2 h. behind a screen in the room, where the light was very obscure, and the cotyledons rose 63° above the horizon; the pot was then replaced on the table, and after 50 m. the cotyledons had fallen 33°. These two pots with seedlings of the same age stood close together, and were exposed to exactly the same amount of light, yet the cotyledons in the one pot were rising, whilst those in the other pot were at the same time sinking. This fact illustrates in a striking manner that their movements are not governed by the actual amount, but by a change in the intensity or degree of the light. A similar experiment was tried with two sets of seedlings, both exposed to a dull light, but different in degree, and the result was the same. The movements of the cotyledons of this Cassia are, however, determined (as in many other cases) largely by habit or inheritance, independently of light; for seedlings which had been moderately illuminated during the day, were kept all night and on the following morning in complete darkness; yet the cotyledons were partially open in the morning and remained open in the dark for about 6 h. The cotyledons in another pot, similarly treated on another occasion, were open at 7 A.M. and remained open in the dark for 4 h. 30 m., after which time they began to close. Yet these same seedlings, when brought in the middle of the day from a moderately bright into only a moderately dull light raised, as we have seen, their cotyledons high above the horizon.

Sensitiveness of Cotyledons to contact.—This subject does not possess much interest, as it is not known that sensitiveness of this kind is of any service to seedling plants. We have observed cases in only four genera, though we have vainly observed the cotyledons of many others. The genus cassia seems to be pre-eminent in this respect: thus, the cotyledons of C. tora, when extended horizontally, were both lightly tapped with a very thin twig for 3 m. and in the course of a few minutes they formed together an angle of 90°, so that each had risen 45°. A single cotyledon of another seedling was tapped in a like manner for 1 m., and it rose 27° in 9 m.; and after eight additional minutes it had risen 10° more; the opposite cotyledon, which was not tapped, hardly moved at all. The cotyledons in all these cases became horizontal again in less than half an hour. The pulvinus is the most sensitive part, for on slightly pricking three cotyledons with a pin in this part, they rose up vertically; but the blade was found also to be sensitive, care having been taken that the pulvinus was not touched. Drops of water placed quietly on these cotyledons produced no effect, but an extremely fine stream of water, ejected from a syringe, caused them to move upwards. When a pot of seedlings was rapidly hit with a stick and thus jarred, the cotyledons rose slightly. When a minute drop of nitric acid was placed on both pulvini of a seedling, the cotyledons rose so quickly that they could easily be seen to move, and almost immediately afterwards they began to fall; but the pulvini had been killed and became brown.

The cotyledons of an unnamed species of Cassia (a large tree from S. Brazil) rose 31° in the course of 26 m. after the pulvini and the blades had both been rubbed during 1 m. with a twig; but when the blade alone was similarly rubbed the cotyledons rose only 8°. The remarkably long and narrow cotyledons, of a third unnamed species from S. Brazil, did not move when their blades were rubbed on six occasions with a pointed stick for 30 s. or for 1 m.; but when the pulvinus was rubbed and slightly pricked with a pin, the cotyledons rose in the course of a few minutes through an angle of 6°o. Several cotyledons of C. neglecta (likewise from S. Brazil) rose in from 5 m. to 15 m. to various angles between 16v and 34°, after being rubbed during 1 m. with a twig. Their sensitiveness is retained to a somewhat advanced age, for the cotyledons of a little plant of C. neglecta, 34 days old and bearing three true leaves, rose when lightly pinched between the finger and thumb. Some seedlings were exposed for 30 m. to a wind (temp. 50° F.) sufficiently strong to keep the cotyledons vibrating, but this to our surprise did not cause any movement. The cotyledons of four seedlings of the Indian C. glauca were either rubbed with a thin twig for 2 m. or were lightly pinched: one rose 34°; a second only 6°; a third 13°; and a fourth 17°. A cotyledon of C. florida similarly treated rose 9°; one of C. corymbosa rose 7½°, and one of the very distinct C. mimosoides only 6°. Those of C. pubescens did not appear to be in the least sensitive; nor were those of C. nodosa, but these latter are rather thick and fleshy, and do not rise at night or go to sleep.

Smithia sensitiva.—This plant belongs to a distinct sub-order of the Leguminosae from Cassia. Both cotyledons of an oldish seedling, with the first true leaf partially unfolded, were rubbed for 1 m. with a fine twig, and in 5 m. each rose 32°; they remained in this position for 15 m., but when looked at again 40 m. after the rubbing, each had fallen 14°. Both cotyledons of another and younger seedling were lightly rubbed in the same manner for 1 m., and after an interval of 32 m. each had risen 30°. They were hardly at all sensitive to a fine jet of water. The cotyledons of S. Pfundii, an African water plant, are thick and fleshy; they are not sensitive and do not go to sleep.

Mimosa pudica and albida.—The blades of several cotyledons of both these plants were rubbed or slightly scratched with a needle during 1 m. or 2 m.; but they did not move in the least. When, however, the pulvini of six cotyledons of M. pudica were thus scratched, two of them were slightly raised. In these two cases perhaps the pulvinus was accidentally pricked, for on pricking the pulvinus of another cotyledon it rose a little. It thus appears that the cotyledons of Mimosa are less sensitive than those of the previously mentioned plants.[29]

[29] The sole notice which we have met with on the sensitiveness of cotyledons, relates to Mimosa; for Aug. P. De Candolle says ('Phys. Vég.,' 1832, tom. ii. p. 865), "les cotyledons du M. pudica tendent à se raprocher par leurs faces supérieures lorsqu'on les irrite."

Oxalis sensitiva.—The blades and pulvini of two cotyledons, standing horizontally, were rubbed or rather tickled for 30 s. with a fine split bristle, and in 10 m. each had risen 48°; when looked at again in 35 m. after being rubbed they had risen 4° more; after 30 additional minutes they were again horizontal. On hitting a pot rapidly with a stick for 1 m., the cotyledons of two seedlings were considerably raised in the course of 11 m. A pot was carried a little distance on a tray and thus jolted; and the cotyledons of four seedlings were all raised in 10 m.; after 17 m. one had risen 56°, a second 45°, a third almost 90°, and a fourth 90°. After an additional interval of 40 m. three of them had re-expanded to a considerable extent. These observations were made before we were aware at what an extraordinarily rapid rate the cotyledons circumnutate, and are therefore liable to error. Nevertheless it is extremely improbable that the cotyledons in the eight cases given, should all have been rising at the time when they were irritated. The cotyledons of Oxalis Valdiviana and rosea were rubbed and did not exhibit any sensitiveness.

Finally, there seems to exist some relation between the habit of cotyledons rising vertically at night or going to sleep, and their sensitiveness, especially that of their pulvini, to a touch; for all the above-named plants sleep at night. On the other hand, there are many plants the cotyledons of which sleep, and are not in the least sensitive. As the cotyledons of several species of Cassia are easily affected both by slightly diminished light and by contact, we thought that these two kinds of sensitiveness might be connected; but this is not necessarily the case, for the cotyledons of Oxalis sensitiva did not rise when kept on one occasion for 1½ h., and on a second occasion for nearly 4 h., in a dark closet. Some other cotyledons, as those of Githago segetum, are much affected by a feeble light, but do not move when scratched by a needle. That with the same plant there is some relation between the sensitiveness of its cotyledons and leaves seems highly probable, for the above described Smithia and Oxalis have been called sensitiva, owing to their leaves being sensitive; and though the leaves of the several species of Cassia are not sensitive to a touch, yet if a branch be shaken or syringed with water, they partially assume their nocturnal dependent position. But the relation between the sensitiveness to contact of the cotyledons and of the leaves of the same plant is not very close, as may be inferred from the cotyledons of Mimosa pudica being only slightly sensitive, whilst the leaves are well known to be so in the highest degree. Again, the leaves of Neptunia oleracea are very sensitive to a touch, whilst the cotyledons do not appear to be so in any degree.

CHAPTER III.
SENSITIVENESS OF THE APEX OF THE RADICLE TO CONTACT AND TO OTHER IRRITANTS.

Manner in which radicles bend when they encounter an obstacle in the soil—Vicia faba, tips of radicles highly sensitive to contact and other irritants—Effects of too high a temperature—Power of discriminating between objects attached on opposite sides—Tips of secondary radicles sensitive—Pisum, tips of radicles sensitive—Effects of such sensitiveness in overcoming geotropism—Secondary radicles—Phaseolus, tips of radicles hardly sensitive to contact, but highly sensitive to caustic and to the removal of a slice—Tropaeolum—Gossypium—Cucurbita—Raphanus—Æsculus, tip not sensitive to slight contact, highly sensitive to caustic—Quercus, tip highly sensitive to contact—Power of discrimination—Zea, tip highly sensitive, secondary radicles—Sensitiveness of radicles to moist air—Summary of chapter.

In order to see how the radicles of seedlings would pass over stones, roots, and other obstacles, which they must incessantly encounter in the soil, germinating beans (Vicia faba) were so placed that the tips of the radicles came into contact, almost rectangularly or at a high angle, with underlying plates of glass. In other cases the beans were turned about whilst their radicles were growing, so that they descended nearly vertically on their own smooth, almost flat, broad upper surfaces. The delicate root-cap, when it first touched any directly opposing surface, was a little flattened transversely; the flattening soon became oblique, and in a few hours quite disappeared, the apex now pointing at right angles, or at nearly right angles, to its former course. The radicle then seemed to glide in its new direction over the surface which had opposed it, pressing on it with very little force. How far such abrupt changes in its former course are aided by the circumnutation of the tip must be left doubtful. Thin slips of wood were cemented on more or less steeply inclined glass-plates, at right angles to the radicles which were gliding down them. Straight lines had been painted along the growing terminal part of some of these radicles, before they met the opposing slip of wood; and the lines became sensibly curved in 2 h. after the apex had come into contact with the slips. In one case of a radicle, which was growing rather slowly, the root-cap, after encountering a rough slip of wood at right angles, was at first slightly flattened transversely: after an interval of 2 h. 30 m. the flattening became oblique; and after an additional 3 hours the flattening had wholly disappeared, and the apex now pointed at right angles to its former course. It then continued to grow in its new direction alongside the slip of wood, until it came to the end of it, round which it bent rectangularly. Soon afterwards when coming to the edge of the plate of glass, it was again bent at a large angle, and descended perpendicularly into the damp sand.

When, as in the above cases, radicles encountered an obstacle at right angles to their course, the terminal growing part became curved for a length of between .3 and .4 of an inch (8–10 mm.), measured from the apex. This was well shown by the black lines which had been previously painted on them. The first and most obvious explanation of the curvature is, that it results merely from the mechanical resistance to the growth of the radicle in its original direction. Nevertheless, this explanation did not seem to us satisfactory. The radicles did not present the appearance of having been subjected to a sufficient pressure to account for their curvature; and Sachs has shown[1] that the growing part is more rigid than the part immediately above which has ceased to grow, so that the latter might have been expected to yield and become curved as soon as the apex encountered an unyielding object; whereas it was the stiff growing part which became curved. Moreover, an object which yields with the greatest ease will deflect a radicle: thus, as we have seen, when the apex of the radicle of the bean encountered the polished surface of extremely thin tin-foil laid on soft sand, no impression was left on it, yet the radicle became deflected at right angles. A second explanation occurred to us, namely, that even the gentlest pressure might check the growth of the apex, and in this case growth could continue only on one side, and thus the radicle would assume a rectangular form; but this view leaves wholly unexplained the curvature of the upper part, extending for a length of 8–10 mm.

[1] 'Arbeiten Bot. Inst. Würzburg,' Heft iii. 1873, p. 398.

We were therefore led to suspect that the apex was sensitive to contact, and that an effect was transmitted from it to the upper part of the radicle, which was thus excited to bend away from the touching object. As a little loop of fine thread hung on a tendril or on the petiole of a leaf-climbing plant, causes it to bend, we thought that any small hard object affixed to the tip of a radicle, freely suspended and growing in damp air, might cause it to bend, if it were sensitive, and yet would not offer any mechanical resistance to its growth. Full details will be given of the experiments which were tried, as the result proved remarkable. The fact of the apex of a radicle being sensitive to contact has never been observed, though, as we shall hereafter see, Sachs discovered that the radicle a little above the apex is sensitive, and bends like a tendril towards the touching object. But when one side of the apex is pressed by any object, the growing part bends away from the object; and this seems a beautiful adaptation for avoiding obstacles in the soil, and, as we shall see, for following the lines of least resistance. Many organs, when touched, bend in one fixed direction, such as the stamens of Berberis, the lobes of Dionaea, etc.; and many organs, such as tendrils, whether modified leaves or flower-peduncles, and some few stems, bend towards a touching object; but no case, we believe, is known of an organ bending away from a touching object.

Sensitiveness of the Apex of the Radicle of Vicia faba.—Common beans, after being soaked in water for 24 h., were pinned with the hilum downwards (in the manner followed by Sachs), inside the cork lids of glass-vessels, which were half filled with water; the sides and the cork were well moistened, and light was excluded. As soon as the beans had protruded radicles, some to a length of less than a tenth of an inch, and others to a length of several tenths, little squares or oblongs of card were affixed to the short sloping sides of their conical tips. The squares therefore adhered obliquely with reference to the longitudinal axis of the radicle; and this is a very necessary precaution, for if the bits of card accidentally became displaced, or were drawn by the viscid matter employed so as to adhere parallel to the side of the radicle, although only a little way above the conical apex, the radicle did not bend in the peculiar manner which we are here considering. Squares of about the 1/20th of an inch (i.e. about 1½ mm.), or oblong bits of nearly the same size, were found to be the most convenient and effective. We employed at first ordinary thin card, such as visiting cards, or bits of very thin glass, and various other objects; but afterwards sand-paper was chiefly employed, for it was almost as stiff as thin card, and the roughened surface favoured its adhesion. At first we generally used very thick gum-water; and this of course, under the circumstances, never dried in the least; on the contrary, it sometimes seemed to absorb vapour, so that the bits of card became separated by a layer of fluid from the tip. When there was no such absorption and the card was not displaced, it acted well and caused the radicle to bend to the opposite side. I should state that thick gum-water by itself induces no action. In most cases the bits of card were touched with an extremely small quantity of a solution of shellac in spirits of wine, which had been left to evaporate until it was thick; it then set hard in a few seconds, and fixed the bits of card well. When small drops of the shellac were placed on the tips without any card, they set into hard little beads, and these acted like any other hard object, causing the radicles to bend to the opposite side. Even extremely minute beads of the shellac occasionally acted in a slight degree, as will hereafter be described. But that it was the cards which chiefly acted in our many trials, was proved by coating one side of the tip with a little bit of goldbeaters' skin (which by itself hardly acts), and then fixing a bit of card to the skin with shellac which never came into contact with the radicle: nevertheless the radicle bent away from the attached card in the ordinary manner.

Some preliminary trials were made, presently to be described, by which the proper temperature was determined, and then the following experiments were made. It should be premised that the beans were always fixed to the cork-lids, for the convenience of manipulation, with the edge from which the radicle and plumule protrudes, outwards; and it must be remembered that owing to what we have called Sachs' curvature, the radicles, instead of growing perpendicularly downwards, often bend somewhat, even as much as about 45° inwards, or under the suspended bean. Therefore when a square of card was fixed to the apex in front, the bowing induced by it coincided with Sachs' curvature, and could be distinguished from it only by being more strongly pronounced or by occurring more quickly. To avoid this source of doubt, the squares were fixed either behind, causing a curvature in direct opposition to that of Sachs', or more commonly to the right or left sides. For the sake of brevity, we will speak of the bits of card, etc., as fixed in front, or behind, or laterally. As the chief curvature of the radicle is at a little distance from the apex, and as the extreme terminal and basal portions are nearly straight, it is possible to estimate in a rough manner the amount of curvature by an angle; and when it is said that the radicle became deflected at any angle from the perpendicular, this implies that the apex was turned upwards by so many degrees from the downward direction which it would naturally have followed, and to the side opposite to that to which the card was affixed. That the reader may have a clear idea of the kind of movement excited by the bits of attached card, we append here accurate sketches of three germinating beans thus treated, and selected out of several specimens to show the gradations in the degrees of curvature. We will now give in detail a series of experiments, and afterwards a summary of the results.

Fig. 65. Vicia faba: A, radicle beginning to bend from the attached little square of card; B, bent at a rectangle; C, bent into a circle or loop, with the tip beginning to bend downwards through the action of geotropism.

In the first 12 trials, little squares or oblongs of sanded card, 1.8 mm. in length, and 1.5 or only 0.9 mm. in breadth (i.e. .071 of an inch in length and .059 or .035 of an inch in breadth) were fixed with shellac to the tips of the radicles. In the subsequent trials the little squares were only occasionally measured, but were of about the same size.

(1.) A young radicle, 4 mm. in length, had a card fixed behind: after 9 h. deflected in the plane in which the bean is flattened, 50° from the perpendicular and from the card, and in opposition to Sachs' curvature: no change next morning, 23 h. from the time of attachment.

(2.) Radicle 5.5 mm. in length, card fixed behind: after 9 h. deflected in the plane of the bean 20° from the perpendicular and from the card, and in opposition to Sachs' curvature: after 23 h. no change.

(3.) Radicle 11 mm. in length, card fixed behind: after 9 h. deflected in the plane of the bean 40° from the perpendicular and from the card, and in opposition to Sachs' curvature. The tip of the radicle more curved than the upper part, but in the same plane. After 23 h. the extreme tip was slightly bent towards the card; the general curvature of the radicle remaining the same.

(4.) Radicle 9 mm. long, card fixed behind and a little laterally: after 9 h. deflected in the plane of the bean only about 7° or 8° from the perpendicular and from the card, in opposition to Sachs' curvature. There was in addition a slight lateral curvature directed partly from the card. After 23 h. no change.

(5.) Radicle 8 mm. long, card affixed almost laterally: after 9 h. deflected 30° from the perpendicular, in the plane of the bean and in opposition to Sachs' curvature; also deflected in a plane at right angles to the above one, 20° from the perpendicular: after 23 h. no change.

(6.) Radicle 9 mm. long, card affixed in front: after 9 h. deflected in the plane of the bean about 40° from the vertical, away from the card and in the direction of Sachs' curvature. Here therefore we have no evidence of the card being the cause of the deflection, except that a radicle never moves spontaneously, as far as we have seen, as much as 40° in the course of 9 h. After 23 h. no change.

(7.) Radicle 7 mm. long, card affixed to the back: after 9 h. the terminal part of the radicle deflected in the plane of the bean 20° from the vertical, away from the card and in opposition to Sachs' curvature. After 22 h. 30 m. this part of the radicle had become straight.

(8.) Radicle 12 mm. long, card affixed almost laterally: after 9 h. deflected laterally in a plane at right angles to that of the bean between 40° and 50° from the vertical and from the card. In the plane of the bean itself the deflection amounted to 8° or 9° from the vertical and from the card, in opposition to Sachs' curvature. After 22 h. 30 m. the extreme tip had become slightly curved towards the card.

(9.) Card fixed laterally: after 11 h. 30 m. no effect, the radicle being still almost vertical.

(10.) Card fixed almost laterally: after 11 h. 30 m. deflected 90° from the vertical and from the card, in a plane intermediate between that of the bean itself and one at right angles to it. Radicle consequently partially deflected from Sachs' curvature.

(11.) Tip of radicle protected with goldbeaters' skin, with a square of card of the usual dimensions affixed with shellac: after 11 h. greatly deflected in the plane of the bean, in the direction of Sachs' curvature, but to a much greater degree and in less time than ever occurs spontaneously.

(12.) Tip of radicle protected as in last case: after 11 h. no effect, but after 24 h. 40 m. radicle clearly deflected from the card. This slow action was probably due to a portion of the goldbeaters' skin having curled round and lightly touched the opposite side of the tip and thus irritated it.

(13.) A radicle of considerable length had a small square of card fixed with shellac to its apex laterally: after only 7 h. 15 m. a length of .4 of an inch from the apex, measured along the middle, was considerably curved from the side bearing the card.

(14.) Case like the last in all respects, except that a length of only .25 of an inch of the radicle was thus deflected.

(15.) A small square of card fixed with shellac to the apex of a young radicle; after 9 h. 15 m. deflected through 90° from the perpendicular and from the card. After 24 h. deflection much decreased, and after an additional day, reduced to 23° from the perpendicular.

(16.) Square of card fixed with shellac behind the apex of a radicle, which from its position having been changed during growth had become very crooked; but the terminal portion was straight, and this became deflected to about 45° from the perpendicular and from the card, in opposition to Sachs' curvature.

(17.) Square of card affixed with shellac: after 8 h. radicle curved at right angles from the perpendicular and from the card. After 15 additional hours curvature much decreased.

(18.) Square of card affixed with shellac: after 8 h. no effect; after 23 h. 3 m. from time of affixing, radicle much curved from the square. (19.) Square of card affixed with shellac: after 24 h. no effect, but the radicle had not grown well and seemed sickly.

(20.) Square of card affixed with shellac: after 24 h. no effect.

(21, 22.) Squares of card affixed with shellac: after 24 h. radicles of both curved at about 45° from the perpendicular and from the cards.

(23.) Square of card fixed with shellac to young radicle: after 9 h. very slightly curved from the card; after 24 h. tip curved towards card. Refixed new square laterally, after 9 h. distinctly curved from the card, and after 24 h. curved at right angles from the perpendicular and from the card.

(24.) A rather large oblong piece of card fixed with shellac to apex: after 24 h. no effect, but the card was found not to be touching the apex. A small square was now refixed with shellac; after 16 h. slight deflection from the perpendicular and from the card. After an additional day the radicle became almost straight.

(25.) Square of card fixed laterally to apex of young radicle; after 9 h. deflection from the perpendicular considerable; after 24 h. deflection reduced. Refixed a fresh square with shellac: after 24 h. deflection about 40° from the perpendicular and from the card.

(26.) A very small square of card fixed with shellac to apex of young radicle: after 9 h. the deflection from the perpendicular and from the card amounted to nearly a right angle; after 24 h. deflection much reduced; after an additional 24 h. radicle almost straight.

(27.) Square of card fixed with shellac to apex of young radicle: after 9 h. deflection from the card and from the perpendicular a right angle; next morning quite straight. Refixed a square laterally with shellac; after 9 h. a little deflection, which after 24 h. increased to nearly 20° from the perpendicular and from the card.

(28.) Square of card fixed with shellac; after 9 h. some deflection; next morning the card dropped off; refixed it with shellac; it again became loose and was refixed; and now on the third trial the radicle was deflected after 14 h. at right angles from the card.

(29.) A small square of card was first fixed with thick gum-water to the apex. It produced a slight effect but soon fell off. A similar square was now affixed laterally with shellac: after 9 h. the radicle was deflected nearly 45° from the perpendicular and from the card. After 36 additional hours angle of deflection reduced to about 30°.

(30.) A very small piece, less than 1/20th of an inch square, of thin tin-foil fixed with shellac to the apex of a young radicle; after 24 h. no effect. Tin-foil removed, and a small square of sanded card fixed with shellac; after 9 h. deflection at nearly right angles from the perpendicular and from the card. Next morning deflection reduced to about 40° from the perpendicular.

(31.) A splinter of thin glass gummed to apex, after 9 h. no effect, but it was then found not to be touching the apex of the radicle. Next morning a square of card was fixed with shellac to it, and after 9 h. radicle greatly deflected from the card. After two additional days the deflection had decreased and was only 35° from the perpendicular.

(32.) Small square of sanded card, attached with thick gum-water laterally to the apex of a long straight radicle: after 9 h. greatly deflected from the perpendicular and from the card. Curvature extended for a length of .22 of an inch from the apex. After 3 additional hours terminal portion deflected at right angles from the perpendicular. Next morning the curved portion was .36 in length.

(33.) Square of card gummed to apex: after 15 h. deflected at nearly 90° from the perpendicular and from the card.

(34.) Small oblong of sanded card gummed to apex: after 15 h. deflected 90° from the perpendicular and from the card: in the course of the three following days the terminal portion became much contorted and ultimately coiled into a helix.

(35.) Square of card gummed to apex: after 9 h. deflected from card: after 24 h. from time of attachment greatly deflected obliquely and partly in opposition to Sachs' curvature.

(36.) Small piece of card, rather less than 1/20th of an inch square, gummed to apex: in 9 h. considerably deflected from card and in opposition to Sachs' curvature; after 24 h. greatly deflected in the same direction. After an additional day the extreme tip was curved towards the card.

(37.) Square of card, gummed to apex in front, caused after 8 h. 30 m. hardly any effect; refixed fresh square laterally, after 15 h. deflected almost 90° from the perpendicular and from the card. After 2 additional days deflection much reduced.

(38.) Square of card gummed to apex: after 9 h. much deflection, which after 24 h. from time of fixing increased to nearly 90°. After an additional day terminal portion was curled into a loop, and on the following day into a helix.

(39.) Small oblong piece of card gummed to apex, nearly in front, but a little to one side; in 9 h. slightly deflected in the direction of Sachs' curvature, but rather obliquely, and to side opposite to card. Next day more curved in the same direction, and after 2 additional days coiled into a ring.

(40.) Square of card gummed to apex: after 9 h. slightly curved from card; next morning radicle straight, and apex had grown beyond the card. Refixed another square laterally with shellac; in 9 h. deflected laterally, but also in the direction of Sachs' curvature. After 2 additional days' curvature considerably increased in the same direction.

(41.) Little square of tin-foil fixed with gum to one side of apex of a young and short radicle: after 15 h. no effect, but tin-foil had become displaced. A little square of card was now gummed to one side of apex, which after 8 h. 40 m. was slightly deflected; in 24 h. from the time of attachment deflected at 90° from the perpendicular and from the card; after 9 additional hours became hooked, with the apex pointing to the zenith. In 3 days from the time of attachment the terminal portion of the radicle formed a ring or circle.

(42.) A little square of thick letter-paper gummed to the apex of a radicle, which after 9 h. was deflected from it. In 24 h. from time when the paper was affixed the deflection much increased, and after 2 additional days it amounted to 50° from the perpendicular and from the paper.

(43.) A narrow chip of a quill was fixed with shellac to the apex of a radicle. After 9 h. no effect; after 24 h. moderate deflection, but now the quill had ceased to touch the apex. Removed quill and gummed a little square of card to apex, which after 8 h. caused slight deflection. On the fourth day from the first attachment of any object, the extreme tip was curved towards the card.

(44.) A rather long and narrow splinter of extremely thin glass, fixed with shellac to apex, it caused in 9 h. slight deflection, which disappeared in 24 h.; the splinter was then found not touching the apex. It was twice refixed, with nearly similar results, that is, it caused slight deflection, which soon disappeared. On the fourth day from the time of first attachment the tip was bent towards the splinter.

From these experiments it is clear that the apex of the radicle of the bean is sensitive to contact, and that it causes the upper part to bend away from the touching object. But before giving a summary of the results, it will be convenient briefly to give a few other observations. Bits of very thin glass and little squares of common card were affixed with thick gum-water to the tips of the radicles of seven beans, as a preliminary trial. Six of these were plainly acted on, and in two cases the radicles became coiled up into complete loops. One radicle was curved into a semi-circle in so short a period as 6 h. 10 m. The seventh radicle which was not affected was apparently sickly, as it became brown on the following day; so that it formed no real exception. Some of these trials were made in the early spring during cold weather in a sitting-room, and others in a greenhouse, but the temperature was not recorded. These six striking cases almost convinced us that the apex was sensitive, but of course we determined to make many more trials. As we had noticed that the radicles grew much more quickly when subjected to considerable heat, and as we imagined that heat would increase their sensitiveness, vessels with germinating beans suspended in damp air were placed on a chimney-piece, where they were subjected during the greater part of the day to a temperature of between 69° and 72° F.; some, however, were placed in the hot-house where the temperature was rather higher. Above two dozen beans were thus tried; and when a square of glass or card did not act, it was removed, and a fresh one affixed, this being often done thrice to the same radicle. Therefore between five and six dozen trials were altogether made. But there was moderately distinct deflection from the perpendicular and from the attached object in only one radicle out of this large number of cases. In five other cases there was very slight and doubtful deflection. We were astonished at this result, and concluded that we had made some inexplicable mistake in the first six experiments. But before finally relinquishing the subject, we resolved to make one other trial for it occurred to us that sensitiveness is easily affected by external conditions, and that radicles growing naturally in the earth in the early spring would not be subjected to a temperature nearly so high as 70° F. We therefore allowed the radicles of 12 beans to grow at a temperature of between 55° and 60° F. The result was that in every one of these cases (included in the above-described experiments) the radicle was deflected in the course of a few hours from the attached object. All the above recorded successful trials, and some others presently to be given, were made in a sitting-room at the temperatures just specified. It therefore appears that a temperature of about, or rather above, 70° F. destroys the sensitiveness of the radicles, either directly, or indirectly through abnormally accelerated growth; and this curious fact probably explains why Sachs, who expressly states that his beans were kept at a high temperature, failed to detect the sensitiveness of the apex of the radicle.

But other causes interfere with this sensibility. Eighteen radicles were tried with little squares of sanded card, some affixed with shellac and some with gum-water, during the few last days of 1878, and few first days of the next year. They were kept in a room at the proper temperature during the day, but were probably too cold at night, as there was a hard frost at the time. The radicles looked healthy but grew very slowly. The result was that only 6 out of the 18 were deflected from the attached cards, and this only to a slight degree and at a very slow rate. These radicles therefore presented a striking contrast with the 44 above described. On March 6th and 7th, when the temperature of the room varied between 53° and 59° F., eleven germinating beans were tried in the same manner, and now every one of the radicles became curved away from the cards, though one was only slightly deflected. Some horticulturists believe that certain kinds of seeds will not germinate properly in the middle of the winter, although kept at a right temperature. If there really is any proper period for the germination of the bean, the feeble degree of sensibility of the above radicles may have resulted from the trial having been made in the middle of the winter, and not simply from the nights being too cold. Lastly, the radicles of four beans, which from some innate cause germinated later than all the others of the same lot, and which grew slowly though appearing healthy, were similarly tried, and even after 24 h. they were hardly at all deflected from the attached cards. We may therefore infer that any cause which renders the growth of the radicles either slower or more rapid than the normal rate, lessens or annuls the sensibility of their tips to contact. It deserves particular attention that when the attached objects failed to act, there was no bending of any kind, excepting Sachs' curvature. The force of our evidence would have been greatly weakened if occasionally, though rarely, the radicles had become curved in any direction independently of the attached objects. In the foregoing numbered paragraphs, however, it may be observed that the extreme tip sometimes becomes, after a considerable interval of time, abruptly curved towards the bit of card; but this is a totally distinct phenomenon, as will presently be explained.

A Summary of the Results of the foregoing Experiments on the Radicles of Vicia faba.—Altogether little squares (about 1/20th of an inch), generally of sanded paper as stiff as thin card (between .15 and .20 mm. in thickness), sometimes of ordinary card, or little fragments of very thin glass etc., were affixed at different times to one side of the conical tips of 55 radicles. The 11 last-mentioned cases, but not the preliminary ones, are here included. The squares, etc., were most commonly affixed with shellac, but in 19 cases with thick gum-water. When the latter was used, the squares were sometimes found, as previously stated, to be separated from the apex by a layer of thick fluid, so that there was no contact, and consequently no bending of the radicle; and such few cases were not recorded. But in every instance in which shellac was employed, unless the square fell off very soon, the result was recorded. In several instances when the squares became displaced, so as to stand parallel to the radicle, or were separated by fluid from the apex, or soon fell off, fresh squares were attached, and these cases (described under the numbered paragraphs) are here included. Out of 55 radicles experimented on under the proper temperature, 52 became bent, generally to a considerable extent from the perpendicular, and away from the side to which the object was attached. Of the three failures, one can be accounted for, as the radicle became sickly on the following day; and a second was observed only during 11 h. 30 m. As in several cases the terminal growing part of the radicle continued for some time to bend from the attached object, it formed itself into a hook, with the apex pointing to the zenith, or even into a ring, and occasionally into a spire or helix. It is remarkable that these latter cases occurred more frequently when objects were attached with thick gum-water, which never became dry, than when shellac was employed. The curvature was often well-marked in from 7 h. to 11 h.; and in one instance a semicircle was formed in 6 h. 10 m, from the time of attachment. But in order to see the phenomenon as well displayed as in the above described cases, it is indispensable that the bits of card, etc., should be made to adhere closely to one side of the conical apex; that healthy radicles should be selected and kept at not too high or too low a temperature, and apparently that the trials should not be made in the middle of the winter.

In ten instances, radicles which had curved away from a square of card or other object attached to their tips, straightened themselves to a certain extent, or even completely, in the course of from one to two days from the time of attachment. This was more especially apt to occur when the curvature was slight. But in one instance (No. 27) a radicle which in 9 h. had been deflected about 90° from the perpendicular, became quite straight in 24 h. from the period of attachment. With No. 26, the radicle was almost straight in 48 h. We at first attributed the straightening process to the radicles becoming accustomed to a slight stimulus, in the same manner as a tendril or sensitive petiole becomes accustomed to a very light loop of thread, and unbends itself though the loop remains still suspended; but Sachs states[2] that radicles of the bean placed horizontally in damp air after curving downwards through geotropism, straighten themselves a little by growth along their lower or concave sides. Why this should occur is not clear: but perhaps it likewise occurred in the above ten cases. There is another occasional movement which must not be passed over: the tip of the radicle, for a length of from 2 to 3 mm., was found in six instances, after an interval of about 24 or more hours, bent towards the bit of still attached card,—that is, in a direction exactly opposite to the previously induced curvature of the whole growing part for a length of from 7 to 8 mm. This occurred chiefly when the first curvature was small, and when an object had been affixed more than once to the apex of the same radicle. The attachment of a bit of card by shellac to one side of the tender apex may sometimes mechanically prevent its growth; or the application of thick gum-water more than once to the same side may injure it; and then checked growth on this side with continued growth on the opposite and unaffected side would account for the reversed curvature of the apex.

[2] 'Arbeiten Bot. Instit., Würzburg,' Heft iii. p. 456.

Various trials were made for ascertaining, as far as we could, the nature and degree of irritation to which the apex must be subjected, in order that the terminal growing part should bend away, as if to avoid the cause of irritation. We have seen in the numbered experiments, that a little square of rather thick letter-paper gummed to the apex induced, though slowly, considerable deflection. Judging from several cases in which various objects had been affixed with gum, and had soon become separated from the apex by a layer of fluid, as well as from some trials in which drops of thick gum-water alone had been applied, this fluid never causes bending. We have also seen in the numbered experiments that narrow splinters of quill and of very thin glass, affixed with shellac, caused only a slight degree of deflection, and this may perhaps have been due to the shellac itself. Little squares of goldbeaters' skin, which is excessively thin, were damped, and thus made to adhere to one side of the tips of two radicles; one of these, after 24 h., produced no effect; nor did the other in 8 h., within which time squares of card usually act; but after 24 h. there was slight deflection.

An oval bead, or rather cake, of dried shellac, 1.01 mm. in length and 0.63 in breadth, caused a radicle to become deflected at nearly right angles in the course of only 6 h.; but after 23 h. it had nearly straightened itself. A very small quantity of dissolved shellac was spread over a bit of card, and the tips of 9 radicles were touched laterally with it; only two of them became slightly deflected to the side opposite to that bearing the speck of dried shellac, and they afterwards straightened themselves. These specks were removed, and both together weighed less than 1/100th of a grain; so that a weight of rather less than 1/200th of a grain (0.32 mg.) sufficed to excite movement in two out of the nine radicles. Here then we have apparently reached nearly the minimum weight which will act.

A moderately thick bristle (which on measurement was found rather flattened, being 0.33 mm. in one diameter, and 0.20 mm. in the other) was cut into lengths of about 1/20th of an inch. These after being touched with thick gum-water, were placed on the tips of eleven radicles. Three of them were affected; one being deflected in 8 h. 15 m. to an angle of about 90° from the perpendicular; a second to the same amount when looked at after 9 h.; but after 24 h. from the time of first attachment the deflection had decreased to only 19°; the third was only slightly deflected after 9 h., and the bit of bristle was then found not touching the apex; it was replaced, and after 15 additional hours the deflection amounted to 26° from the perpendicular. The remaining eight radicles were not at all acted on by the bits of bristle, so that we here appear to have nearly reached the minimum of size of an object which will act on the radicle of the bean. But it is remarkable that when the bits of bristle did act, that they should have acted so quickly and efficiently.

As the apex of a radicle in penetrating the ground must be pressed on all sides, we wished to learn whether it could distinguish between harder or more resisting, and softer substances. A square of the sanded paper, almost as stiff as card, and a square of extremely thin paper (too thin for writing on), of exactly the same size (about 1/20th of an inch), were fixed with shellac on opposite sides of the apices of 12 suspended radicles. The sanded card was between 0.15 and 0.20 mm. (or between 0.0059 and 0.0079 of an inch), and the thin paper only 0.045 mm. (or 0.00176 of an inch) in thickness. In 8 out of the 12 cases there could be no doubt that the radicle was deflected from the side to which the card-like paper was attached, and towards the opposite side, bearing the very thin paper. This occurred in some instances in 9 h., but in others not until 24 h. had elapsed. Moreover, some of the four failures can hardly be considered as really failures: thus, in one of them, in which the radicle remained quite straight, the square of thin paper was found, when both were removed from the apex, to have been so thickly coated with shellac that it was almost as stiff as the card: in the second case, the radicle was bent upwards into a semicircle, but the deflection was not directly from the side bearing the card, and this was explained by the two squares having become cemented laterally together, forming a sort of stiff gable, from which the radicle was deflected: in the third case, the square of card had been fixed by mistake in front, and though there was deflection from it, this might have been due to Sachs' curvature: in the fourth case alone no reason could be assigned why the radicle had not been at all deflected. These experiments suffice to prove that the apex of the radicle possesses the extraordinary power of discriminating between thin card and very thin paper, and is deflected from the side pressed by the more resisting or harder substance.

Some trials were next made by irritating the tips without any object being left in contact with them. Nine radicles, suspended over water, had their tips rubbed, each six times with a needle, with sufficient force to shake the whole bean; the temperature was favourable, viz. about 63° F. In 7 out of these cases no effect whatever was produced; in the eighth case the radicle became slightly deflected from, and in the ninth case slightly deflected towards, the rubbed side; but these two latter opposed curvatures were probably accidental, as radicles do not always grow perfectly straight downwards. The tips of two other radicles were rubbed in the same manner for 15 seconds with a little round twig, two others for 30 seconds, and two others for 1 minute, but without any effect being produced. We may therefore conclude from these 15 trials that the radicles are not sensitive to temporary contact, but are acted on only by prolonged, though very slight, pressure.

We then tried the effects of cutting off a very thin slice parallel to one of the sloping sides of the apex, as we thought that the wound would cause prolonged irritation, which might induce bending towards the opposite side, as in the case of an attached object. Two preliminary trials were made: firstly, slices were cut from the radicles of 6 beans suspended in damp air, with a pair of scissors, which, though sharp, probably caused considerable crushing, and no curvature followed. Secondly, thin slices were cut with a razor obliquely off the tips of three radicles similarly suspended; and after 44 h. two were found plainly bent from the sliced surface; and the third, the whole apex of which had been cut off obliquely by accident, was curled upwards over the bean, but it was not clearly ascertained whether the curvature had been at first directed from the cut surface. These results led us to pursue the experiment, and 18 radicles, which had grown vertically downwards in damp air, had one side of their conical tips sliced off with a razor. The tips were allowed just to enter the water in the jars, and they were exposed to a temperature 14°–16° C. (57°–61° F.). The observations were made at different times. Three were examined 12 h. after being sliced, and were all slightly curved from the cut surface; and the curvature increased considerably after an additional 12 h. Eight were examined after 19 h.; four after 22 h. 30 m.; and three after 25 h. The final result was that out of the 18 radicles thus tried, 13 were plainly bent from the cut surface after the above intervals of time; and one other became so after an additional interval of 13 h. 30 m. So that only 4 out of the 18 radicles were not acted on. To these 18 cases the 3 previously mentioned ones should be added. It may, therefore, be concluded that a thin slice removed by a razor from one side of the conical apex of the radicle causes irritation, like that from an attached object, and induces curvature from the injured surface.

Lastly, dry caustic (nitrate of silver) was employed to irritate one side of the apex. If one side of the apex or of the whole terminal growing part of a radicle, is by any means killed or badly injured, the other side continues to grow; and this causes the part to bend over towards the injured side.[3] But in the following experiments we endeavoured, generally with success, to irritate the tips on one side, without badly injuring them. This was effected by first drying the tip as far as possible with blotting-paper, though it still remained somewhat damp, and then touching it once with quite dry caustic. Seventeen radicles were thus treated, and were suspended in moist air over water at a temperature of 58° F. They were examined after an interval of 21 h. or 24h. The tips of two were found blackened equally all round, so that they could tell nothing and were rejected, 15 being left. Of these, 10 were curved from the side which had been touched, where there was a minute brown or blackish mark. Five of these radicles, three of which were already slightly deflected, were allowed to enter the water in the jar, and were re-examined after an additional interval of 27 h. (i.e. in 48 h. after the application of the caustic), and now four of them had become hooked, being bent from the discoloured side, with their points directed to the zenith; the fifth remained unaffected and straight. Thus 11 radicles out of the 15 were acted on. But the curvature of the four just described was so plain, that they alone would have sufficed to show that the radicles of the bean bend away from that side of the apex which has been slightly irritated by caustic.

[3] Ciesielski found this to be the case ('Untersuchungen über die Abwartskrümmung der Wurzel,' 1871, p. 28) after burning with heated platinum one side of a radicle. So did we when we painted longitudinally half of the whole length of 7 radicles, suspended over water, with a thick layer of grease, which is very injurious or even fatal to growing parts; for after 48 hours five of these radicles were curved towards the greased side, two remaining straight.

The Power of an Irritant on the apex of the Radicle of the Bean, compared with that of Geotropism.—We know that when a little square of card or other object is fixed to one side of the tip of a vertically dependent radicle, the growing part bends from it often into a semicircle, in opposition to geotropism, which force is conquered by the effect of the irritation from the attached object. Radicles were therefore extended horizontally in damp air, kept at the proper low temperature for full sensitiveness, and squares of card were affixed with shellac on the lower sides of their tips, so that if the squares acted, the terminal growing part would curve upwards. Firstly, eight beans were so placed that their short, young, horizontally extended radicles would be simultaneously acted on both by geotropism and by Sachs' curvature, if the latter came into play; and they all eight became bowed downwards to the centre of the earth in 20 h., excepting one which was only slightly acted on. Two of them were a little bowed downwards in only 5 h.! Therefore the cards, affixed to the lower sides of their tips, seemed to produce no effect; and geotropism easily conquered the effects of the irritation thus caused. Secondly, 5 oldish radicles, 1½ inch in length, and therefore less sensitive than the above-mentioned young ones, were similarly placed and similarly treated. From what has been seen on many other occasions, it may be safely inferred that if they had been suspended vertically they would have bent away from the cards; and if they had been extended horizontally, without cards attached to them, they would have quickly bent vertically downwards through geotropism; but the result was that two of these radicles were still horizontal after 23 h.; two were curved only slightly, and the fifth as much as 40° beneath the horizon. Thirdly, 5 beans were fastened with their flat surfaces parallel to the cork-lid, so that Sachs' curvature would not tend to make the horizontally extended radicles turn either upwards or downwards, and little squares of card were affixed as before, to the lower sides of their tips. The result was that all five radicles were bent down, or towards the centre of the earth, after only 8 h. 20 m. At the same time and within the same jars, 3 radicles of the same age, with squares affixed to one side, were suspended vertically; and after 8 h. 20 m. they were considerably deflected from the cards, and therefore curved upwards in opposition to geotropism. In these latter cases the irritation from the squares had over-powered geotropism; whilst in the former cases, in which the radicles were extended horizontally, geotropism had overpowered the irritation. Thus within the same jars, some of the radicles were curving upwards and others downwards at the same time—these opposite movements depending on whether the radicles, when the squares were first attached to them, projected vertically down, or were extended horizontally. This difference in their behaviour seems at first inexplicable, but can, we believe, be simply explained by the difference between the initial power of the two forces under the above circumstances, combined with the well-known principle of the after-effects of a stimulus. When a young and sensitive radicle is extended horizontally, with a square attached to the lower side of the tip, geotropism acts on it at right angles, and, as we have seen, is then evidently more efficient than the irritation from the square; and the power of geotropism will be strengthened at each successive period by its previous action—that is, by its after-effects. On the other hand, when a square is affixed to a vertically dependent radicle, and the apex begins to curve upwards, this movement will be opposed by geotropism acting only at a very oblique angle, and the irritation from the card will be strengthened by its previous action. We may therefore conclude that the initial power of an irritant on the apex of the radicle of the bean, is less than that of geotropism when acting at right angles, but greater than that of geotropism when acting obliquely on it.

Sensitiveness of the tips of the Secondary Radicles of the Bean to contact.—All the previous observations relate to the main or primary radicle. Some beans suspended to cork-lids, with their radicles dipping into water, had developed secondary or lateral radicles, which were afterwards kept in very damp air, at the proper low temperature for full sensitiveness. They projected, as usual, almost horizontally, with only a slight downward curvature, and retained this position during several days. Sachs has shown[4] that these secondary roots are acted on in a peculiar manner by geotropism, so that if displaced they reassume their former sub-horizontal position, and do not bend vertically downwards like the primary radicle. Minute squares of the stiff sanded paper were affixed by means of shellac (but in some instances with thick gum-water) to the tips of 39 secondary radicles of different ages, generally the uppermost ones. Most of the squares were fixed to the lower sides of the apex, so that if they acted the radicle would bend upwards; but some were fixed laterally, and a few on the upper side. Owing to the extreme tenuity of these radicles, it was very difficult to attach the square to the actual apex. Whether owing to this or some other circumstance, only nine of the squares induced any curvature. The curvature amounted in some cases to about 45° above the horizon, in others to 90°, and then the tip pointed to the zenith. In one instance a distinct upward curvature was observed in 8 h. 15 m., but usually not until 24 h. had elapsed. Although only 9 out of 39 radicles were affected, yet the curvature was so distinct in several of them, that there could be no doubt that the tip is sensitive to slight contact, and that the growing part bends away from the touching object. It is possible that some secondary radicles are more sensitive than others; for Sachs has proved[5] the interesting fact that each individual secondary radicle possesses its own peculiar constitution.

[4] 'Arbeiten Bot. Inst., Würzburg,' Heft iv. 1874, p. 605–617.

[5] 'Arbeiten Bot. Instit., Würzburg,' Heft, iv. 1874, p. 620.

Sensitiveness to contact of the Primary Radicle, a little above the apex, in the Bean (Vicia faba) and Pea (Pisum sativum).—The sensitiveness of the apex of the radicle in the previously described cases, and the consequent curvature of the upper part from the touching object or other source of irritation, is the more remarkable, because Sachs[6] has shown that pressure at the distance of a few millimeters above the apex causes the radicle to bend, like a tendril, towards the touching object. By fixing pins so that they pressed against the radicles of beans suspended vertically in damp air, we saw this kind of curvature; but rubbing the part with a twig or needle for a few minutes produced no effect. Haberlandt remarks,[7] that these radicles in breaking through the seed-coats often rub and press against the ruptured edges, and consequently bend round them. As little squares of the card-like paper affixed with shellac to the tips were highly efficient in causing the radicles to bend away from them, similar pieces (of about 1/20th inch square, or rather less) were attached in the same manner to one side of the radicle at a distance of 3 or 4 mm. above the apex. In our first trial on 15 radicles no effect was produced. In a second trial on the same number, three became abruptly curved (but only one strongly) towards the card within 24 h. From these cases we may infer that the pressure from a bit of card affixed with shellac to one side above the apex, is hardly a sufficient irritant; but that it occasionally causes the radicle to bend like a tendril towards this side.

[6] Ibid. Heft iii. 1873, p. 437.

[7] 'Die Schutzeinrichtungen der Keimpflanze,' 1877, p. 25.

We next tried the effect of rubbing several radicles at a distance of 4 mm. from the apex for a few seconds with lunar caustic (nitrate of silver); and although the radicles had been wiped dry and the stick of caustic was dry, yet the part rubbed was much injured and a slight permanent depression was left. In such cases the opposite side continues to grow, and the radicle necessarily becomes bent towards the injured side. But when a point 4 mm.

from the apex was momentarily touched with dry caustic, it was only faintly discoloured, and no permanent injury was caused. This was shown by several radicles thus treated straightening themselves after one or two days; yet at first they became curved towards the touched side, as if they had been there subjected to slight continued pressure. These cases deserve notice, because when one side of the apex was just touched with caustic, the radicle, as we have seen, curved itself in an opposite direction, that is, away from the touched side.

The radicle of the common pea at a point a little above the apex is rather more sensitive to continued pressure than that of the bean, and bends towards the pressed side.[8] We experimented on a variety (*Yorkshire Hero*) which has a much wrinkled tough skin, too large for the included cotyledons; so that out of 30 peas which had been soaked for 24 h. and allowed to germinate on damp sand, the radicles of three were unable to escape, and were crumpled up in a strange manner within the skin; four other radicles were abruptly bent round the edges of the ruptured skin against which they had pressed. Such abnormalities would probably never, or very rarely, occur with forms developed in a state of nature and subjected to natural selection. One of the four radicles just mentioned in doubling backwards came into contact with the pin by which the pea was fixed to the cork-lid; and now it bent at right angles round the pin, in a direction quite different from that of the first curvature due to contact with the ruptured skin; and it thus afforded a good illustration of the tendril-like sensitiveness of the radicle a little above the apex.

[8] Sachs, 'Arbeiten Bot. Institut., Würzburg,' Heft iii. p. 438.

Little squares of the card-like paper were next affixed to radicles of the pea at 4 mm. above the apex, in the same manner as with the bean. Twenty-eight radicles suspended vertically over water were thus treated on different occasions, and 13 of them became curved towards the cards. The greatest degree of curvature amounted to 62° from the perpendicular; but so large an angle was only once formed. On one occasion a slight curvature was perceptible after 5 h. 45 m., and it was generally well-marked after 14 h. There can therefore be no doubt that with the pea, irritation from a bit of card attached to one side of the radicle above the apex suffices to induce curvature.

Squares of card were attached to one side of the tips of 11 radicles within the same jars in which the above trials were made, and five of them became plainly, and one slightly, curved away from this side. Other analogous cases will be immediately described. The fact is here mentioned because it was a striking spectacle, showing the difference in the sensitiveness of the radicle in different parts, to behold in the same jar one set of radicles curved away from the squares on their tips, and another set curved towards the squares attached a little higher up. Moreover, the kind of curvature in the two cases is different. The squares attached above the apex cause the radicle to bend abruptly, the part above and beneath remaining nearly straight; so that here there is little or no transmitted effect. On the other hand, the squares attached to the apex affect the radicle for a length of from about 4 to even 8 mm., inducing in most cases a symmetrical curvature; so that here some influence is transmitted from the apex for this distance along the radicle.

Pisum sativum (var. Yorkshire Hero): Sensitiveness of the apex of the Radicle.—Little squares of the same card-like paper were affixed (April 24th) with shellac to one side of the apex of 10 vertically suspended radicles: the temperature of the water in the bottom of the jars was 60°–61° F. Most of these radicles were acted on in 8 h. 30 m.; and eight of them became in the course of 24 h. conspicuously, and the remaining two slightly, deflected from the perpendicular and from the side bearing the attached squares. Thus all were acted on; but it will suffice to describe two conspicuous cases. In one the terminal portion of the radicle was bent at right angles (A, Fig. 66) after 24h.; and in the other (B) it had by this time become hooked, with the apex pointing to the zenith. The two bits of card here used were .07 inch in length and .04 inch in breadth. Two other radicles, which after 8 h. 30 m. were moderately deflected, became straight again after 24h. Another trial was made in the same manner with 15 radicles; but from circumstances, not worth explaining, they were only once and briefly examined after the short interval of 5 h. 30 m.; and we merely record in our notes "almost all bent slightly from the perpendicular, and away from the squares; the deflection amounting in one or two instances to nearly a rectangle." These two sets of cases, especially the first one, prove that the apex of the radicle is sensitive to slight contact and that the upper part bends from the touching object. Nevertheless, on June 1st and 4th, 8 other radicles were tried in the same manner at a temperature of 58°–60° F., and after 24 h. only 1 was decidedly bent from the card, 4 slightly, 2 doubtfully, and 1 not in the least. The amount of curvature was unaccountably small; but all the radicles which were at all bent, were bent away from the cards.

Fig. 66. Pisum sativum: deflection produced within 24 hours in the growth of vertically dependent radicles, by little squares of card affixed with shellac to one side of apex: A, bent at right angles; B, hooked.

We now tried the effects of widely different temperatures on the sensitiveness of these radicles with squares of card attached to their tips. Firstly, 13 peas, most of them having very short and young radicles, were placed in an ice-box, in which the temperature rose during three days from 44° to 47° F. They grew slowly, but 10 out of the 13 became in the course of the three days very slightly curved from the squares; the other 3 were not affected; so that this temperature was too low for any high degree of sensitiveness or for much movement. Jars with 13 other radicles were next placed on a chimney-piece, where they were subjected to a temperature of between 68° and 72° F., and after 24 h., 4 were conspicuously curved from the cards, 2 slightly, and 7 not at all; so that this temperature was rather too high. Lastly 12 radicles were subjected to a temperature varying between 72° and 85° F., and none of them were in the least affected by the squares. The above several trials, especially the first recorded one, indicate that the most favourable temperature for the sensitiveness of the radicle of the pea is about 60° F.

The tips of 6 vertically dependent radicles were touched once with dry caustic, in the manner described under Vicia faba. After 24 h. four of them were bent from the side bearing a minute black mark; and the curvature increased in one case after 38 h., and in another case after 48 h., until the terminal part projected almost horizontally. The two remaining radicles were not affected.

With radicles of the bean, when extended horizontally in damp air, geotropism always conquered the effects of the irritation caused by squares of card attached to the lower sides of their tips. A similar experiment was tried on 13 radicles of the pea; the squares being attached with shellac, and the temperature between 58°–60° F. The result was somewhat different; for these radicles are either less strongly acted on by geotropism, or, what is more probable, are more sensitive to contact. After a time geotropism always prevailed, but its action was often delayed; and in three instances there was a most curious struggle between geotropism and the irritation caused by the cards. Four of the 13 radicles were a little curved downwards within 6 or 8 h., always reckoning from the time when the squares were first attached, and after 23 h. three of them pointed vertically downwards, and the fourth at an angle of 45° beneath the horizon. These four radicles therefore did not seem to have been at all affected by the attached squares. Four others were not acted on by geotropism within the first 6 or 8 h., but after 23 h. were much bowed down. Two others remained almost horizontal for 23 h., but afterwards were acted on. So that in these latter six cases the action of geotropism was much delayed. The eleventh radicle was slightly curved down after 8 h., but when looked at again after 23 h. the terminal portion was curved upwards; if it had been longer observed, the tip no doubt would have been found again curved down, and it would have formed a loop as in the following case. The twelfth radicle after 6 h. was slightly curved downwards; but when looked at again after 21 h., this curvature had disappeared and the apex pointed upwards; after 30 h. the radicle formed a hook, as shown at A (Fig. 67); which hook after 45 h. was converted into a loop (B). The thirteenth radicle after 6 h. was slightly curved downwards, but within 21 h. had curved considerably up, and then down again at an angle of 45° beneath the horizon, afterwards becoming perpendicular. In these three last cases geotropism and the irritation caused by the attached squares alternately prevailed in a highly remarkable manner; geotropism being ultimately victorious.

Fig. 67. Pisum sativum: a radicle extended horizontally in damp air with a little square of card affixed to the lower side of its tip, causing it to bend upwards in opposition to geotropism. The deflection of the radicle after 21 hours is shown at A, and of the same radicle after 45 hours at B, now forming a loop.

Similar experiments were not always quite so successful as in the above cases. Thus 6 radicles, horizontally extended with attached squares, were tried on June 8th at a proper temperature, and after 7 h. 30 m. none were in the least curved upwards and none were distinctly geotropic; whereas of 6 radicles without any attached squares, which served as standards of comparison or controls, 3 became slightly and 3 almost rectangularly geotropic within the 7 h. 30 m.; but after 23 h. thc two lots were equally geotropic. On July 10th another trial was made with 6 horizontally extended radicles, with squares attached in the same manner beneath their tips; and after 7 h. 30 m., 4 were slightly geotropic, 1 remained horizontal, and 1 was curved upwards in opposition to gravity or geotropism. This latter radicle after 48 h. formed a loop, like that at B (Fig. 67).

An analogous trial was now made, but instead of attaching squares of card to the lower sides of the tips, these were touched with dry caustic. The details of the experiment will be given in the chapter on Geotropism, and it will suffice here to say that 10 peas, with radicles extended horizontally and not cauterised, were laid on and under damp friable peat; these, which served as standards or controls, as well as 10 others which had been touched on the upper side with the caustic, all became strongly geotropic in 24 h. Nine radicles, similarly placed, had their tips touched on the lower side with the caustic; and after 24 h., 3 were slightly geotropic, 2 remained horizontal, and 4 were bowed upwards in opposition to gravity and to geotropism. This upward curvature was distinctly visible in 8 h. 45m. after the lower sides of the tips had been cauterised.

Little squares of card were affixed with shellac on two occasions to the tips of 22 young and short secondary radicles, which had been emitted from the primary radicle whilst growing in water, but were now suspended in damp air. Besides the difficulty of attaching the squares to such finely pointed objects as were these radicles, the temperature was too high,—varying on the first occasion from 72° to 77° F., and on the second being almost steadily 78° F.; and this probably lessened the sensitiveness of the tips. The result was that after an interval of 8 h. 30 m., 6 of the 22 radicles were bowed upwards (one of them greatly) in opposition to gravity, and 2 laterally; the remaining 14 were not affected. Considering the unfavourable circumstances, and bearing in mind the case of the bean, the evidence appears sufficient to show that the tips of the secondary radicles of the pea are sensitive to slight contact.

Phaseolus multiflorus: Sensitiveness of the apex of the Radicle.—Fifty-nine radicles were tried with squares of various sizes of the same card-like paper, also with bits of thin glass and rough cinders, affixed with shellac to one side of the apex. Rather large drops of the dissolved shellac were also placed on them and allowed to set into hard beads. The specimens were subjected to various temperatures between 60° and 72° F., more commonly at about the latter. But out of this considerable number of trials only 5 radicles were plainly bent, and 8 others slightly or even doubtfully, from the attached objects; the remaining 46 not being at all affected. It is therefore clear that the tips of the radicles of this Phaseolus are much less sensitive to contact than are those of the bean or pea. We thought that they might be sensitive to harder pressure, but after several trials we could not devise any method for pressing harder on one side of the apex than on the other, without at the same time offering mechanical resistance to its growth. We therefore tried other irritants.

The tips of 13 radicles, dried with blotting-paper, were thrice touched or just rubbed on one side with dry nitrate of silver. They were rubbed thrice, because we supposed from the foregoing trials, that the tips were not highly sensitive. After 24 h. the tips were found greatly blackened; 6 were blackened equally all round, so that no curvature to any one side could be expected; 6 were much blackened on one side for a length of about 1/10th of an inch, and this length became curved at right angles towards the blackened surface, the curvature afterwards increasing in several instances until little hooks were formed. It was manifest that the blackened side was so much injured that it could not grow, whilst the opposite side continued to grow. One alone out of these 13 radicles became curved from the blackened side, the curvature extending for some little distance above the apex.

After the experience thus gained, the tips of six almost dry radicles were once touched with the dry caustic on one side; and after an interval of 10 m. were allowed to enter water, which was kept at a temperature of 65°–67° F. The result was that after an interval of 8 h. a minute blackish speck could just be distinguished on one side of the apex of five of these radicles, all of which became curved towards the opposite side—in two cases at about an angle of 45°—in two other cases at nearly a rectangle—and in the fifth case at above a rectangle, so that the apex was a little hooked; in this latter case the black mark was rather larger than in the others. After 24 h. from the application of the caustic, the curvature of three of these radicles (including the hooked one) had diminished; in the fourth it remained the same, and in the fifth it had increased, the tip being now hooked. It has been said that after 8 h. black specks could be seen on one side of the apex of five of the six radicles; on the sixth the speck, which was extremely minute, was on the actual apex and therefore central; and this radicle alone did not become curved. It was therefore again touched on one side with caustic, and after 15 h. 30 m. was found curved from the perpendicular and from the blackened side at an angle of 34°, which increased in nine additional hours to 54°.

It is therefore certain that the apex of the radicle of this Phaseolus is extremely sensitive to caustic, more so than that of the bean, though the latter is far more sensitive to pressure. In the experiments just given, the curvature from the slightly cauterised side of the tip, extended along the radicle for a length of nearly 10 mm.; whereas in the first set of experiments, when the tips of several were greatly blackened and injured on one side, so that their growth was arrested, a length of less than 3 mm. became curved towards the much blackened side, owing to the continued growth of the opposite side. This difference in the results is interesting, for it shows that too strong an irritant does not induce any transmitted effect, and does not cause the adjoining, upper and growing part of the radicle to bend. We have analogous cases with Drosera, for a strong solution of carbonate of ammonia when absorbed by the glands, or too great heat suddenly applied to them, or crushing them, does not cause the basal part of the tentacles to bend, whilst a weak solution of the carbonate, or a moderate heat, or slight pressure always induced such bending. Similar results were observed with Dionaea and Pinguicula.

The effect of cutting off with a razor a thin slice from one side of the conical apex of 14 young and short radicles was next tried. Six of them after being operated on were suspended in damp air; the tips of the other eight, similarly suspended, were allowed to enter water at a temperature of about 65° F. It was recorded in each case which side of the apex had been sliced off, and when they were afterwards examined the direction of the curvature was noted, before the record was consulted. Of the six radicles in damp air, three had their tips curved after an interval of 10 h. 15 m. directly away from the sliced surface, whilst the other three were not affected and remained straight; nevertheless, one of them after 13 additional hours became slightly curved from the sliced surface. Of the eight radicles with their tips immersed in water, seven were plainly curved away from the sliced surfaces after 10 h. 15 m.; and with respect to the eighth which remained quite straight, too thick a slice had been accidentally removed, so that it hardly formed a real exception to the general result. When the seven radicles were looked at again, after an interval of 23 h. from the time of slicing, two had become distorted; four were deflected at an angle of about 70° from the perpendicular and from the cut surface; and one was deflected at nearly 90°, so that it projected almost horizontally, but with the extreme tip now beginning to bend downwards through the action of geotropism. It is therefore manifest that a thin slice cut off one side of the conical apex, causes the upper growing part of the radicle of this Phaseolus to bend, through the transmitted effects of the irritation, away from the sliced surface.

Tropaeolum majus: Sensitiveness of the apex of the Radicle to contact.—Little squares of card were attached with shellac to one side of the tips of 19 radicles, some of which were subjected to 78° F., and others to a much lower temperature. Only 3 became plainly curved from the squares, 5 slightly, 4 doubtfully, and 7 not at all. These seeds were, as we believed, old, so we procured a fresh lot, and now the results were widely different. Twenty-three were tried in the same manner; five of the squares produced no effect, but three of these cases were no real exceptions, for in two of them the squares had slipped and were parallel to the apex, and in the third the shellac was in excess and had spread equally all round the apex. One radicle was deflected only slightly from the perpendicular and from the card; whilst seventeen were plainly deflected. The angles in several of these latter cases varied between 40° and 65° from the perpendicular; and in two of them it amounted after 15 h. or 16 h. to about 90°. In one instance a loop was nearly completed in 16 h. There can, therefore, be no doubt that the apex is highly sensitive to slight contact, and that the upper part of the radicle bends away from the touching object.

Gossypium herbaceum: Sensitiveness of the apex of the Radicle.—Radicles were experimented on in the same manner as before, but they proved ill-fitted for our purpose, as they soon became unhealthy when suspended in damp air. Of 38 radicles thus suspended, at temperatures varying from 66° to 69° F., with squares of card attached to their tips, 9 were plainly and 7 slightly or even doubtfully deflected from the squares and from the perpendicular; 22 not being affected. We thought that perhaps the above temperature was not high enough, so 19 radicles with attached squares, likewise suspended in damp air, were subjected to a temperature of from 74° to 79° F., but not one of them was acted on, and they soon became unhealthy. Lastly, 19 radicles were suspended in water at a temperature from 70° to 75° F., with bits of glass or squares of the card attached to their tips by means of Canada-balsam or asphalte, which adhered rather better than shellac beneath the water. The radicles did not keep healthy for long. The result was that 6 were plainly and 2 doubtfully deflected from the attached objects and the perpendicular; 11 not being affected. The evidence consequently is hardly conclusive, though from the two sets of cases tried under a moderate temperature, it is probable that the radicles are sensitive to contact; and would be more so under favourable conditions.

Fifteen radicles which had germinated in friable peat were suspended vertically over water. Seven of them served as controls, and they remained quite straight during 24 h. The tips of the other eight radicles were just touched with dry caustic on one side. After only 5 h. 10 m. five of them were slightly curved from the perpendicular and from the side bearing the little blackish marks. After 8 h. 40 m., 4 out of these 5 were deflected at angles between 15° and 65° from the perpendicular. On the other hand, one which had been slightly curved after 5 h. 10 m., now became straight. After 24 h. the curvature in two cases had considerably increased; also in four other cases, but these latter radicles had now become so contorted, some being turned upwards, that it could no longer be ascertained whether they were still curved from the cauterised side. The control specimens exhibited no such irregular growth, and the two sets presented a striking contrast. Out of the 8 radicles which had been touched with caustic, two alone were not affected, and the marks left on their tips by the caustic were extremely minute. These marks in all cases were oval or elongated; they were measured in three instances, and found to be of nearly the same size, viz. 2/3 of a mm. in length. Bearing this fact in mind, it should be observed that the length of the curved part of the radicle, which had become deflected from the cauterised side in the course of 8 h. 40 m. was found to be in three cases 6, 7, and 9 mm.

Cucurbita ovifera: Sensitiveness of the apex of the Radicle.—The tips proved ill-fitted for the attachment of cards, as they are extremely fine and flexible. Moreover, owing to the hypocotyls being soon developed and becoming arched, the whole radicle is quickly displaced and confusion is thus caused. A large number of trials were made, but without any definite result, excepting on two occasions, when out of 23 radicles 10 were deflected from the attached squares of card, and 13 were not acted on. Rather large squares, though difficult to affix, seemed more efficient than very small ones.

We were much more successful with caustic; but in our first trial, 15 radicles were too much cauterised, and only two became curved from the blackened side; the others being either killed on one side, or blackened equally all round. In our next trial the dried tips of 11 radicles were touched momentarily with dry caustic, and after a few minutes were immersed in water. The elongated marks thus caused were never black, only brown, and about ½ mm. in length, or even less. In 4 h. 30 m. after the cauterisation, 6 of them were plainly curved from the side with the brown mark, 4 slightly, and 1 not at all. The latter proved unhealthy, and never grew; and the marks on 2 of the 4 slightly curved radicles were excessively minute, one being distinguishable only with the aid of a lens. Of 10 control specimens tried in the same jars at the same time, not one was in the least curved. In 8 h. 40 m. after the cauterisation, 5 of the radicles out of the 10 (the one unhealthy one being omitted) were deflected at about 90°, and 3 at about 45° from the perpendicular and from the side bearing the brown mark. After 24 h. all 10 radicles had increased immensely in length; in 5 of them the curvature was nearly the same, in 2 it had increased, and in 3 it had decreased. The contrast presented by the 10 controls, after both the 8 h. 40 m. and the 24 h. intervals, was very great; for they had continued to grow vertically downwards, excepting two which, from some unknown cause, had become somewhat tortuous.

In the chapter on Geotropism we shall see that 10 radicles of this plant were extended horizontally on and beneath damp friable peat, under which conditions they grow better and more naturally than in damp air; and their tips were slightly cauterised on the lower side, brown marks about ½ mm. in length being thus caused. Uncauterised specimens similarly placed became much bent downwards through geotropism in the course of 5 or 6 hours. After 8 h. only 3 of the cauterised ones were bowed downwards, and this in a slight degree; 4 remained horizontal; and 3 were curved upwards in opposition to geotropism and from the side bearing the brown mark. Ten other specimens had their tips cauterised at the same time and in the same degree, on the upper side; and this, if it produced any effect, would tend to increase the power of geotropism; and all these radicles were strongly bowed downwards after 8 h. From the several foregoing facts, there can be no doubt that the cauterisation of the tip of the radicle of this Cucurbita on one side, if done lightly enough, causes the whole growing part to bend to the opposite side. Raphanus sativus: Sensitiveness of the apex of the Radicle.—We here encountered many difficulties in our trials, both with squares of card and with caustic; for when seeds were pinned to a cork-lid, many of the radicles, to which nothing had been done, grew irregularly, often curving upwards, as if attracted by the damp surface above; and when they were immersed in water they likewise often grew irregularly. We did not therefore dare to trust our experiments with attached squares of card; nevertheless some of them seemed to indicate that the tips were sensitive to contact. Our trials with caustic generally failed from the difficulty of not injuring too greatly the extremely fine tips. Out of 7 radicles thus tried, one became bowed after 22 h. at an angle of 60°, a second at 40°, and a third very slightly from the perpendicular and from the cauterised side.

Æsculus hippocastanum: Sensitiveness of the apex of the Radicle.—Bits of glass and squares of card were affixed with shellac or gum-water to the tips of 12 radicles of the horse-chestnut; and when these objects fell off, they were refixed; but not in a single instance was any curvature thus caused. These massive radicles, one of which was above 2 inches in length and .3 inch in diameter at its base, seemed insensible to so slight a stimulus as any small attached object. Nevertheless, when the apex encountered an obstacle in its downward course, the growing part became so uniformly and symmetrically curved, that its appearance indicated not mere mechanical bending, but increased growth along the whole convex side, due to the irritation of the apex.

That this is the correct view may be inferred from the effects of the more powerful stimulus of caustic. The bending from the cauterised side occurred much slower than in the previously described species, and it will perhaps be worth while to give our trials in detail.

The seeds germinated in sawdust, and one side of the tips of the radicles were slightly rubbed once with dry nitrate of silver; and after a few minutes were allowed to dip into water. They were subjected to a rather varying temperature, generally between 52° and 58° F. A few cases have not been thought worth recording, in which the whole tip was blackened, or in which the seedling soon became unhealthy.

(1.) The radicle was slightly deflected from the cauterised side in one day (i.e. 24 h.); in three days it stood at 60° from the perpendicular; in four days at 90°; on the fifth day it was curved up about 40° above the horizon; so that it had passed through an angle of 130° in the five days, and this was the greatest amount of curvature observed.

(2.) In two days radicle slightly deflected; after seven days deflected 69° from the perpendicular and from the cauterised side; after eight days the angle amounted to nearly 90°.

(3.) After one day slight deflection, but the cauterised mark was so faint that the same side was again touched with caustic. In four days from the first touch deflection amounted to 78°, which in an additional day increased to 90°.

(4.) After two days slight deflection, which during the next three days certainly increased but never became great; the radicle did not grow well and died on the eighth day.

(5.) After two days very slight deflection; but this on the fourth day amounted to 56° from the perpendicular and from the cauterised side.

(6.) After three days doubtfully, but after four days certainly deflected from the cauterised side. On the fifth day deflection amounted to 45° from the perpendicular, and this on the seventh day increased to about 90°.

(7.) After two days slightly deflected; on the third day the deflection amounted to 25° from the perpendicular, and this did not afterwards increase.

(8.) After one day deflection distinct; on the third day it amounted to 44°, and on the fourth day to 72° from the perpendicular and the cauterised side.

(9.) After two days deflection slight, yet distinct; on the third day the tip was again touched on the same side with caustic and thus killed.

(10.) After one day slight deflection, which after six days increased to 50° from the perpendicular and the cauterised side.

(11.) After one day decided deflection, which after six days increased to 62° from the perpendicular and from the cauterised side.

(12.) After one day slight deflection, which on the second day amounted to 35°, on the fourth day to 50°, and the sixth day to 63° from the perpendicular and the cauterised side.

(13.) Whole tip blackened, but more on one side than the other; on the fourth day slightly, and on the sixth day greatly deflected from the more blackened side; the deflection on the ninth day amounted to 90° from the perpendicular.

(14.) Whole tip blackened in the same manner as in the last case: on the second day decided deflection from the more blackened side, which increased on the seventh day to nearly 90°; on the following day the radicle appeared unhealthy.

(15.) Here we had the anomalous case of a radicle bending slightly *towards* the cauterised side on the first day, and continuing to do so for the next three days, when the deflection amounted to about 90° from the perpendicular. The cause appeared to lie in the tendril-like sensitiveness of the upper part of the radicle, against which the point of a large triangular flap of the seed-coats pressed with considerable force; and this irritation apparently conquered that from the cauterised apex.

These several cases show beyond doubt that the irritation of one side of the apex, excites the upper part of the radicle to bend slowly towards the opposite side. This fact was well exhibited in one lot of five seeds pinned to the cork-lid of a jar; for when after 6 days the lid was turned upside down and viewed from directly above, the little black marks made by the caustic were now all distinctly visible on the upper sides of the tips of the laterally bowed radicles. A thin slice was shaved off with a razor from one side of the tips of 22 radicles, in the manner described under the common bean; but this kind of irritation did not prove very effective. Only 7 out of the 22 radicles became moderately deflected in from 3 to 5 days from the sliced surface, and several of the others grew irregularly. The evidence, therefore, is far from conclusive.

Quercus robur: Sensitiveness of the apex of the Radicle.—The tips of the radicles of the common oak are fully as sensitive to slight contact as are those of any plant examined by us. They remained healthy in damp air for 10 days, but grew slowly. Squares of the card-like paper were fixed with shellac to the tips of 15 radicles, and ten of these became conspicuously bowed from the perpendicular and from the squares; two slightly, and three not at all. But two of the latter were not real exceptions, as they were at first very short, and hardly grew afterwards. Some of the more remarkable cases are worth describing. The radicles were examined on each successive morning, at nearly the same hour, that is, after intervals of 24 h.

No. 1. This radicle suffered from a series of accidents, and acted in an anomalous manner, for the apex appeared at first insensible and afterwards sensitive to contact. The first square was attached on Oct 19th; on the 21st the radicle was not at all curved, and the square was accidentally knocked off; it was refixed on the 22nd, and the radicle became slightly curved from the square, but the curvature disappeared on the 23rd, when the square was removed and refixed. No curvature ensued, and the square was again accidentally knocked off, and refixed. On the morning of the 27th it was washed off by having reached the water in the bottom of the jar. The square was refixed, and on the 29th, that is, ten days after the first square had been attached, and two days after the attachment of the last square, the radicle had grown to the great length of 3.2 inches, and now the terminal growing part had become bent away from the square into a hook (see Fig. 68).

Fig. 68. Quercus robur: radicle with square of card attached to one side of apex, causing it to become hooked. Drawing one-half natural scale.

No. 2. Square attached on the 19th; on the 20th radicle slightly deflected from it and from the perpendicular; on the 21st deflected at nearly right angles; it remained during the next two days in this position, but on the 25th the upward curvature was lessened through the action of geotropism, and still more so on the 26th.

No. 3. Square attached on the 19th; on the 21st a trace of curvature from the square, which amounted on the 22nd to about 40°, and on the 23rd to 53° from the perpendicular.

No. 4. Square attached on the 21st; on the 22nd trace of curvature from the square; on the 23rd completely hooked with the point turned up to the zenith. Three days afterwards (i.e. 26th) the curvature had wholly disappeared and the apex pointed perpendicularly downwards.

No. 5. Square attached on the 21st; on the 22nd decided though slight curvature from the square; on the 23rd the tip had curved up above the horizon, and on the 24th was hooked with the apex pointing almost to the zenith, as in Fig. 68.

No. 6. Square attached on the 21st; on the 22nd slightly curved from the square; 23rd more curved; 25th considerably curved; 27th all curvature lost, and the radicle was now directed perpendicularly downwards.

No. 7. Square attached on the 21st; on the 22nd a trace of curvature from the square, which increased next day, and on the 24th amounted to a right angle.

It is, therefore, manifest that the apex of the radicle of the oak is highly sensitive to contact, and retains its sensitiveness during several days. The movement thus induced was, however, slower than in any of the previous cases, with the exception of that of Æsculus. As with the bean, the terminal growing part, after bending, sometimes straightened itself through the action of geotropism, although the object still remained attached to the tip.

The same remarkable experiment was next tried, as in the case of the bean; namely, little squares of exactly the same size of the card-like sanded paper and of very thin paper (the thicknesses of which have been given under Vicia faba) were attached with shellac on opposite sides (as accurately as could be done) of the tips of 13 radicles, suspended in damp air, at a temperature of 65°–66° F. The result was striking, for 9 out of these 13 radicles became plainly, and 1 very slightly, curved from the thick paper towards the side bearing the thin paper. In two of these cases the apex became completely hooked after two days; in four cases the deflection from the perpendicular and from the side bearing the thick paper, amounted in from two to four days to angles of 90°, 72°, 60°, and 49°, but in two other cases to only 18° and 15°. It should, however, be stated that in the case in which the deflection was 49°, the two squares had accidentally come into contact on one side of the apex, and thus formed a lateral gable; and the deflection was directed in part from this gable and in part from the thick paper. In three cases alone the radicles were not affected by the difference in thickness of the squares of paper attached to their tips, and consequently did not bend away from the side bearing the stiffer paper.

Zea mays: Sensitiveness of the apex of the Radicle to contact.—A large number of trials were made on this plant, as it was the only monocotyledon on which we experimented. An abstract of the results will suffice. In the first place, 22 germinating seeds were pinned to cork-lids without any object being attached to their radicles, some being exposed to a temperature of 65°–66° F., and others to between 74° and 79°; and none of them became curved, though some were a little inclined to one side. A few were selected, which from having germinated on sand were crooked, but when suspended in damp air the terminal part grew straight downwards. This fact having been ascertained, little squares of the card-like paper were affixed with shellac, on several occasions, to the tips of 68 radicles. Of these the terminal growing part of 39 became within 24 h. conspicuously curved away from the attached squares and from the perpendicular; 13 out of the 39 forming hooks with their points directed towards the zenith, and 8 forming loops. Moreover, 7 other radicles out of the 68, were slightly and two doubtfully deflected from the cards. There remain 20 which were not affected; but 10 of these ought not to be counted; for one was diseased, two had their tips quite surrounded by shellac, and the squares on 7 had slipped so as to stand parallel to the apex, instead of obliquely on it. There were therefore only 10 out of the 68 which certainly were not acted on. Some of the radicles which were experimented on were young and short, most of them of moderate length, and two or three exceeded three inches in length. The curvature in the above cases occurred within 24 h., but it was often conspicuous within a much shorter period. For instance, the terminal growing part of one radicle was bent upwards into a rectangle in 8 h. 15 m., and of another in 9 h. On one occasion a hook was formed in 9 h. Six of the radicles in a jar containing nine seeds, which stood on a sand-bath, raised to a temperature varying from 76° to 82° F., became hooked, and a seventh formed a complete loop, when first looked at after 15 hours.

The accompanying figures of four germinating seeds (Fig. 69) show, firstly, a radicle (A) the apex of which has become so much bent away from the attached square as to form a hook. Secondly (B), a hook converted through the continued irritation of the card, aided perhaps by geotropism, into an almost complete circle or loop. The tip in the act of forming a loop generally rubs against the upper part of the radicle, and pushes off the attached square; the loop then contracts or closes, but never disappears; and the apex afterwards grows vertically downwards, being no longer irritated by any attached object. This frequently occurred, and is represented at C. The jar above mentioned with the six hooked radicles and another jar were kept for two additional days, for the sake of observing how the hooks would be modified. Most of them became converted into simple loops, like that figured at C; but in one case the apex did not rub against the upper part of the radicle and thus remove the card; and it consequently made, owing to the continued irritation from the card, two complete loops, that is, a helix of two spires; which afterwards became pressed closely together. Then geotropism prevailed and caused the apex to grow perpendicularly downwards. In another case, shown at (D), the apex in making a second turn or spire, passed through the first loop, which was at first widely open, and in doing so knocked off the card; it then grew perpendicularly downwards, and thus tied itself into a knot, which soon became tight!

Fig. 69. Zea mays: radicles excited to bend away from the little squares of card attached to one side of their tips.

Secondary Radicles of Zea.—A short time after the first radicle has appeared, others protrude from the seed, but not laterally from the primary one. Ten of these secondary radicles, which were directed obliquely downwards, were experimented on with very small squares of card attached with shellac to the lower sides of their tips. If therefore the squares acted, the radicles would bend upwards in opposition to gravity. The jar stood (protected from light) on a sand-bath, which varied between 76° and 82° F. After only 5 h. one appeared to be a little deflected from the square, and after 20 h. formed a loop. Four others were considerably curved from the squares after 20 h., and three of them became hooked, with their tips pointing to the zenith,—one after 29 h. and the two others after 44 h. By this latter time a sixth radicle had become bent at a right angle from the side bearing the square. Thus altogether six out of the ten secondary radicles were acted on, four not being affected. There can, therefore, be no doubt that the tips of these secondary radicles are sensitive to slight contact, and that when thus excited they cause the upper part to bend from the touching object; but generally, as it appears, not in so short a time as in the case of the first-formed radicle.

SENSITIVENESS OF THE TIP OF THE RADICLE TO MOIST AIR.

Sachs made the interesting discovery, a few years ago, that the radicles of many seedling plants bend towards an adjoining damp surface.[9] We shall here endeavour to show that this peculiar form of sensitiveness resides in their tips. The movement is directly the reverse of that excited by the irritants hitherto considered, which cause the growing part of the radicle to bend away from the source of irritation. In our experiments we followed Sachs' plan, and sieves with seeds germinating in damp sawdust were suspended so that the bottom was generally inclined at 40° with the horizon. If the radicles had been acted on solely by geotropism, they would have grown out of the bottom of the sieve perpendicularly downwards; but as they were attracted by the adjoining damp surface they bent towards it and were deflected 50° from the perpendicular. For the sake of ascertaining whether the tip or the whole growing part of the radicle was sensitive to the moist air, a length of from 1 to 2 mm. was coated in a certain number of cases with a mixture of olive-oil and lamp-black. This mixture was made in order to give consistence to the oil, so that a thick layer could be applied, which would exclude, at least to a large extent, the moist air, and would be easily visible. A greater number of experiments than those which were actually tried would have been necessary, had not it been clearly established that the tip of the radicle is the part which is sensitive to various other irritants.

[9] 'Arbeiten des Bot. Institut., in Würzburg,' vol. i. 1872, p. 209.

Phaseolus multiflorus.—Twenty-nine radicles, to which nothing had been done, growing out of a sieve, were observed at the same time with those which had their tips greased, and for an equal length of time. Of the 29, 24 curved themselves so as to come into close contact with the bottom of the sieve. The place of chief curvature was generally at a distance of 5 or 6 mm. from the apex. Eight radicles had their tips greased for a length of 2 mm., and two others for a length of 1½ mm.; they were kept at a temperature of 15°–16° C. After intervals of from 19 h. to 24 h. all were still vertically or almost vertically dependent, for some of them had moved towards the adjoining damp surface by about 10°. They had therefore not been acted on, or only slightly acted on, by the damper air on one side, although the whole upper part was freely exposed. After 48 h. three of these radicles became considerably curved towards the sieve; and the absence of curvature in some of the others might perhaps be accounted for by their not having grown very well. But it should be observed that during the first 19 h. to 24 h. all grew well; two of them having increased 2 and 3 mm. in length in 11 h.; five others increased 5 to 8 mm. in 19 h.; and two, which had been at first 4 and 6 mm. in length, increased in 24 h. to 15 and 20 mm.

The tips of 10 radicles, which likewise grew well, were coated with the grease for a length of only 1 mm., and now the result was somewhat different; for of these 4 curved themselves to the sieve in from 21 h. to 24h., whilst 6 did not do so. Five of the latter were observed for an additional day, and now all excepting one became curved to the sieve.

The tips of 5 radicles were cauterised with nitrate of silver, and about 1 mm. in length was thus destroyed. They were observed for periods varying between 11 h. and 24h., and were found to have grown well. One of them had curved until it came into contact with the sieve; another was curving towards it; whilst the remaining three were still vertically dependent. Of 7 not cauterised radicles observed at the same time, all had come into contact with the sieve.

The tips of 11 radicles were protected by moistened gold-beaters' skin, which adheres closely, for a length varying from 1½ to 2½ mm. After 22 h. to 24 h., 6 of these radicles were clearly bent towards or had come into contact with the sieve; 2 were slightly curved in this direction, and 3 not at all. All had grown well. Of 14 control specimens observed at the same time, all excepting one had closely approached the sieve. It appears from these cases that a cap of goldbeaters' skin checks, though only to a slight degree, the bending of the radicles to an adjoining damp surface. Whether an extremely thin sheet of this substance when moistened allows moisture from the air to pass through it, we do not know. One case indicated that the caps were sometimes more efficient than appears from the above results; for a radicle, which after 23 h. had only slightly approached the sieve, had its cap (1½ mm. in length) removed, and during the next 15½ h. it curved itself abruptly towards the source of moisture, the chief seat of curvature being at a distance of 2 to 3 mm. from the apex.

Vicia faba.—The tips of 13 radicles were coated with the grease for a length of 2 mm.; and it should be remembered that with these radicles the seat of chief curvature is about 4 or 5 mm. from the apex. Four of them were examined after 22h., three after 26 h., and six after 36 h., and none had been attracted towards the damp lower surface of the sieve. In another trial 7 radicles were similarly treated, and 5 of them still pointed perpendicularly downwards after 11 h., whilst 2 were a little curved towards the sieve; by an accident they were not subsequently observed. In both these trials the radicles grew well; 7 of them, which were at first from 4 to 11 mm. in length, were after 11 h. between 7 and 16 mm.; 3 which were at first from 6 to 8 mm. after 26 h. were 11.5 to 18 mm. in length; and lastly, 4 radicles which were at first 5 to 8 mm. after 46 h. were 18 to 23 mm. in length. The control or ungreased radicles were not invariably attracted towards the bottom of the sieve. But on one occasion 12 out of 13, which were observed for periods between 22 h. and 36 h., were thus attracted. On two other occasions taken together, 38 out of 40 were similarly attracted. On another occasion only 7 out of 14 behaved in this manner, but after two more days the proportion of the curved increased to 17 out of 23. On a last occasion only 11 out of 20 were thus attracted. If we add up these numbers, we find that 78 out of 96 of the control specimens curved themselves towards the bottom of the sieve. Of the specimens with greased tips, 2 alone out of the 20 (but 7 of these were not observed for a sufficiently long time) thus curved themselves. We can, therefore, hardly doubt that the tip for a length of 2 mm. is the part which is sensitive to a moist atmosphere, and causes the upper part to bend towards its source.

The tips of 15 radicles were cauterised with nitrate of silver, and they grew as well as those above described with greased tips. After an interval of 24 h., 9 of them were not at all curved towards the bottom of the sieve; 2 were curved towards it at angles of 20° and 12° from their former vertical position, and 4 had come into close contact with it. Thus the destruction of the tip for a length of about 1 mm. prevented the curvature of the greater number of these radicles to the adjoining damp surface. Of 24 control specimens, 23 were bent to the sieve, and on a second occasion 15 out of 16 were similarly curved in a greater or less degree. These control trials are included in those given in the foregoing paragraph.

Avena sativa.—The tips of 13 radicles, which projected between 2 and 4 mm. from the bottom of the sieve, many of them not quite perpendicularly downwards, were coated with the black grease for a length of from 1 to 1½ mm. The sieves were inclined at 30° with the horizon. The greater number of these radicles were examined after 22 h., and a few after 25 h., and within these intervals they had grown so quickly as to have nearly doubled their lengths. With the ungreased radicles the chief seat of curvature is at a distance of not less than between 3.5 and 5.5 mm., and not more than between 7 and 10 mm. from the apex. Out of the 13 radicles with greased tips, 4 had not moved at all towards the sieve; 6 were deflected towards it and from the perpendicular by angles varying between 10° and 35°; and 3 had come into close contact with it. It appears, therefore, at first sight that greasing the tips of these radicles had checked but little their bending to the adjoining damp surface. But the inspection of the sieves on two occasions produced a widely different impression on the mind; for it was impossible to behold the radicles with the black greased tips projecting from the bottom, and all those with ungreased tips, at least 40 to 50 in number, clinging closely to it, and feel any doubt that the greasing had produced a great effect. On close examination only a single ungreased radicle could be found which had not become curved towards the sieve. It is probable that if the tips had been protected by grease for a length of 2 mm. instead of from 1 to 1½ mm., they would not have been affected by the moist air and none would have become curved.

Triticum vulgare.—Analogous trials were made on 8 radicles of the common wheat; and greasing their tips produced much less effect than in the case of the oats. After 22 h., 5 of them had come into contact with the bottom of the sieve; 2 had moved towards it 10° and 15°, and one alone remained perpendicular. Not one of the very numerous ungreased radicles failed to come into close contact with the sieve. These trials were made on Nov. 28th, when the temperature was only 4.8° C. at 10 A.M. We should hardly have thought this case worth notice, had it not been for the following circumstance. In the beginning of October, when the temperature was considerably higher, viz., 12° to 13° C., we found that only a few of the ungreased radicles became bent towards the sieve; and this indicates that sensitiveness to moisture in the air is increased by a low temperature, as we have seen with the radicles of Vicia faba relatively to objects attached to their tips. But in the present instance it is possible that a difference in the dryness of the air may have caused the difference in the results at the two periods.

Finally, the facts just given with respect to Phaseolus multiflorus, Vicia faba, and Avena sativa show, as it seems to us, that a layer of grease spread for a length of 1½ to 2 mm. over the tip of the radicle, or the destruction of the tip by caustic, greatly lessens or quite annuls in the upper and exposed part the power of bending towards a neighbouring source of moisture. We should bear in mind that the part which bends most, lies at some little distance above the greased or cauterised tip; and that the rapid growth of this part, proves that it has not been injured by the tips having been thus treated. In those cases in which the radicles with greased tips became curved, it is possible that the layer of grease was not sufficiently thick wholly to exclude moisture, or that a sufficient length was not thus protected, or, in the case of the caustic, not destroyed. When radicles with greased tips are left to grow for several days in damp air, the grease is drawn out into the finest reticulated threads and dots, with narrow portions of the surface left clean. Such portions would, it is probable, be able to absorb moisture, and thus we can account for several of the radicles with greased tips having become curved towards the sieve after an interval of one or two days. On the whole, we may infer that sensitiveness to a difference in the amount of moisture in the air on the two sides of a radicle resides in the tip, which transmits some influence to the upper part, causing it to bend towards the source of moisture. Consequently, the movement is the reverse of that caused by objects attached to one side of the tip, or by a thin slice being cut off, or by being slightly cauterised. In a future chapter it will be shown that sensitiveness to the attraction of gravity likewise resides in the tip; so that it is the tip which excites the adjoining parts of a horizontally extended radicle to bend towards the centre of the earth.

SECONDARY RADICLES BECOMING VERTICALLY GEOTROPIC BY THE DESTRUCTION OR INJURY OF THE TERMINAL PART OF THE PRIMARY RADICLE.

Sachs has shown that the lateral or secondary radicles of the bean, and probably of other plants, are acted on by geotropism in so peculiar a manner, that they grow out horizontally or a little inclined downwards; and he has further shown[10] the interesting fact, that if the end of the primary radicle be cut off, one of the nearest secondary radicles changes its nature and grows perpendicularly downwards, thus replacing the primary radicle. We repeated this experiment, and planted beans with amputated radicles in friable peat, and saw the result described by Sachs; but generally two or three of the secondary radicles grew perpendicularly downwards. We also modified the experiment, by pinching young radicles a little way above their tips, between the arms of a U-shaped piece of thick leaden wire. The part pinched was thus flattened, and was afterwards prevented from growing thicker. Five radicles had their ends cut off, and served as controls or standards. Eight were pinched; of these 2 were pinched too severely and their ends died and dropped off; 2 were not pinched enough and were not sensibly affected; the remaining 4 were pinched sufficiently to check the growth of the terminal part, but did not appear otherwise injured. When the U-shaped wires were removed, after an interval of 15 days, the part beneath the wire was found to be very thin and easily broken, whilst the part above was thickened. Now in these four cases, one or more of the secondary radicles, arising from the thickened part just above the wire, had grown perpendicularly downwards. In the best case the primary radicle (the part below the wire being 1½ inch in length) was somewhat distorted, and was not half as long as three adjoining secondary radicles, which had grown vertically, or almost vertically, downwards. Some of these secondary radicles adhered together or had become confluent. We learn from these four cases that it is not necessary, in order that a secondary radicle should assume the nature of a primary one, that the latter should be actually amputated; it is sufficient that the flow of sap into it should be checked, and consequently should be directed into the adjoining secondary radicles; for this seems to be the most obvious result of the primary radicle being pinched between the arms of a U-shaped wire.

[10] 'Arbeiten Bot. Institut., Würzburg,' Heft iv. 1874, p. 622.

This change in the nature of secondary radicles is clearly analogous, as Sachs has remarked, to that which occurs with the shoots of trees, when the leading one is destroyed and is afterwards replaced by one or more of the lateral shoots; for these now grow upright instead of sub-horizontally. But in this latter case the lateral shoots are rendered apogeotropic, whereas with radicles the lateral ones are rendered geotropic. We are naturally led to suspect that the same cause acts with shoots as with roots, namely, an increased flow of sap into the lateral ones. We made some trials with Abies communis and pectinata, by pinching with wire the leading and all the lateral shoots excepting one. But we believe that they were too old when experimented on; and some were pinched too severely, and some not enough. Only one case succeeded, namely, with the spruce-fir. The leading shoot was not killed, but its growth was checked; at its base there were three lateral shoots in a whorl, two of which were pinched, one being thus killed; the third was left untouched. These lateral shoots, when operated on (July 14th) stood at an angle of 8° above the horizon; by Sept. 8th the unpinched one had risen 35°; by Oct. 4th it had risen 46°, and by Jan. 26th

48°, and it had now become a little curved inwards. Part of this rise of 48° may be attributed to ordinary growth, for the pinched shoot rose 12° within the same period. It thus follows that the unpinched shoot stood, on Jan. 26th, 56° above the horizon, or 34° from the vertical; and it was thus obviously almost ready to replace the slowly growing, pinched, leading shoot. Nevertheless, we feel some doubt about this experiment, for we have since observed with spruce-firs growing rather unhealthily, that the lateral shoots near the summit sometimes become highly inclined, whilst the leading shoot remains apparently sound.

A widely different agency not rarely causes shoots which naturally would have brown out horizontally to grow up vertically. The lateral branches of the Silver Fir (A. pectinata) are often affected by a fungus, Æcidium elatinum, which causes the branch to enlarge into an oval knob formed of hard wood, in one of which we counted 24 rings of growth. According to De Bary[11], when the mycelium penetrates a bud beginning to elongate, the shoot developed from it grows vertically upwards. Such upright shoots afterwards produce lateral and horizontal branches; and they then present a curious appearance, as if a young fir-tree had grown out of a ball of clay surrounding the branch. These upright shoots have manifestly changed their nature and become apogeotropic; for if they had not been affected by the Æcidium, they would have grown out horizontally like all the other twigs on the same branches. This change can hardly be due to an increased flow of sap into the part; but the presence of the mycelium will have greatly disturbed its natural constitution.

[11] See his valuable article in 'Bot. Zeitung,' 1867, p. 257, on these monstrous growths, which are called in German "Hexenbesen," or "witch-brooms."

According to Mr. Meehan,[12] the stems of three species of Euphorbia and of Portulaca oleracea are "normally prostrate or procumbent;" but when they are attacked by an Æcidium, they "assume an erect habit." Dr. Stahl informs us that he knows of several analogous cases; and these seem to be closely related to that of the Abies. The rhizomes of Sparganium ramosum grow out horizontally in the soil to a considerable length, or are diageotropic; but F. Elfving found that when they were cultivated in water their tips turned upwards, and they became apogeotropic. The same result followed when the stem of the plant was bent until it cracked or was merely much bowed.[13]

[12] 'Proc. Acad. Nat. Sc. Philadelphia,' June 16th, 1874, and July 23rd, 1875.

[13] See F. Elfving's interesting paper in 'Arbeiten Bot. Institut., in Würzburg,' vol. ii. 1880, p. 489. Carl Kraus (Triesdorf) had previously

observed ('Flora,' 1878, p. 324) that the underground shoots of Triticum repens bend vertically up when the parts above ground are removed, and when the rhizomes are kept partly immersed in water.

No explanation has hitherto been attempted of such cases as the foregoing,—namely, of secondary radicles growing vertically downwards, and of lateral shoots growing vertically upwards, after the amputation of the primary radicle or of the leading shoot. The following considerations give us, as we believe, the clue. Firstly, any cause which disturbs the constitution[14] is apt to induce reversion; such as the crossing of two distinct races, or a change of conditions, as when domestic animals become feral. But the case which most concerns us, is the frequent appearance of peloric flowers on the summit of a stem, or in the centre of the inflorescence,—parts which, it is believed, receive the most sap; for when an irregular flower becomes perfectly regular or peloric, this may be attributed, at least partly, to reversion to a primitive and normal type. Even the position of a seed at the end of the capsule sometimes gives to the seedling developed from it a tendency to revert. Secondly, reversions often occur by means of buds, independently of reproduction by seed; so that a bud may revert to the character of a former state many bud-generations ago. In the case of animals, reversions may occur in the individual with advancing age. Thirdly and lastly, radicles when they first protrude from the seed are always geotropic, and plumules or shoots almost always apogeotropic. If then any cause, such as an increased flow of sap or the presence of mycelium, disturbs the constitution of a lateral shoot or of a secondary radicle, it is apt to revert to its primordial state; and it becomes either apogeotropic or geotropic, as the case may be, and consequently grows either vertically upwards or downwards. It is indeed possible, or even probable, that this tendency to reversion may have been increased, as it is manifestly of service to the plant.

[14] The facts on which the following conclusions are founded are given in 'The Variation of Animals and Plants under Domestication,' 2nd edit. 1875. On the causes leading to reversion see chap. xii. vol. ii. and p. 59, chap. xiv. On peloric flowers, chap. xiii. p. 32; and see p. 337 on their position on the plant. With respect to seeds, p. 340. On reversion by means of buds, p. 438, chap. xi. vol. i.

A SUMMARY OF CHAPTER.

A part or organ may be called sensitive, when its irritation excites movement in an adjoining part. Now it has been shown in this chapter, that the tip of the radicle of the bean is in this sense sensitive to the contact of any small object attached to one side by shellac or gum-water; also to a slight touch with dry caustic, and to a thin slice cut off one side. The radicles of the pea were tried with attached objects and caustic, both of which acted. With Phaseolus multiflorus the tip was hardly sensitive to small squares of attached card, but was sensitive to caustic and to slicing. The radicles of Tropaeolum were highly sensitive to contact; and so, as far as we could judge, were those of Gossypium herbaceum, and they were certainly sensitive to caustic. The tips of the radicles of Cucurbita ovifera were likewise highly sensitive to caustic, though only moderately so to contact. Raphanus sativus offered a somewhat doubtful case. With Æsculus the tips were quite indifferent to bodies attached to them, though sensitive to caustic. Those of Quercus robur and Zea mays were highly sensitive to contact, as were the radicles of the latter to caustic. In several of these cases the difference in sensitiveness of the tip to contact and to caustic was, as we believe, merely apparent; for with Gossypium, Raphanus, and Cucurbita, the tip was so fine and flexible that it was very difficult to attach any object to one of its sides. With the radicles of Æsculus, the tips were not at all sensitive to small bodies attached to them; but it does not follow from this fact that they would not have been sensitive to somewhat greater continued pressure, if this could have been applied.

The peculiar form of sensitiveness which we are here considering, is confined to the tip of the radicle for a length of from 1 mm. to 1.5 mm. When this part is irritated by contact with any object, by caustic, or by a thin slice being cut off, the upper adjoining part of the radicle, for a length of from 6 or 7 to even 12 mm., is excited to bend away from the side which has been irritated. Some influence must therefore be transmitted from the tip along the radicle for this length. The curvature thus caused is generally symmetrical. The part which bends most apparently coincides with that of the most rapid growth. The tip and the basal part grow very slowly and they bend very little.

Considering the widely separated position in the vegetable series of the several above-named genera, we may conclude that the tips of the radicles of all, or almost all, plants are similarly sensitive, and transmit an influence causing the upper part to bend. With respect to the tips of the secondary radicles, those of Vicia faba, Pisum sativum, and Zea mays were alone observed, and they were found similarly sensitive.

In order that these movements should be properly displayed, it appears necessary that the radicles should grow at their normal rate. If subjected to a high temperature and made to grow rapidly, the tips seem either to lose their sensitiveness, or the upper part to lose the power of bending. So it appears to be if they grow very slowly from not being vigorous, or from being kept at too low a temperature; also when they are forced to germinate in the middle of the winter.

The curvature of the radicle sometimes occurs within from 6 to 8 hours after the tip has been irritated, and almost always within 24 h., excepting in the case of the massive radicles of Æsculus. The curvature often amounts to a rectangle,—that is, the terminal part bends upwards until the tip, which is but little curved, projects almost horizontally. Occasionally the tip, from the continued irritation of the attached object, continues to bend up until it forms a hook with the point directed towards the zenith, or a loop, or even a spire. After a time the radicle apparently becomes accustomed to the irritation, as occurs in the case of tendrils, for it again grows downwards, although the bit of card or other object may remain attached to the tip. It is evident that a small object attached to the free point of a vertically suspended radicle can offer no mechanical resistance to its growth as a whole, for the object is carried downwards as the radicle elongates, or upwards as the radicle curves upwards. Nor can the growth of the tip itself be mechanically checked by an object attached to it by gum-water, which remains all the time perfectly soft. The weight of the object, though quite insignificant, is opposed to the upward curvature. We may therefore conclude that it is the irritation due to contact which excites the movement. The contact, however, must be prolonged, for the tips of 15 radicles were rubbed for a short time, and this did not cause them to bend. Here then we have a case of specialised sensibility, like that of the glands of Drosera; for these are exquisitely sensitive to the slightest pressure if prolonged, but not to two or three rough touches.

When the tip of a radicle is lightly touched on one side with dry nitrate of silver, the injury caused is very slight, and the adjoining upper part bends away from the cauterised point, with more certainty in most cases than from an object attached on one side. Here it obviously is not the mere touch, but the effect produced by the caustic, which induces the tip to transmit some influence to the adjoining part, causing it to bend away. If one side of the tip is badly injured or killed by the caustic, it ceases to grow, whilst the opposite side continues growing; and the result is that the tip itself bends towards the injured side and often becomes completely hooked; and it is remarkable that in this case the adjoining upper part does not bend. The stimulus is too powerful or the shock too great for the proper influence to be transmitted from the tip. We have strictly analogous cases with Drosera, Dionaea and Pinguicula, with which plants a too powerful stimulus does not excite the tentacles to become incurved, or the lobes to close, or the margin to be folded inwards.

With respect to the degree of sensitiveness of the apex to contact under favourable conditions, we have seen that with Vicia faba a little square of writing-paper affixed with shellac sufficed to cause movement; as did on one occasion a square of merely damped goldbeaters' skin, but it acted very slowly. Short bits of moderately thick bristle (of which measurements have been given) affixed with gum-water acted in only three out of eleven trials, and beads of dried shellac under 1/200th of a grain in weight acted only twice in nine cases; so that here we have nearly reached the minimum of necessary irritation. The apex, therefore, is much less sensitive to pressure than the glands of Drosera, for these are affected by far thinner objects than bits of bristle, and by a very much less weight than 1/200th of a grain. But the most interesting evidence of the delicate sensitiveness of the tip of the radicle, was afforded by its power of discriminating between equal-sized squares of card-like and very thin paper, when these were attached on opposite sides, as was observed with the radicles of the bean and oak.

When radicles of the bean are extended horizontally with squares of card attached to the lower sides of their tips, the irritation thus caused was always conquered by geotropism, which then acts under the most favourable conditions at right angles to the radicle. But when objects were attached to the radicles of any of the above-named genera, suspended vertically, the irritation conquered geotropism, which latter power at first acted obliquely on the radicle; so that the immediate irritation from the attached object, aided by its after-effects, prevailed and caused the radicle to bend upwards, until sometimes the point was directed to the zenith. We must, however, assume that the after-effects of the irritation of the tip by an attached object come into play, only after movement has been excited. The tips of the radicles of the pea seem to be more sensitive to contact than those of the bean, for when they were extended horizontally with squares of card adhering to their lower sides, a most curious struggle occasionally arose, sometimes one and sometimes the other force prevailing, but ultimately geotropism was always victorious; nevertheless, in two instances the terminal part became so much curved upwards that loops were subsequently formed. With the pea, therefore, the irritation from an attached object, and from geotropism when acting at right angles to the radicle, are nearly balanced forces. Closely similar results were observed with the horizontally extended radicles of *Cucurbita ovifera*, when their tips were slightly cauterised on the lower side.

Finally, the several co-ordinated movements by which radicles are enabled to perform their proper functions are admirably perfect. In whatever direction the primary radicle first protrudes from the seed, geotropism guides it perpendicularly downwards; and the capacity to be acted on by the attraction of gravity resides in the tip. But Sachs has proved[15] that the secondary radicles, or those emitted by the primary one, are acted on by geotropism in such a manner that they tend to bend only obliquely downwards. If they had been acted on like the primary radicle, all the radicles would have penetrated the ground in a close bundle. We have seen that if the end of the primary radicle is cut off or injured, the adjoining secondary radicles become geotropic and grow vertically downwards. This power must often be of great service to the plant, when the primary radicle has been destroyed by the larvae of insects, burrowing animals, or any other accident. The tertiary radicles, or those emitted by the secondary ones, are not influenced, at least in the case of the bean, by geotropism; so they grow out freely in all directions. From this manner of growth of the various kinds of radicles, they are distributed, together with their absorbent hairs, throughout the surrounding soil, as Sachs has remarked, in the most advantageous manner; for the whole soil is thus closely searched.

[15] 'Arbeiten Bot. Institut, Würzburg,' Heft iv. 1874, pp. 605–631.

Geotropism, as was shown in the last chapter, excites the primary radicle to bend downwards with very little force, quite insufficient to penetrate the ground. Such penetration is effected by the pointed apex (protected by the root-cap) being pressed down by the longitudinal expansion or growth of the terminal rigid portion, aided by its transverse expansion, both of which forces act powerfully. It is, however, indispensable that the seeds should be at first held down in some manner. When they lie on the bare surface they are held down by the attachment of the root-hairs to any adjoining objects; and this apparently is effected by the conversion of their outer surfaces into a cement. But many seeds get covered up by various accidents, or they fall into crevices or holes. With some seeds their own weight suffices. The circumnutating movement of the terminal growing part both of the primary and secondary radicles is so feeble that it can aid them very little in penetrating the ground, excepting when the superficial layer is very soft and damp. But it must aid them materially when they happen to break obliquely into cracks, or into burrows made by earth-worms or larvae. This movement, moreover, combined with the sensitiveness of the tip to contact, can hardly fail to be of the highest importance; for as the tip is always endeavouring to bend to all sides it will press on all sides, and will thus be able to discriminate between the harder and softer adjoining surfaces, in the same manner as it discriminated between the attached squares of card-like and thin paper. Consequently it will tend to bend from the harder soil, and will thus follow the lines of least resistance. So it will be if it meets with a stone or the root of another plant in the soil, as must incessantly occur. If the tip were not sensitive, and if it did not excite the upper part of the root to bend away, whenever it encountered at right angles some obstacle in the ground, it would be liable to be doubled up into a contorted mass. But we have seen with radicles growing down inclined plates of glass, that as soon as the tip merely touched a slip of wood cemented across the plate, the whole terminal growing part curved away, so that the tip soon stood at right angles to its former direction; and thus it would be with an obstacle encountered in the ground, as far as the pressure of the surrounding soil would permit. We can also understand why thick and strong radicles, like those of Æsculus, should be endowed with less sensitiveness than more delicate ones; for the former would be able by the force of their growth to overcome any slight obstacle.

After a radicle, which has been deflected by some stone or root from its natural downward course, reaches the edge of the obstacle, geotropism will direct it to grow again straight downward; but we know that geotropism acts with very little force, and here another excellent adaptation, as Sachs has remarked,[16] comes into play. For the upper part of the radicle, a little above the apex, is, as we have seen, likewise sensitive; and this sensitiveness causes the radicle to bend like a tendril towards the touching object, so that as it rubs over the edge of an obstacle, it will bend downwards; and the curvature thus induced is abrupt, in which respect it differs from that caused by the irritation of one side of the tip. This downward bending coincides with that due to geotropism, and both will cause the root to resume its original course.

[16] 'Arbeiten Bot. Inst., Würzburg,' Heft iii. p. 456.

As radicles perceive an excess of moisture in the air on one side and bend towards this side, we may infer that they will act in the same manner with respect to moisture in the earth. The sensitiveness to moisture resides in the tip, which determines the bending of the upper part. This capacity perhaps partly accounts for the extent to which drain-pipes often become choked with roots.

Considering the several facts given in this chapter, we see that the course followed by a root through the soil is governed by extraordinarily complex and diversified agencies,—by geotropism acting in a different manner on the primary, secondary, and tertiary radicles,—by sensitiveness to contact, different in kind in the apex and in the part immediately above the apex, and apparently by sensitiveness to the varying dampness of different parts of the soil. These several stimuli to movement are all more powerful than geotropism, when this acts obliquely on a radicle, which has been deflected from its perpendicular downward course. The roots, moreover, of most plants are excited by light to bend either to or from it; but as roots are not naturally exposed to the light it is doubtful whether this sensitiveness, which is perhaps only the indirect result of the radicles being highly sensitive to other stimuli, is of any service to the plant. The direction which the apex takes at each successive period of the growth of a root, ultimately determines its whole course; it is therefore highly important that the apex should pursue from the first the most advantageous direction; and we can thus understand why sensitiveness to geotropism, to contact and to moisture, all reside in the tip, and why the tip determines the upper growing part to bend either from or to the exciting cause. A radicle may be compared with a burrowing animal such as a mole, which wishes to penetrate perpendicularly down into the ground. By continually moving his head from side to side, or circumnutating, he will feel any stone or other obstacle, as well as any difference in the hardness of the soil, and he will turn from that side; if the earth is damper on one than on the other side he will turn thitherward as a better hunting-ground. Nevertheless, after each interruption, guided by the sense of gravity, he will be able to recover his downward course and to burrow to a greater depth.

CHAPTER IV.
THE CIRCUMNUTATING MOVEMENTS OF THE SEVERAL PARTS OF MATURE PLANTS.

Circumnutation of stems: concluding remarks on—Circumnutation of stolons: aid thus afforded in winding amongst the stems of surrounding plants—Circumnutation of flower-stems—Circumnutation of Dicotyledonous leaves—Singular oscillatory movement of leaves of Dionaea—Leaves of Cannabis sink at night—Leaves of Gymnosperms—Of Monocotyledons—Cryptogams—Concluding remarks on the circumnutation of leaves; generally rise in the evening and sink in the morning.

We have seen in the first chapter that the stems of all seedlings, whether hypocotyls or epicotyls, as well as the cotyledons and the radicles, are continually circumnutating—that is they grow first on one side and then on another, such growth being probably preceded by increased turgescence of the cells. As it was unlikely that plants should change their manner of growth with advancing age, it seemed probable that the various organs of all plants at all ages, as long as they continued to grow, would be found to circumnutate, though perhaps to an extremely small extent. As it was important for us to discover whether this was the case, we determined to observe carefully a certain number of plants which were growing vigorously, and which were not known to move in any manner. We commenced with stems. Observations of this kind are tedious, and it appeared to us that it would be sufficient to observe the stems in about a score of genera, belonging to widely distinct families and inhabitants of various countries. Several plants were selected which, from being woody, or for other reasons, seemed the least likely to circumnutate. The observations and the diagrams were made in the manner described in the Introduction. Plants in pots were subjected to a proper temperature, and whilst being observed, were kept either in darkness or were feebly illuminated from above. They are arranged in the order adopted by Hooker in Le Maout and Decaisne's 'System of Botany.' The number of the family to which each genus belongs is appended, as this serves to show the place of each in the series.

(1.) Iberis umbellata (Cruciferae, Fam. 14).—The movement of the stem of a young plant, 4 inches in height, consisting of four internodes (the hypocotyl included) besides a large bud on the summit, was traced, as here shown, during 24 h. (Fig. 70). As far as we could judge the uppermost inch alone of the stem circumnutated, and this in a simple manner. The movement was slow, and the rate very unequal at different times. In part of its course an irregular ellipse, or rather triangle, was completed in 6 h. 30 m.

Fig. 70. Iberis umbellata: circumnutation of stem of young plant, traced from 8.30 A.M. Sept. 13th to same hour on following morning. Distance of summit of stem beneath the horizontal glass 7.6 inches. Diagram reduced to half of original size. Movement as here shown magnified between 4 and 5 times.

(2.) Brassica oleracea (Cruciferae).—A very young plant, bearing three leaves, of which the longest was only three-quarters of an inch in length, was placed under a microscope, furnished with an eye-piece micrometer, and the tip of the largest leaf was found to be in constant movement. It crossed five divisions of the micrometer, that is, 1/100th of an inch, in 6 m. 20 s. There could hardly be a doubt that it was the stem which chiefly moved, for the tip did not get quickly out of focus; and this would have occurred had the movement been confined to the leaf, which moves up or down in nearly the same vertical plane.

(3.) Linum usitatissimum (Lineae, Fam. 39).—The stems of this plant, shortly before the flowering period, are stated by Fritz Müller ('Jenaische Zeitschrift,' B. v. p. 137) to revolve, or circumnutate.

(4.) Pelargonium zonale (Geraniaceae, Fam. 47).—A young plant, 7½ inches in height, was observed in the usual manner; but, in order to see the bead at the end of the glass filament and at the same time the mark beneath, it was necessary to cut off three leaves on one side. We do not know whether it was owing to this cause, or to the plant having previously become bent to one side through heliotropism, but from the morning of the 7th of March to 10.30 P.M. on the 8th, the stem moved a considerable distance in a zigzag line in the same general direction. During the night of the 8th it moved to some distance at right angles to its former course, and next morning (9th) stood for a time almost still. At noon on the 9th a new tracing was begun (see Fig. 71), which was continued till 8 A.M. on the 11th. Between noon on the 9th and 5 P.M. on the 10th (i.e. in the course of 29 h.), the stem described a circle. This plant therefore circumnutates, but at a very slow rate, and to a small extent.

Fig. 71. Pelargonium zonale: circumnutation of stem of young plant, feebly illuminated from above. Movement of bead magnified about 11 times; traced on a horizontal glass from noon on March 9th to 8 A.M. on the 11th.

(5.) Tropaeolum majus (?) (dwarfed var. called Tom Thumb); (Geraniaceae, Fam. 47).—The species of this genus climb by the aid of their sensitive petioles, but some of them also twine round supports; but even these latter species do not begin to circumnutate in a conspicuous manner whilst young. The variety here treated of has a rather thick stem, and is so dwarf that apparently it does not climb in any manner. We therefore wished to ascertain whether the stem of a young plant, consisting of two internodes, together 3.2 inches in height, circumnutated. It was observed during 25 h., and we see in Fig. 72 that the stem moved in a zigzag course, indicating circumnutation.

Fig. 72. Tropaeolum majus (?): circumnutation of stem of young plant, traced on a horizontal glass from 9 A.M. Dec. 26th to 10 A.M. on 27th. Movement of bead magnified about 5 times, and here reduced to half of original scale.

Fig. 73. Trifolium resupinatum: circumnutation of stem, traced on vertical glass from 9.30 A.M. to 4.30 P.M. Nov. 3rd. Tracing not greatly magnified, reduced to half of original size. Plant feebly illuminated from above.

(6.) Trifolium resupinatum (Leguminosae, Fam. 75).—When we treat of the sleep of plants, we shall see that the stems in several Leguminous genera, for instance, those of Hedysarum, Mimosa, Melilotus, etc., which are not climbers, circumnutate in a conspicuous manner. We will here give only a single instance (Fig. 73), showing the circumnutation of the stem of a large plant of a clover, Trifolium resupinatum. In the course of 7 h. the stem changed its course greatly eight times and completed three irregular circles or ellipses. It therefore circumnutated rapidly. Some of the lines run at right angles to one another.

Fig. 74. Rubus (hybrid): circumnutation of stem, traced on horizontal glass, from 4 P.M. March 14th to 8.30 A.M. 16th. Tracing much magnified, reduced to half of original size. Plant illuminated feebly from above.

(7.) Rubus idæus (hybrid) (Rosaceae, Fam. 76).—As we happened to have a young plant, 11 inches in height and growing vigorously, which had been raised from a cross between the raspberry (Rubus idæus) and a North American Rubus, it was observed in the usual manner. During the morning of March 14th the stem almost completed a circle, and then moved far to the right. At 4 P.M. it reversed its course, and now a fresh tracing was begun, which was continued during 40½ h., and is given in Fig. 74. We here have well-marked circumnutation.

(8.) Deutzia gracilis (Saxifrageae, Fam. 77).—A shoot on a bush about 18 inches in height was observed. The bead changed its course greatly eleven times in the course of 10 h. 30 m. (Fig. 75), and there could be no doubt about the circumnutation of the stem.

Fig. 75. Deutzia gracilis: circumnutation of stem, kept in darkness, traced on horizontal glass, from 8.30 A.M. to 7 P.M. March 20th. Movement of bead originally magnified about 20 times, here reduced to half scale.

(9.) Fuchsia (greenhouse var., with large flowers, probably a hybrid) (Onagrarieae, Fam. 100).—A young plant, 15 inches in height, was observed during nearly 48 h. The accompanying figure (Fig. 76) gives the necessary particulars, and shows that the stem circumnutated, though rather slowly.

Fig. 76. Fuchsia (garden var.): circumnutation of stem, kept in darkness, traced on horizontal glass, from 8.30 A.M. to 7 P.M. March 20th. Movement of bead originally magnified about 40 times, here reduced to half scale.

(10.) Cereus speciocissimus (garden var., sometimes called Phyllocactus multiflorus) (Cacteæ, Fam. 109).—This plant, which was growing vigorously from having been removed a few days before from the greenhouse to the hot-house, was observed with especial interest, as it seemed so little probable that the stem would circumnutate. The branches are flat, or flabelliform; but some of them are triangular in section, with the three sides hollowed out. A branch of this latter shape, 9 inches in length and 1½ in diameter, was chosen for observation, as less likely to circumnutate than a flabelliform branch. The movement of the bead at the end of the glass filament, affixed to the summit of the branch, was traced (A, Fig. 77) from 9.23 A.M. to 4.30 P.M. on Nov. 23rd, during which time it changed its course greatly six times. On the 24th another tracing was made (see B), and the bead on this day changed its course oftener, making in 8 h. what may be considered as four ellipses, with their longer axes differently directed. The position of the stem and its commencing course on the following morning are likewise shown. There can be no doubt that this branch, though appearing quite rigid, circumnutated; but the extreme amount of movement during the time was very small, probably rather less than the 1/20th of an inch.

Fig 77. Cereus speciocissimus: circumnutation of stem, illuminated from above, traced on a horizontal glass, in A from 9 A.M. to 4.30 P.M. on Nov. 23rd; and in B from 8.30 A.M. on the 24th to 8 A.M. on the 25th. Movement of the bead in B magnified about 38 times.

(11.) Hedera helix (Araliaceae, Fam. 114).—The stem is known to be apheliotropic, and several seedlings growing in a pot in the greenhouse became bent in the middle of the summer at right angles from the light. On

Sept. 2nd some of these stems were tied up so as to stand vertically, and were placed before a north-east window; but to our surprise they were now decidedly heliotropic, for during 4 days they curved themselves towards the light, and their course being traced on a horizontal glass, was strongly zigzag. During the 6 succeeding days they circumnutated over the same small space at a slow rate, but there could be no doubt about their circumnutation. The plants were kept exactly in the same place before the window, and after an interval of 15 days the stems were again observed during 2 days and their movements traced, and they were found to be still circumnutating, but on a yet smaller scale.

(12.) Gazania ringens (Compositæ, Fam. 122).—The circumnutation of the stem of a young plant, 7 inches in height, as measured to the tip of the highest leaf, was traced during 33 h., and is shown in the accompanying figure (Fig. 78). Two main lines may be observed running at nearly right angles to two other main lines; but these are interrupted by small loops.

Fig. 78. Gazania ringens: circumnutation of stem traced from 9 A.M. March 21st to 6 P.M. on 22nd; plant kept in darkness. Movement of bead at the close of the observations magnified 34 times, here reduced to half the original scale.

(13.) Azalea Indica (Ericineae, Fam. 128).—A bush 21 inches in height was selected for observation, and the circumnutation of its leading shoot was traced during 26 h. 40 m., as shown in the following figure (Fig. 79).

(14.) Plumbago Capensis (Plumbagineae, Fam. 134).—A small lateral branch which projected from a tall freely growing bush, at an angle of 35° above the horizon, was selected for observation. For the first 11 h. it moved to a considerable distance in a nearly straight line to one side, owing probably to its having been previously deflected by the light whilst standing in the greenhouse. At 7.20 P.M. on March 7th a fresh tracing was begun and continued for the next 43 h. 40 m. (see Fig. 80). During the first 2 h. it followed nearly the same direction as before, and then changed it a little; during the night it moved at nearly right angles to its previous course. Next day (8th) it zigzagged greatly, and on the 9th moved irregularly round and round a small circular space. By 3 P.M. on the 9th the figure had become so complicated that no more dots could be made; but the shoot continued during the evening of the 9th, the whole of the 10th, and the morning of the 11th to circumnutate over the same small space, which was only about the 1/26th of an inch (.97 mm.) in diameter. Although this branch circumnutated to a very small extent, yet it changed its course frequently. The movements ought to have been more magnified.

Fig. 79. Azalea Indica: circumnutation of stem, illuminated from above, traced on horizontal glass, from 9.30 A.M. March 9th to 12.10 P.M. on the

10th. But on the morning of the 10th only four dots were made between 8.30 A.M. and 12.10 P.M., both hours included, so that the circumnutation is not fairly represented in this part of the diagram. Movement of the bead here magnified about 30 times.

Fig. 80. Plumbago Capensis: circumnutation of tip of a lateral branch, traced on horizontal glass, from 7.20 P.M. on March 7th to 3 P.M. on the 9th. Movement of bead magnified 13 times. Plant feebly illuminated from above.

(15.) Aloysia citriodora (Verbenaceae, Fam. 173).—The following figure (Fig. 81) gives the movements of a shoot during 31 h. 40 m., and shows that it circumnutated. The bush was 15 inches in height.

Fig. 81. Aloysia citriodora: circumnutation of stem, traced from 8.20 A.M. on March 22nd to 4 P.M. on 23rd. Plant kept in darkness. Movement magnified about 40 times.

(16.) Verbena melindres (?) (a scarlet-flowered herbaceous var.) (Verbenaceae).—A shoot 8 inches in height had been laid horizontally, for the sake of observing its apogeotropism, and the terminal portion had grown vertically upwards for a length of 1½ inch. A glass filament, with a bead at the end, was fixed upright to the tip, and its movements were traced during 41 h. 30 m. on a vertical glass (Fig. 82). Under these circumstances the lateral movements were chiefly shown; but as the lines from side to side are not on the same level, the shoot must have moved in a plane at right angles to that of the lateral movement, that is, it must have circumnutated. On the next day (6th) the shoot moved in the course of 16 h. four times to the right, and four times to the left; and this apparently represents the formation of four ellipses, so that each was completed in 4 h. (17.) Ceratophyllum demersum (Ceratophylleae, Fam. 220).—An interesting account of the movements of the stem of this water-plant has been published by M. E. Rodier.[1] The movements are confined to the young internodes, becoming less and less lower down the stem; and they are extraordinary from their amplitude. The stems sometimes moved through an angle of above 200° in 6 h., and in one instance through 220° in 3 h. They generally bent from right to left in the morning, and in an opposite direction in the afternoon; but the movement was sometimes temporarily reversed or quite arrested. It was not affected by light. It does not appear that M. Rodier made any diagram on a horizontal plane representing the actual course pursued by the apex, but he speaks of the "branches executing round their axes of growth a movement of torsion." From the particulars above given, and remembering in the case of twining plants and of tendrils, how difficult it is not to mistake their bending to all points of the compass for true torsion, we are led to believe that the stems of this Ceratophyllum circumnutate, probably in the shape of narrow ellipses, each completed in about 26 h. The following statement, however, seems to indicate something different from ordinary circumnutation, but we cannot fully understand it. M. Rodier says: "Il est alors facile de voir que le mouvement de flexion se produit d'abord dans les mérithalles supérieurs, qu'il se propage ensuite, en s'amoindrissant du haut en bas; tandis qu'au contraire le movement de redressement commence par la partie inférieur pour se terminer a la partie supérieure qui, quelquefois, peu de temps avant de se relever tout à fait, forme avec l'axe un angle très aigu."

[1] 'Comptes Rendus,' April 30th, 1877. Also a second notice published separately in Bourdeaux, Nov. 12th, 1877.

Fig. 82. Verbena melindres: circumnutation of stem in darkness, traced on vertical glass, from 5.30 P.M. on June 5th to 11 A.M. June 7th. Movement of bead magnified 9 times.

(18.) Coniferæ.—Dr. Maxwell Masters states ('Journal Linn. Soc.,' Dec. 2nd, 1879) that the leading shoots of many Coniferæ during the season of their active growth exhibit very remarkable movements of revolving nutation, that is, they circumnutate. We may feel sure that the lateral shoots whilst growing would exhibit the same movement if carefully observed.

(19.) Lilium auratum (Fam. Liliaceae).—The circumnutation of the stem of a plant 24 inches in height is represented in the above figure (Fig. 83).

Fig. 83. Lilium auratum: circumnutation of a stem in darkness, traced on a horizontal glass, from 8 A.M. on March 14th to 8.35 A.M. on 16th. But it should be noted that our observations were interrupted between 6 P.M. on the 14th and 12.15 P.M. on the 15th, and the movements during this interval of 18 h. 15 m. are represented by a long broken line. Diagram reduced to half original scale.

Fig. 84. Cyperus alternifolius: circumnutation of stem, illuminated from above, traced on horizontal glass, from 9.45 A.M. March 9th to 9 P.M. on 10th. The stem grew so rapidly whilst being observed, that it was not possible to estimate how much its movements were magnified in the tracing.

(20.) Cyperus alternifolius (Fam. Cyperaceae.)—A glass filament, with a bead at the end, was fixed across the summit of a young stem 10 inches in height, close beneath the crown of elongated leaves. On March 8th, between 12.20 and 7.20 P.M. the stem described an ellipse, open at one end. On the following day a new tracing was begun (Fig. 84), which plainly shows that the stem completed three irregular figures in the course of 35 h. 15 m.

Concluding Remarks on the Circumnutation of Stems.—Any one who will inspect the diagrams now given, and will bear in mind the widely separated position of the plants described in the series,—remembering that we have good grounds for the belief that the hypocotyls and epicotyls of all seedlings circumnutate,—not forgetting the number of plants distributed in the most distinct families which climb by a similar movement,—will probably admit that the growing stems of all plants, if carefully observed, would be found to circumnutate to a greater or less extent. When we treat of the sleep and other movements of plants, many other cases of circumnutating stems will be incidentally given. In looking at the diagrams, we should remember that the stems were always growing, so that in each case the circumnutating apex as it rose will have described a spire of some kind. The dots were made on the glasses generally at intervals of an hour, or hour and a half, and were then joined by straight lines. If they had been made at intervals of 2 or 3 minutes, the lines would have been more curvilinear, as in the case of the tracks left on the smoked glass-plates by the tips of the circumnutating radicles of seedling plants. The diagrams generally approach in form to a succession of more or less irregular ellipses or ovals, with their longer axes directed to different points of the compass during the same day or on succeeding days. The stems therefore, sooner or later, bend to all sides; but after a stem has bent in any one direction, it commonly bends back at first in nearly, though not quite, the opposite direction; and this gives the tendency to the formation of ellipses, which are generally narrow, but not so narrow as those described by stolons and leaves. On the other hand, the figures sometimes approach in shape to circles. Whatever the figure may be, the course pursued is often interrupted by zigzags, small triangles, loops, or ellipses. A stem may describe a single large ellipse one day, and two on the next. With different plants the complexity, rate, and amount of movement differ much. The stems, for instance, of Iberis and Azalea described only a single large ellipse in 24 h.; whereas those of the Deutzia made four or five deep zigzags or narrow ellipses in 11½ h., and those of the Trifolium three triangular or quadrilateral figures in 7 h.

CIRCUMNUTATION OF STOLONS OR RUNNERS.

Stolons consist of much elongated, flexible branches, which run along the surface of the ground and form roots at a distance from the parent-plant. They are therefore of the same homological nature as stems; and the three following cases may be added to the twenty previously given cases.

Fragaria (cultivated garden var.): Rosaceae.—A plant growing in a pot had emitted a long stolon; this was supported by a stick, so that it projected for the length of several inches horizontally. A glass filament bearing two minute triangles of paper was affixed to the terminal bud, which was a little upturned; and its movements were traced during 21 h., as shown in Fig. 85. In the course of the first 12 h. it moved twice up and twice down in somewhat zigzag lines, and no doubt travelled in the same manner during the night. On the following morning after an interval of 20 h. the apex stood a little higher than it did at first, and this shows that the stolon had not been acted on within this time by geotropism;[2] nor had its own weight caused it to bend downwards.

[2] Dr. A. B. Frank states ('Die Naturliche wagerechte Richtung von Pflanzentheilen,' 1870, p. 20) that the stolons of this plant are acted on by geotropism, but only after a considerable interval of time.

Fig. 85. Fragaria: circumnutation of stolon, kept in darkness, traced on vertical glass, from 10.45 A.M. May 18th to 7.45 A.M. on 19th.

On the following morning (19th) the glass filament was detached and refixed close behind the bud, as it appeared possible that the circumnutation of the terminal bud and of the adjoining part of the stolon might be different. The movement was now traced during two consecutive days (Fig. 86). During the first day the filament travelled in the course of 14 h. 30 m. five times up and four times down, besides some lateral movement. On the 20th the course was even more complicated, and can hardly be followed in the figure; but the filament moved in 16 h. at least five times up and five times down, with very little lateral deflection. The first and last dots made on this second day, viz., at 7 A.M. and 11 P.M., were close together, showing that the stolon had not fallen or risen. Nevertheless, by comparing its position on the morning of the 19th and 21st, it is obvious that the stolon had sunk; and this may be attributed to slow bending down either from its own weight or from geotropism.

Fig. 86. Fragaria: circumnutation of the same stolon as in the last figure, observed in the same manner, and traced from 8 A.M. May 19th to 8 A.M. 21st.

During a part of the 20th an orthogonal tracing was made by applying a cube of wood to the vertical glass and bringing the apex of the stolon at successive periods into a line with one edge; a dot being made each time on the glass. This tracing therefore represented very nearly the actual amount of movement of the apex; and in the course of 9 h. the distance of the extreme dots from one another was .45 inch. By the same method it was ascertained that the apex moved between 7 A.M. on the 20th and 8 A.M. on the 21st a distance of .82 inch.

A younger and shorter stolon was supported so that it projected at about 45° above the horizon, and its movement was traced by the same orthogonal method. On the first day the apex soon rose above the field of vision. By the next morning it had sunk, and the course pursued was now traced during 14 h. 30 m. (Fig. 87). The amount of movement was almost the same, from side to side as up and down; and differed in this respect remarkably from the movement in the previous cases. During the latter part of the day, viz., between 3 and 10.30 P.M., the actual distance travelled by the apex amounted to 1.15 inch; and in the course of the whole day to at least 2.67 inches. This is an amount of movement almost comparable with that of some climbing plants. The same stolon was observed on the following day, and now it moved in a somewhat less complex manner, in a plane not far from vertical. The extreme amount of actual movement was 1.55 inch in one direction, and .6 inch in another direction at right angles. During neither of these days did the stolon bend downwards through geotropism or its own weight.

Fig. 87. Fragaria: circumnutation of another and younger stolon, traced from 8 A.M. to 10.30 P.M. Figure reduced to one-half of original scale.

Four stolons still attached to the plant were laid on damp sand in the back of a room, with their tips facing the north-east windows. They were thus placed because De Vries says[3] that they are apheliotropic when exposed to the light of the sun; but we could not perceive any effect from the above feeble degree of illumination. We may add that on another occasion, late in the summer, some stolons, placed upright before a south-west window on a cloudy day, became distinctly curved towards the light, and were therefore heliotropic. Close in front of the tips of the prostrate stolons, a crowd of very thin sticks and the dried haulms of grasses were driven into the sand, to represent the crowded stems of surrounding plants in a state of nature. This was done for the sake of observing how the growing stolons would pass through them. They did so easily in the course of 6 days, and their circumnutation apparently facilitated their passage. When the tips encountered sticks so close together that they could not pass between them, they rose up and passed over them. The sticks and haulms were removed after the passage of the four stolons, two of which were found to have assumed a permanently sinuous shape, and two were still straight. But to this subject we shall recur under Saxifraga.

[3] 'Arbeiten Bot Inst., Würzburg,' 1872, p. 434.

Saxifraga sarmentosa (Saxifrageae).—A plant in a suspended pot had emitted long branched stolons, which depended like threads on all sides. Two were tied up so as to stand vertically, and their upper ends became gradually bent downwards, but so slowly in the course of several days, that the bending was probably due to their weight and not to geotropism. A glass filament with little triangles of paper was fixed to the end of one of these stolons, which was 17½ inches in length, and had already become much bent down, but still projected at a considerable angle above the horizon. It moved only slightly three times from side to side and then upwards; on the following day the movement was even less. As this stolon was so long we thought that its growth was nearly completed, so we tried another which was thicker and shorter, viz., 10 1/4 inches in length. It moved greatly, chiefly upwards, and changed its course five times in the course of the day. During the night it curved so much upwards in opposition to gravity, that the movement could no longer be traced on the vertical glass, and a horizontal one had to be used. The movement was followed during the next 25 h., as shown in Fig. 88. Three irregular ellipses, with their longer axes somewhat differently directed, were almost completed in the first 15 h. The extreme actual amount of movement of the tip during the 25 h. was .75 inch. Several stolons were laid on a flat surface of damp sand, in the same manner as with those of the strawberry. The friction of the sand did not interfere with their circumnutation; nor could we detect any evidence of their being sensitive to contact. In order to see how in a state of nature they would act, when encountering a stone or other obstacle on the ground, short pieces of smoked glass, an inch in height, were stuck upright into the sand in front of two thin lateral branches. Their tips scratched the smoked surface in various directions; one made three upward and two downward lines, besides a nearly horizontal one; the other curled quite away from the glass; but ultimately both surmounted the glass and pursued their original course. The apex of a third thick stolon swept up the glass in a curved line, recoiled and again came into contact with it; it then moved to the right, and after ascending, descended vertically; ultimately it passed round one end of the glass instead of over it.

Fig. 88. Saxifraga sarmentosa: circumnutation of an inclined stolon, traced in darkness on a horizontal glass, from 7.45 A.M. April 18th to 9 A.M. on 19th. Movement of end of stolon magnified 2.2 times.

Many long pins were next driven rather close together into the sand, so as to form a crowd in front of the same two thin lateral branches; but these easily wound their way through the crowd. A thick stolon was much delayed in its passage; at one place it was forced to turn at right angles to its former course; at another place it could not pass through the pins, and the hinder part became bowed; it then curved upwards and passed through an opening between the upper part of some pins which happened to diverge; it then descended and finally emerged through the crowd. This stolon was rendered permanently sinuous to a slight degree, and was thicker where sinuous than elsewhere, apparently from its longitudinal growth having been checked.

Cotyledon umbilicus (Crassulaceæ).—A plant growing in a pan of damp moss had emitted 2 stolons, 22 and 20 inches in length. One of these was supported, so that a length of 4½ inches projected in a straight and horizontal line, and the movement of the apex was traced. The first dot was made at 9.10 A.M.; the terminal portion soon began to bend downwards and continued to do so until noon. Therefore a straight line, very nearly as long as the whole figure here given (Fig. 89), was first traced on the glass; but the upper part of this line has not been copied in the diagram. The curvature occurred in the middle of the penultimate internode; and its chief seat was at the distance of 1 1/4 inch from the apex; it appeared due to the weight of the terminal portion, acting on the more flexible part of the internode, and not to geotropism. The apex after thus sinking down from 9.10 A.M. to noon, moved a little to the left; it then rose up and circumnutated in a nearly vertical plane until 10.35 P.M. On the following day (26th) it was observed from 6.40 A.M. to 5.20 P.M., and within this time it moved twice up and twice down. On the morning of the 27th the apex stood as high as it did at 11.30 A.M. on the 25th. Nor did it sink down during the 28th, but continued to circumnutate about the same place.

Fig. 89. Cotyledon umbilicus: circumnutation of stolon, traced from 11.15 A.M. Aug. 25th to 11 A.M. 27th. Plant illuminated from above. The terminal internode was .25 inch in length, the penultimate 2.25 and the third 3.0 inches in length. Apex of stolon stood at a distance of 5.75 inches from the vertical glass; but it was not possible to ascertain how much the tracing was magnified, as it was not known how great a length of the internode circumnutated.

Another stolon, which resembled the last in almost every respect, was observed during the same two days, but only two inches of the terminal portion was allowed to project freely and horizontally. On the 25th it continued from 9.10 A.M. to 1.30 P.M. to bend straight downwards, apparently owing to its weight (Fig. 90); but after this hour until 10.35 P.M. it zigzagged. This fact deserves notice, for we here probably see the combined effects of the bending down from weight and of circumnutation. The stolon, however, did not circumnutate when it first began to bend down, as may be observed in the present diagram, and as was still more evident in the last case, when a longer portion of the stolon was left unsupported. On the following day (26th) the stolon moved twice up and twice down, but still continued to fall; in the evening and during the night it travelled from some unknown cause in an oblique direction.

Fig. 90. Cotyledon umbilicus: circumnutation and downward movement of another stolon, traced on vertical glass, from 9.11 A.M. Aug. 25th to 11

A.M. 27th. Apex close to glass, so that figure but little magnified, and here reduced to two-thirds of original size.

We see from these three cases that stolons or runners circumnutate in a very complex manner. The lines generally extend in a vertical plane, and this may probably be attributed to the effect of the weight of the unsupported end of the stolon; but there is always some, and occasionally a considerable, amount of lateral movement. The circumnutation is so great in amplitude that it may almost be compared with that of climbing plants. That the stolons are thus aided in passing over obstacles and in winding between the stems of the surrounding plants, the observations above given render almost certain. If they had not circumnutated, their tips would have been liable to have been doubled up, as often as they met with obstacles in their path; but as it is, they easily avoid them. This must be a considerable advantage to the plant in spreading from its parent-stock; but we are far from supposing that the power has been gained by the stolons for this purpose, for circumnutation seems to be of universal occurrence with all growing parts; but it is not improbable that the amplitude of the movement may have been specially increased for this purpose.

CIRCUMNUTATION OF FLOWER-STEMS.

We did not think it necessary to make any special observations on the circumnutation of flower-stems, these being axial in their nature, like stems or stolons; but some were incidentally made whilst attending to other subjects, and these we will here briefly give. A few observations have also been made by other botanists. These taken together suffice to render it probable that all peduncles and sub-peduncles circumnutate whilst growing.

Oxalis carnosa.—The peduncle which springs from the thick and woody stem of this plant bears three or four sub-peduncles. A filament with little triangles of paper was fixed within the calyx of a flower which stood upright. Its movements were observed for 48 h.; during the first half of this time the flower was fully expanded, and during the second half withered. The figure here given (Fig. 91) represents 8 or 9 ellipses. Although the main peduncle circumnutated, and described one large and two smaller ellipses in the course of 24 h., yet the chief seat of movement lies in the sub-peduncles, which ultimately bend vertically downwards, as will be described in a future chapter. The peduncles of Oxalis acetosella likewise bend downwards, and afterwards, when the pods are nearly mature, upwards; and this is effected by a circumnutating movement.

Fig. 91. Oxalis carnosa: flower-stem, feebly illuminated from above, its circumnutation traced from 9 A.M. April 13th to 9 A.M. 15th. Summit of flower 8 inches beneath the horizontal glass. Movement probably magnified about 6 times.

It may be seen in the above figure that the flower-stem of O. carnosa circumnutated during two days about the same spot. On the other hand, the flower-stem of O. sensitiva undergoes a strongly marked, daily, periodical change of position, when kept at a proper temperature. In the middle of the day it stands vertically up, or at a high angle; in the afternoon it sinks, and in the evening projects horizontally, or almost horizontally, rising again during the night. This movement continues from the period when the flowers are in bud to when, as we believe, the pods are mature: and it ought perhaps to have been included amongst the so-called sleep-movements of plants. A tracing was not made, but the angles were measured at successive periods during one whole day; and these showed that the movement was not continuous, but that the peduncle oscillated up and down. We may therefore conclude that it circumnutated. At the base of the peduncle there is a mass of small cells, forming a well-developed pulvinus, which is exteriorly coloured purple and hairy. In no other genus, as far as we know, is the peduncle furnished with a pulvinus. The peduncle of O. Ortegesii behaved differently from that of O. sensitiva, for it stood at a less angle above the horizon in the middle of the day, then in the morning or evening. By 10.20 P.M. it had risen greatly. During the middle of the day it oscillated much up and down.

Trifolium subterraneum.—A filament was fixed vertically to the uppermost part of the peduncle of a young and upright flower-head (the stem of the plant having been secured to a stick); and its movements were traced during 36 h. Within this time it described (see Fig. 92) a figure which represents four ellipses; but during the latter part of the time the peduncle began to bend downwards, and after 10.30 P.M. on the 24th it curved so rapidly down, that by 6.45 A.M. on the 25th it stood only 19° above the horizon. It went on circumnutating in nearly the same position for two days. Even after the flower-heads have buried themselves in the ground they continue, as will hereafter be shown, to circumnutate. It will also be seen in the next chapter that the sub-peduncles of the separate flowers of *Trifolium repens* circumnutate in a complicated course during several days. I may add that the gynophore of *Arachis hypogoea*, which looks exactly like a peduncle, circumnutates whilst growing vertically downwards, in order to bury the young pod in the ground.

Fig. 92. Trifolium subterraneum: main flower-peduncle, illuminated from above, circumnutation traced on horizontal glass, from 8.40 A.M. July 23rd to 10.30 P.M. 24th.

The movements of the flowers of Cyclamen Persicum were not observed; but the peduncle, whilst the pod is forming, increases much in length, and bows itself down by a circumnutating movement. A young peduncle of Maurandia semperflorens, 1½ inch in length, was carefully observed during a whole day, and it made 4½ narrow, vertical, irregular and short ellipses, each at an average rate of about 2 h. 25 m. An adjoining peduncle described during the same time similar, though fewer, ellipses.[4] According to Sachs[5] the flower-stems, whilst growing, of many plants, for instance, those of Brassica napus, revolve or circumnutate; those of Allium porrum bend from side to side, and, if this movement had been traced on a horizontal glass, no doubt ellipses would have been formed. Fritz Müller has described[6] the spontaneous revolving movements of the flower-stems of an Alisma, which he compares with those of a climbing plant.

[4] 'The Movements and Habits of Climbing Plants,' 2nd edit., 1875, p. 68.

[5] 'Text-Book of Botany,' 1875, p. 766. Linnæus and Treviranus (according to Pfeffer, 'Die Periodischen Bewegungen,' etc., p. 162) state that the flower-stalks of many plants occupy different positions by night and day, and we shall see in the chapter on the Sleep of Plants that this implies circumnutation.

[6] 'Jenaische Zeitsch.,' B. v. p. 133.

We made no observations on the movements of the different parts of flowers. Morren, however, has observed[7] in the stamens of Sparmannia and Cereus a "fremissement spontané," which, it may be suspected, is a circumnutating movement. The circumnutation of the gynostemium of Stylidium, as described by Gad,[8] is highly remarkable, and apparently aids in the fertilisation of the flowers. The gynostemium, whilst spontaneously moving, comes into contact with the viscid labellum, to which it adheres, until freed by the increasing tension of the parts or by being touched.

[7] 'N. Mem. de l'Acad. R. de Bruxelles,' tom. xiv. 1841, p. 3.

[8] 'Sitzungbericht des bot. Vereins der P. Brandenburg,' xxi. p. 84.

We have now seen that the flower-stems of plants belonging to such widely different families as the Cruciferae, Oxalidæ, Leguminosae, Primulaceae, Scrophularineae, Alismaceae, and Liliaceae, circumnutate; and that there are indications of this movement in many other families. With these facts before us, bearing also in mind that the tendrils of not a few plants consist of modified peduncles, we may admit without much doubt that all growing flower-stems circumnutate.

CIRCUMNUTATION OF LEAVES: DICOTYLEDONS.

Several distinguished botanists, Hofmeister, Sachs, Pfeffer, De Vries, Batalin, Millardet, etc., have observed, and some of them with the greatest care, the periodical movements of leaves; but their attention has been chiefly, though not exclusively, directed to those which move largely and are commonly said to sleep at night. From considerations hereafter to be given, plants of this nature are here excluded, and will be treated of separately. As we wished to ascertain whether all young and growing leaves circumnutated, we thought that it would be sufficient if we observed between 30 and 40 genera, widely distributed throughout the vegetable series, selecting some unusual forms and others on woody plants. All the plants were healthy and grew in pots. They were illuminated from above, but the light perhaps was not always sufficiently bright, as many of them were observed under a skylight of ground-glass. Except in a few specified cases, a fine glass filament with two minute triangles of paper was fixed to the leaves, and their movements were traced on a vertical glass (when not stated to the contrary) in the manner already described. I may repeat that the broken lines represent the nocturnal course. The stem was always secured to a stick, close to the base of the leaf under observation. The arrangement of the species, with the number of the Family appended, is the same as in the case of stems.

Fig. 93. Sarracenia purpurea: circumnutation of young pitcher, traced from 8 A.M. July 3rd to 10.15 A.M. 4th. Temp. 17°–18° C. Apex of pitcher 20 inches from glass, so movement greatly magnified.

(1.) Sarracenia purpurea (Sarraceneae, Fam. 11).—A young leaf, or pitcher, 8½ inches in height, with the bladder swollen but with the hood not as yet open, had a filament fixed transversely across its apex; it was observed for 48 h., and during the whole of this time it circumnutated in a nearly similar manner, but to a very small extent. The tracing given (Fig. 93) relates only to the movement during the first 26 h.

(2) Glaucium luteum (Papaveraceae, Fam. 12).—A young plant, bearing only 8 leaves, had a filament attached to the youngest leaf but one, which was 3 inches in length, including the petiole. The circumnutating movement was traced during 47 h. On both days the leaf descended from before 7 A.M. until about 11 A.M., and then ascended slightly during the rest of the day and the early part of the night. During the latter part of the night it fell greatly. It did not ascend so much during the second as during the first day, and it descended considerably lower on the second night than on the first. This difference was probably due to the illumination from above having been insufficient during the two days of observation. Its course during the two days is shown in Fig. 94.

Fig. 94. Glaucium luteum: circumnutation of young leaf, traced from 9.30 A.M. June 14th to 8.30 A.M. 16th. Tracing not much magnified, as apex of leaf stood only 5½ inches from the glass.

(3.) Crambe maritima (Cruciferae, Fam. 14).—A leaf 9½ inches in length on a plant not growing vigorously was first observed. Its apex was in constant movement, but this could hardly be traced, from being so small in extent. The apex, however, certainly changed its course at least 6 times in the course of 14 h. A more vigorous young plant, bearing only 4 leaves, was then selected, and a filament was affixed to the midrib of the third leaf from the base, which, with the petiole, was 5 inches in length. The leaf stood up almost vertically, but the tip was deflected, so that the filament projected almost horizontally, and its movements were traced during 48 h. on a vertical glass as shown in the accompanying figure (Fig. 95). We here plainly see that the leaf was continually circumnutating; but the proper periodicity of its movements was disturbed by its being only dimly illuminated from above through a double skylight. We infer that this was the case, because two leaves on plants growing out of doors, had their angles above the horizon measured in the middle of the day and at 9 to about 10 P.M. on successive nights, and they were found at this latter hour to have risen by an average angle of 9° above their mid-day position: on the following morning they fell to their former position. Now it may be observed in the diagram that the leaf rose during the second night, so that it stood at 6.40 A.M. higher than at 10.20 P.M. on the preceding night; and this may be attributed to the leaf adjusting itself to the dim light, coming exclusively from above.

Fig. 95. Crambe maritima: circumnutation of leaf, disturbed by being insufficiently illuminated from above, traced from 7.50 A.M. June 23rd to 8 A.M. 25th. Apex of leaf 15 1/4 inches from the vertical glass, so that the tracing was much magnified, but is here reduced to one-fourth of original scale.

(4.) Brassica oleracea (Cruciferae).—Hofmeister and Batalin[9] state that the leaves of the cabbage rise at night, and fall by day. We covered a young plant, bearing 8 leaves, under a large bell-glass, placing it in the same position with respect to the light in which it had long remained, and a filament was fixed at the distance of .4 of an inch from the apex of a young leaf nearly 4 inches in length. Its movements were then traced during three days, but the tracing is not worth giving. The leaf fell during the whole morning, and rose in the evening and during the early part of the night. The ascending and descending lines did not coincide, so that an irregular ellipse was formed each 24 h. The basal part of the midrib did not move, as was ascertained by measuring at successive periods the angle which it formed with the horizon, so that the movement was confined to the terminal portion of the leaf, which moved through an angle of 11° in the course of 24 h., and the distance travelled by the apex, up and down, was between .8 and .9 of an inch.

[9] 'Flora,' 1873, p. 437.

In order to ascertain the effect of darkness, a filament was fixed to a leaf 5½ inches in length, borne by a plant which after forming a head had produced a stem. The leaf was inclined 44° above the horizon, and its movements were traced on a vertical glass every hour by the aid of a taper. During the first day the leaf rose from 8 A.M. to 10.40 P.M. in a slightly zigzag course, the actual distance travelled by the apex being .67 of an inch. During the night the leaf fell, whereas it ought to have risen; and by 7 A.M. on the following morning it had fallen .23 of an inch, and it continued falling until 9.40 A.M. It then rose until 10.50 P.M., but the rise was interrupted by one considerable oscillation, that is, by a fall and re-ascent. During the second night it again fell, but only to a very short distance, and on the following morning re-ascended to a very short distance. Thus the normal course of the leaf was greatly disturbed, or rather completely inverted, by the absence of light; and the movements were likewise greatly diminished in amplitude.

We may add that, according to Mr. A. Stephen Wilson,[10] the young leaves of the Swedish turnip, which is a hybrid between B. oleracea and rapa, draw together in the evening so much "that the horizontal breadth diminishes about 30 per cent. of the daylight breadth." Therefore the leaves must rise considerably at night.

[10] 'Trans. Bot. Soc. Edinburgh,' vol. xiii. p. 32. With respect to the origin of the Swedish turnip, see Darwin, 'Animals and Plants under Domestication,' 2nd edit. vol. i. p. 344.

(5.) Dianthus caryophyllus (Caryophylleae, Fam. 26).—The terminal shoot of a young plant, growing very vigorously, was selected for observation. The young leaves at first stand up vertically and close together, but they soon bend outwards and downwards, so as to become horizontal, and often at the same time a little to one side. A filament was fixed to the tip of a young leaf whilst still highly inclined, and the first dot was made on the vertical glass at 8.30 A.M. June 13th, but it curved downwards so quickly that by 6.40 A.M. on the following morning it stood only a little above the horizon. In Fig. 96 the long, slightly zigzag line representing this rapid downward course, which was somewhat inclined to the left, is not given; but the figure shows the highly tortuous and zigzag course, together with some loops, pursued during the next 2½ days. As the leaf continued to move all the time to the left, it is evident that the zigzag line represents many circumnutations.

Fig. 96. Dianthus caryophyllus: circumnutation of young leaf, traced from 10.15 P.M. June 13th to 10.35 P.M. 16th. Apex of leaf stood, at the close of our observations, 8 3/4 inches from the vertical glass, so tracing not greatly magnified. The leaf was 5 1/4 inches long. Temp. 15½°–17½° C.

(6.) Camellia Japonica (Camelliaceae, Fam. 32).—A youngish leaf, which together with its petiole was 2 3/4 inches in length and which arose from a side branch on a tall bush, had a filament attached to its apex. This leaf sloped downwards at an angle of 40° beneath the horizon. As it was thick and rigid, and its petiole very short, much movement could not be expected. Nevertheless, the apex changed its course completely seven times in the course of 11½ h., but moved to only a very small distance. On the next day the movement of the apex was traced during 26 h. 20 m. (as shown in Fig. 97), and was nearly of the same nature, but rather less complex. The movement seems to be periodical, for on both days the leaf circumnutated in the forenoon, fell in the afternoon (on the first day until between 3 and 4 P.M., and on the second day until 6 P.M.), and then rose, falling again during the night or early morning.

Fig. 97. Camellia Japonica: circumnutation of leaf, traced from 6.40 A.M. June 14th to 6.50 A.M. 15th. Apex of leaf 12 inches from the vertical glass, so figure considerably magnified. Temp. 16°–16½° C.

In the chapter on the Sleep of Plants we shall see that the leaves in several Malvaceous genera sink

Fig. 98. Pelargonium zonale: circumnutation and downward movement of young leaf, traced from 9.30 A.M. June 14th to 6.30 P.M. 16th. Apex of leaf 9 1.4 inches from the vertical glass, so figure moderately magnified. Temp. 15°–16½° C.

at night; and as they often do not then occupy a vertical position, especially if they have not been well illuminated during the day, it is doubtful whether some of these cases ought not to have been included in the present chapter.

(7.) Pelargonium zonale (Geraniaceae, Fam. 47).—A young leaf, 1 1/4 inch in breadth, with its petiole 1 inch long, borne on a young plant, was observed in the usual manner during 61 h.; and its course is shown in the preceding figure (Fig. 98). During the first day and night the leaf moved downwards, but circumnutated between 10 A.M. and 4.30 P.M. On the second day it sank and rose again, but between 10 A.M. and 6 P.M. it circumnutated on an extremely small scale. On the third day the circumnutation was more plainly marked.

(8.) Cissus discolor (Ampelideae, Fam. 67).—A leaf, not nearly full-grown, the third from the apex of a shoot on a cut-down plant, was observed during 31 h. 30 m. (see Fig. 99). The day was cold (15°–16° C.), and if the plant had been observed in the hot-house, the circumnutation, though plain enough as it was, would probably have been far more conspicuous.

Fig. 99. Cissus discolor: circumnutation of leaf, traced from 10.35 A.M. May 28th to 6 P.M. 29th. Apex of leaf 8 3/4 inches from the vertical glass.

(9.) Vicia faba (Leguminosae, Fam. 75).—A young leaf, 3.1 inches in length, measured from base of petiole to end of leaflets, had a filament affixed to the midrib of one of the two terminal leaflets, and its movements were traced during 51½ h. The filament fell all morning (July 2nd) till 3 P.M., and then rose greatly till 10.35 P.M.; but the rise this day was so great, compared with that which subsequently occurred, that it was probably due in part to the plant being illuminated from above. The latter part of the course on July 2nd is alone given in the following figure (Fig. 100). On the next day (July 3rd) the leaf again fell in the morning, then circumnutated in a conspicuous manner, and rose till late at night; but the movement was not traced after 7.15 P.M., as by that time the filament pointed towards the upper edge of the glass. During the latter part of the night or early morning it again fell in the same manner as before.

As the evening rise and the early morning fall were unusually large, the angle of the petiole above the horizon was measured at the two periods, and the leaf was found to have risen 19° between 12.20 P.M. and 10.45 P.M., and to have fallen 23° 30 seconds between the latter hour and 10.20 A.M. on the following morning.

Fig. 100. Vicia faba: circumnutation of leaf, traced from 7.15 P.M. July 2nd to 10.15 A.M. 4th. Apex of the two terminal leaflets 7 1/4 inches from the vertical glass. Figure here reduced to two-thirds of original scale. Temp. 17°–18° C.

The main petiole was now secured to a stick close to the base of the two terminal leaflets, which were 1.4 inch in length; and the movements of one of them were traced during 48 h. (see Fig. 101). The course pursued is closely analogous to that of the whole leaf. The zigzag line between 8.30 A.M. and 3.30 P.M. on the second day represents 5 very small ellipses, with their longer axes differently directed. From these observations it follows that both the whole leaf and the terminal leaflets undergo a well-marked daily periodical movement, rising in the evening and falling during the latter part of the night or early morning; whilst in the middle of the day they generally circumnutate round the same small space.

Fig 101. Vicia faba: circumnutation of one of the two terminal leaflets, the main petiole having been secured, traced from 10.40 A.M. July 4th to 10.30

A.M. 6th. Apex of leaflet 6 5/8 inches from the vertical glass. Tracing here reduced to one-half of original scale. Temp. 16°–18° C.

(10.) Acacia retinoides (Leguminosae).—The movement of a young phyllode, 2 3/8 inches in length, and inclined at a considerable angle above the horizon, was traced during 45 h. 30 m.; but in the figure here given (Fig. 102), its circumnutation is shown during only 21 h. 30 m. During part of this time (viz., 14 h. 30 m.) the phyllode described a figure representing 5 or 6 small ellipses. The actual amount of movement in a vertical direction was .3 inch. The phyllode rose considerably between 1.30 P.M. and 4 P.M., but there was no evidence on either day of a regular periodic movement.

Fig. 102. Acacia retinoides: circumnutation of a young phyllode, traced from 10.45 A.M. July 18th to 8.15 A.M. 19th. Apex of phyllode 9 inches from the vertical glass; temp. 16½°–17½° C.

(11.) Lupinus speciosus (Leguminosae).—Plants were raised from seed purchased under this name. This is one of the species in this large genus, the leaves of which do not sleep at night. The petioles rise direct from the ground, and are from 5 to 7 inches in length. A filament was fixed to the midrib of one of the longer leaflets, and the movement of the whole leaf was traced, as shown in Fig. 103. In the course of 6 h. 30 m. the filament went four times up and three times down. A new tracing was then begun (not here given), and during 12½ h. the leaf moved eight times up and seven times down; so that it described 7½ ellipses in this time, and this is an extraordinary rate of movement. The summit of the petiole was then secured to a stick, and the separate leaflets were found to be continually circumnutating.

Fig. 103. Lupinus speciosus: circumnutation of leaf, traced on vertical glass, from 10.15 A.M. to 5.45 P.M.; i.e., during 6 h. 30 m.

(12.) Echeveria stolonifera (Crassulaceæ, Fam. 84).—The older leaves of this plant are so thick and fleshy, and the young ones so short and broad, that it seemed very improbable that any circumnutation could be detected. A filament was fixed to a young upwardly inclined leaf, .75 inch in length and .28 in breadth, which stood on the outside of a terminal rosette of leaves, produced by a plant growing very vigorously. Its movement was traced during 3 days, as here shown (Fig. 104). The course was chiefly in an upward direction, and this may be attributed to the elongation of the leaf through growth; but we see that the lines are strongly zigzag, and that occasionally there was distinct circumnutation, though on a very small scale.

Fig. 104. Echeveria stolonifera: circumnutation of leaf, traced from 8.20 A.M. June 25th to 8.45 A.M. 28th. Apex of leaf 12 1/4 inches from the glass, so that the movement was much magnified; temp. 23°–24½° C.

(13.) Bryophyllum (vel Calanchæ) calycinum (Crassulaceæ).—Duval-Jouve ('Bull. Soc. Bot. de France,' Feb. 14th, 1868) measured the distance between the tips of the upper pair of leaves on this plant, with the result shown in the following Table. It should be noted that the measurements on Dec. 2nd were made on a different pair of leaves:—

8 A.M. 2 P.M. 7 P.M. Nov. 16. . . . . .15 mm.. . . . . .25 mm. . . ..(?) " 19. . . . . .48 " . . . . 60 ". . . . . . . 48 mm. Dec. 2. . . . . .22 ". . . . . 43 ". . . . . . . .28 "

We see from this Table that the leaves stood considerably further apart at 2 P.M. than at either 8 A.M. or 7 P.M.; and this shows that they rise a little in the evening and fall or open in the forenoon.

(14.) Drosera rotundifolia (Droseraceae, Fam. 85).—The movements of a young leaf, having a long petiole but with its tentacles (or gland-bearing hairs) as yet unfolded, were traced during 47 h. 15 m. The figure (Fig. 105) shows that it circumnutated largely, chiefly in a vertical direction, making two ellipses each day. On both days the leaf began to descend after 12 or 1 o'clock, and continued to do so all night, though to a very unequal distance on the two occasions. We therefore thought that the movement was periodic; but on observing three other leaves during several successive days and nights, we found this to be an error; and the case is given merely as a caution. On the third morning the above leaf occupied almost exactly the same position as on the first morning; and the tentacles by this time had unfolded sufficiently to project at right angles to the blade or disc.

Fig. 105. Drosera rotundifolia: circumnutation of young leaf, with filament fixed to back of blade, traced from 9.15 A.M. June 7th to 8.30 A.M. June 9th. Figure here reduced to one-half original scale.

The leaves as they grow older generally sink more and more downwards. The movement of an oldish leaf, the glands of which were still secreting freely, was traced for 24 h., during which time it continued to sink a little in a slightly zigzag line. On the following morning, at 7 A.M., a drop of a solution of carbonate of ammonia (2 gr. to 1 oz. of water) was placed on the disc, and this blackened the glands and induced inflection of many of the tentacles. The weight of the drop caused the leaf at first to sink a little; but immediately afterwards it began to rise in a somewhat zigzag course, and continued to do so till 3 P.M. It then circumnutated about the same spot on a very small scale for 21 h.; and during the next 21 h. it sank in a zigzag line to nearly the same level which it had held when the ammonia was first administered. By this time the tentacles had re-expanded, and the glands had recovered their proper colour. We thus learn that an old leaf circumnutates on a small scale, at least whilst absorbing carbonate of ammonia; for it is probable that this absorption may stimulate growth and thus re-excite circumnutation. Whether the rising of the glass filament which was attached to the back of the leaf, resulted from its margin becoming slightly inflected (as generally occurs), or from the rising of the petiole, was not ascertained.

In order to learn whether the tentacles or gland-bearing hairs circumnutate, the back of a young leaf, with the innermost tentacles as yet incurved, was firmly cemented with shellac to a flat stick driven into compact damp argillaceous sand. The plant was placed under a microscope with the stage removed and with an eye-piece micrometer, of which each division equalled 1/500 of an inch. It should be stated that as the leaves grow older the tentacles of the exterior rows bend outwards and downwards, so as ultimately to become deflected considerably beneath the horizon. A tentacle in the second row from the margin was selected for observation, and was found to be moving outwards at a rate of 1/500 of an inch in 20 m., or 1/100 of inch in 1 h. 40 m.; but as it likewise moved from side to side to an extent of above 1/500 of inch, the movement was probably one of modified circumnutation. A tentacle on an old leaf was next observed in the same manner. In 15 m. after being placed under the microscope it had moved about 1/1000 of an inch. During the next 7½ h. it was looked at repeatedly, and during this whole time it moved only another 1/1000 of an inch; and this small movement may have been due to the settling of the damp sand (on which the plant rested), though the sand had been firmly pressed down. We may therefore conclude that the tentacles when old do not circumnutate; yet this tentacle was so sensitive, that in 23 seconds after its gland had been merely touched with a bit of raw meat, it began to curl inwards. This fact is of some importance, as it apparently shows that the inflection of the tentacles from the stimulus of absorbed animal matter (and no doubt from that of contact with any object) is not due to modified circumnutation.

(15.) Dionoea muscipula (Droseraceae).—It should be premised that the leaves at an early stage of their development have the two lobes pressed closely together. These are at first directed back towards the centre of the plant; but they gradually rise up and soon stand at right angles to the petiole, and ultimately in nearly a straight line with it. A young leaf, which with the petiole was only 1.2 inch in length, had a filament fixed externally along the midrib of the still closed lobes, which projected at right angles to the petiole. In the evening this leaf completed an ellipse in the course of 2 h. On the following day (Sept. 25th) its movements were traced during 22 h.; and we see in Fig. 106 that it moved in the same general direction, due to the straightening of the leaf, but in an extremely zigzag line. This line represents several drawn-out or modified ellipses. There can therefore be no doubt that this young leaf circumnutated.

Fig. 106. Dionaea muscipula: circumnutation of a young and expanding leaf, traced on a horizontal glass in darkness, from noon Sept. 24th to 10 A.M. 25th. Apex of leaf 13½ inches from the glass, so tracing considerably magnified.

A rather old, horizontally extended leaf, with a filament attached along the under side of the midrib, was next observed during 7 h. It hardly moved, but when one of its sensitive hairs was touched, the blades closed, though not very quickly. A new dot was now made on the glass, but in the course of 14 h. 20 m. there was hardly any change in the position of the filament. We may therefore infer that an old and only moderately sensitive leaf does not circumnutate plainly; but we shall soon see that it by no means follows that such a leaf is absolutely motionless. We may further infer that the stimulus from a touch does not re-excite plain circumnutation.

Another full-grown leaf had a filament attached externally along one side of the midrib and parallel to it, so that the filament would move if the lobes closed. It should be first stated that, although a touch on one of the sensitive hairs of a vigorous leaf causes it to close quickly, often almost instantly, yet when a bit of damp meat or some solution of carbonate of ammonia is placed on the lobes, they close so slowly that generally 24 h. is required for the completion of the act. The above leaf was first observed for 2 h. 30 m., and did not circumnutate, but it ought to have been observed for a longer period; although, as we have seen, a young leaf completed a fairly large ellipse in 2 h. A drop of an infusion of raw meat was then placed on the leaf, and within 2 h. the glass filament rose a little; and this implies that the lobes had begun to close, and perhaps the petiole to rise. It continued to rise with extreme slowness for the next 8 h. 30 m. The position of the pot was then (7.15 P.M., Sept. 24th) slightly changed and an additional drop of the infusion given, and a new tracing was begun (Fig. 107). By 10.50 P.M. the filament had risen only a little more, and it fell during the night. On the following morning the lobes were closing more quickly, and by 5 P.M. it was evident to the eye that they had closed considerably; by 8.48. P.M. this was still plainer, and by 10.45 P.M. the marginal spikes were interlocked. The leaf fell a little during the night, and next morning (25th) at 7 A.M. the lobes were completely shut. The course pursued, as may be seen in the figure, was strongly zigzag, and this indicates that the closing of the lobes was combined with the circumnutation of the whole leaf; and there cannot be much doubt, considering how motionless the leaf was during 2 h. 30 m. before it received the infusion, that the absorption of the animal matter had excited it to circumnutate. The leaf was occasionally observed for the next four days, but was kept in rather too cool a place; nevertheless, it continued to circumnutate to a small extent, and the lobes remained closed.

Fig. 107. Dionoea muscipula: closure of the lobes and circumnutation of a full-grown leaf, whilst absorbing an infusion of raw meat, traced in darkness, from 7.15 P.M. Sept. 24th to 9 A.M. 26th. Apex of leaf 8½ inches from the vertical glass. Figure here reduced to two-thirds of original scale.

It is sometimes stated in botanical works that the lobes close or sleep at night; but this is an error. To test the statement, very long glass filaments were fixed inside the two lobes of three leaves, and the distances between their tips were measured in the middle of the day and at night; but no difference could be detected.

The previous observations relate to the movements of the whole leaf, but the lobes move independently of the petiole, and seem to be continually opening and shutting to a very small extent. A nearly full-grown leaf (afterwards proved to be highly sensitive to contact) stood almost horizontally, so that by driving a long thin pin through the foliaceous petiole close to the blade, it was rendered motionless. The plant, with a little triangle of paper attached to one of the marginal spikes, was placed under a microscope with an eye-piece micrometer, each division of which equalled 1/500 of an inch. The apex of the paper-triangle was now seen to be in constant slight movement; for in 4 h. it crossed nine divisions, or 9/500 of an inch, and after ten additional hours it moved back and had crossed 5/500 in an opposite direction. The plant was kept in rather too cool a place, and on the following day it moved rather less, namely, 1/500 in 3 h., and 2/500 in an opposite direction during the next 6 h. The two lobes, therefore, seem to be constantly closing or opening, though to a very small distance; for we must remember that the little triangle of paper affixed to the marginal spike increased its length, and thus exaggerated somewhat the movement. Similar observations, with the important difference that the petiole was left free and the plant kept under a high temperature, were made on a leaf, which was healthy, but so old that it did not close when its sensitive hairs were repeatedly touched, though judging from other cases it would have slowly closed if it had been stimulated by animal matter. The apex of the triangle was in almost, though not quite, constant movement, sometimes in one direction and sometimes in an opposite one; and it thrice crossed five divisions of the micrometer (i.e. 1/100 of an inch) in 30 m. This movement on so small a scale is hardly comparable with ordinary circumnutation; but it may perhaps be compared with the zigzag lines and little loops, by which the larger ellipses made by other plants are often interrupted.

In the first chapter of this volume, the remarkable oscillatory movements of the circumnutating hypocotyl of the cabbage have been described. The leaves of Dionaea present the same phenomenon, which is a wonderful one, as viewed under a low power (2-inch object-glass), with an eye-piece micrometer of which each division (1/500 of an inch) appeared as a rather wide space. The young unexpanded leaf, of which the circumnutating movements were traced (Fig. 106), had a glass filament fixed perpendicularly to it; and the movement of the apex was observed in the hot-house (temp. 84° to 86° F.), with light admitted only from above, and with any lateral currents of air excluded. The apex sometimes crossed one or two divisions of the micrometer at an imperceptibly slow rate, but generally it moved onwards by rapid starts or jerks of 2/1000 or 3/1000, and in one instance of 4/1000 of an inch. After each jerk forwards, the apex drew itself backwards with comparative slowness for part of the distance which had just been gained; and then after a very short time made another jerk forwards. Four conspicuous jerks forwards, with slower retreats, were seen on one occasion to occur in exactly one minute, besides some minor oscillations. As far as we could judge, the advancing and retreating lines did not coincide, and if so, extremely minute ellipses were each time described. Sometimes the apex remained quite motionless for a short period. Its general course during the several hours of observation was in two opposite directions, so that the leaf was probably circumnutating.

An older leaf with the lobes fully expanded, and which was afterwards proved to be highly sensitive to contact, was next observed in a similar manner, except that the plant was exposed to a lower temperature in a room. The apex oscillated forwards and backwards in the same manner as before; but the jerks forward were less in extent, viz. about 1/1000 inch; and there were longer motionless periods. As it appeared possible that the movements might be due to currents of air, a wax taper was held close to the leaf during one of the motionless periods, but no oscillations were thus caused. After 10 m., however, vigorous oscillations commenced, perhaps owing to the plant having been warmed and thus stimulated. The candle was then removed and before long the oscillations ceased; nevertheless, when looked at again after an interval of 1 h. 30 m., it was again oscillating. The plant was taken back into the hot-house, and on the following morning was seen to be oscillating, though not very vigorously. Another old but healthy leaf, which was not in the least sensitive to a touch, was likewise observed during two days in the hot-house, and the attached filament made many little jerks forwards of about 2/1000 or only 1/1000 of an inch.

Finally, to ascertain whether the lobes independently of the petiole oscillated, the petiole of an old leaf was cemented close to the blade with shellac to the top of a little stick driven into the soil. But before this was done the leaf was observed, and found to be vigorously oscillating or jerking; and after it had been cemented to the stick, the oscillations of about 2/1000 of an inch still continued. On the following day a little infusion of raw meat was placed on the leaf, which caused the lobes to close together very slowly in the course of two days; and the oscillations continued during this whole time and for the next two days. After nine additional days the leaf began to open and the margins were a little everted, and now the apex of the glass filament remained for long periods motionless, and then moved backwards and forwards for a distance of about 1/1000 of an inch slowly, without any jerks. Nevertheless, after warming the leaf with a taper held close to it, the jerking movement recommenced.

This same leaf had been observed 2½ months previously, and was then found to be oscillating or jerking. We may therefore infer that this kind of movement goes on night and day for a very long period; and it is common to young unexpanded leaves and to leaves so old as to have lost their sensitiveness to a touch, but which were still capable of absorbing nitrogenous matter. The phenomenon when well displayed, as in the young leaf just described, is a very interesting one. It often brought before our minds the idea of effort, or of a small animal struggling to escape from some constraint.

(16.) Eucalyptus resinifera (Myrtaceae, Fam. 94).—A young leaf, two inches in length together with the petiole, produced by a lateral shoot from a cut-down tree, was observed in the usual manner. The blade had not as yet assumed its vertical position. On June 7th only a few observations were made, and the tracing merely showed that the leaf had moved three times upwards and three downwards. On the following day it was observed more frequently; and two tracings were made (see A and B, Fig. 108), as a single one would have been too complicated. The apex changed its course 13 times in the course of 16 h., chiefly up and down, but with some lateral movement. The actual amount of movement in any one direction was small.

Fig. 108. Eucalyptus resinifera: circumnutation of a leaf, traced, A, from 6.40 A.M. to 1 P.M. June 8th; B, from 1 P.M. 8th to 8.30 A.M. 9th. Apex of leaf 14½ inches from the horizontal glass, so figures considerably magnified.

(17.) Dahlia (garden var.) (Compositæ, Fam. 122).—A fine young leaf 5 3/4 inches in length, produced by a young plant 2 feet high, growing vigorously in a large pot, was directed at an angle of about 45° beneath the horizon. On June 18th the leaf descended from 10 A.M. till 11.35 A.M. (see Fig. 109); it then ascended greatly till 6 P.M., this ascent being probably due to the light coming only from above. It zigzagged between 6 P.M. and 10.35 P.M., and ascended a little during the night. It should be remarked that the vertical distances in the lower part of the diagram are much exaggerated, as the leaf was at first deflected beneath the horizon, and after it had sunk downwards, the filament pointed in a very oblique line towards the glass. Next day the leaf descended from 8.20 A.M. till 7.15 P.M., then zigzagged and ascended greatly during the night. On the morning of the 20th the leaf was probably beginning to descend, though the short line in the diagram is horizontal. The actual distances travelled by the apex of the leaf were considerable, but could not be calculated with safety. From the course pursued on the second day, when the plant had accommodated itself to the light from above, there cannot be much doubt that the leaves undergo a daily periodic movement, sinking during the day and rising at night.

Fig. 109. Dahlia: circumnutation of leaf, traced from 10 A.M. June 18th to 8.10 A.M. 20th, but with a break of 1 h. 40 m. on the morning of the 19th, as, owing to the glass filament pointing too much to one side, the pot had to be slightly moved; therefore the relative position of the two tracings is somewhat arbitrary. The figure here given is reduced to one-fifth of the original scale. Apex of leaf 9 inches from the glass in the line of its inclination, and 4 3/4 in a horizontal line.

(18.) Mutisia clematis (Compositæ).—The leaves terminate in tendrils and circumnutate like those of other tendril-bearers; but this plant is here mentioned, on account of an erroneous statement[11] which has been published, namely, that the leaves sink at night and rise during the day. The leaves which behaved in this manner had been kept for some days in a northern room and had not been sufficiently illuminated. A plant therefore was left undisturbed in the hot-house, and three leaves had their angles measured at noon and at 10 P.M. All three were inclined a little beneath the horizon at noon, but one stood at night 2°, the second 21°, and the third 10° higher than in the middle of the day; so that instead of sinking they rise a little at night.

[11] 'The Movements and Habits of Climbing Plants,' 1875, p. 118.

(19.) Cyclamen Persicum (Primulaceae, Fam. 135).—A young leaf, 1.8 of an inch in length, petiole included, produced by an old root-stock, was observed during three days in the usual manner (Fig. 110). On the first day the leaf fell more than afterwards, apparently from adjusting itself to the light from above. On all three days it fell from the early morning to about 7 P.M., and from that hour rose during the night, the course being slightly zigzag. The movement therefore is strictly periodic. It should be noted that the leaf would have sunk each evening a little lower down than it did, had not the glass filament rested between 5 and 6 P.M. on the rim of the pot. The amount of movement was considerable; for if we assume that the whole leaf to the base of the petiole became bent, the tracing would be magnified rather less than five times, and this would give to the apex a rise and fall of half an inch, with some lateral movement. This amount, however, would not attract attention without the aid of a tracing or measurement of some kind.

(20.) Allamanda Schottii (Apocyneae, Fam. 144).—The young leaves of this shrub are elongated, with the blade bowed so much downwards as almost to form a semicircle. The chord—that is, a line drawn from the apex of the blade to the base of the petiole—of a young leaf, 4 3/4 inches in length, stood at 2.50 P.M. on Dec. 5th at an angle of 13° beneath the horizon, but by 9.30 P.M. the blade had straightened itself so much, which implies the raising of the apex, that the chord now stood at 37° above the horizon, and had therefore risen 50°. On the next day similar angular measurements of the same leaf were made; and at noon the chord stood 36° beneath the horizon, and 9.30 P.M. 3½° above it, so had risen 39½°. The chief cause of the rising movement lies in the straightening of the blade, but the short petiole rises between 4° and 5°. On the third night the chord stood at 35° above the horizon, and if the leaf occupied the same position at noon, as on the previous day, it had risen 71°. With older leaves no such change of curvature could be detected. The plant was then brought into the house and kept in a north-east room, but at night there was no change in the curvature of the young leaves; so that previous exposure to a strong light is apparently requisite for the periodical change of curvature in the blade, and for the slight rising of the petiole.

Fig. 110. Cyclamen Persicum: circumnutation of leaf, traced from 6.45 A.M. June 2nd to 6.40 A.M. 5th. Apex of leaf 7 inches from the vertical glass.

(21.) Wigandia (Hydroleaceae, Fam. 149).—Professor Pfeffer informs us that the leaves of this plant rise in the evening; but as we do not know whether or not the rising is great, this species ought perhaps to be classed amongst sleeping plants.

Fig. 111. Petunia violacea: downward movement and circumnutation of a very young leaf, traced from 10 A.M. June 2nd to 9.20 A.M. June 6th. N.B.—At 6.40 A.M. on the 5th it was necessary to move the pot a little, and a new tracing was begun at the point where two dots are not joined in the diagram. Apex of leaf 7 inches from the vertical glass. Temp. generally 17½° C.

(22.) Petunia violacea (Solaneae, Fam. 157).—A very young leaf, only 3/4 inch in length, highly inclined upwards, was observed for four days. During the whole of this time it bent outwards and downwards, so as to become more and more nearly horizontal. The strongly marked zigzag line in the figure on p. 248 (Fig. 111), shows that this was effected by modified circumnutation; and during the latter part of the time there was much ordinary circumnutation on a small scale. The movement in the diagram is magnified between 10 and 11 times. It exhibits a clear trace of periodicity, as the leaf rose a little each evening; but this upward tendency appeared to be almost conquered by the leaf striving to become more and more horizontal as it grew older. The angles which two older leaves formed together, were measured in the evening and about noon on 3 successive days, and each night the angle decreased a little, though irregularly.

Fig. 112. Acanthus mollis: circumnutation of young leaf, traced from 9.20 A.M. June 14th to 8.30 A.M. 16th. Apex of leaf 11 inches from the vertical glass, so movement considerably magnified. Figure here reduced to one-half of original scale. Temp. 15°–16½° C.

(23.) Acanthus mollis (Acanthaceae, Fam. 168).—The younger of two leaves, 2 1/4 inches in length, petiole included, produced by a seedling plant, was observed during 47 h. Early on each of the three mornings, the apex of the leaf fell; and it continued to fall till 3 P.M., on the two afternoons when observed. After 3 P.M. it rose considerably, and continued to rise on the second night until the early morning. But on the first night it fell instead of rising, and we have little doubt that this was owing to the leaf being very young and becoming through epinastic growth more and more horizontal; for it may be seen in the diagram (Fig. 112), that the leaf stood on a higher level on the first than on the second day. The leaves of an allied species ('A. spinosus') certainly rose every night; and the rise between noon and 10.15 P.M., when measured on one occasion, was 10°. This rise was chiefly or exclusively due to the straightening of the blade, and not to the movement of the petiole. We may therefore conclude that the leaves of Acanthus circumnutate periodically, falling in the morning and rising in the afternoon and night.

(24.) Cannabis sativa (Cannabineae, Fam. 195).—We have here the rare case of leaves moving downwards in the evening, but not to a sufficient degree to be called sleep.[12] In the early morning, or in the latter part of the night, they move upwards. For instance, all the young leaves near the summits of several stems stood almost horizontally at 8 A.M. May 29th and at 10.30 P.M. were considerably declined. On a subsequent day two leaves stood at 2 P.M. at 21° and 12° beneath the horizon, and at 10 P.M. at 38° beneath it. Two other leaves on a younger plant were horizontal at 2 P.M., and at 10 P.M. had sunk to 36° beneath the horizon. With respect to this downward movement of the leaves, Kraus believes that it is due to their epinastic growth. He adds, that the leaves are relaxed during the day, and tense at night, both in sunny and rainy weather.

[12] We were led to observe this plant by Dr. Carl Kraus' paper, 'Beiträge zur Kentniss der Bewegungen Wachsender Laubblätter,' Flora, 1879, p. 66. We regret that we cannot fully understand parts of this paper.

(25.) Pinus pinaster (Coniferæ, Fam. 223).—The leaves on the summits of the terminal shoots stand at first in a bundle almost upright, but they soon diverge and ultimately become almost horizontal. The movements of a young leaf, nearly one inch in length, on the summit of a seedling plant only 3 inches high, were traced from the early morning of June 2nd to the evening of the 7th. During these five days the leaf diverged, and its apex descended at first in an almost straight line; but during the two latter days it zigzagged so much that it was evidently circumnutating. The same little plant, when grown to a height of 5 inches, was again observed during four days. A filament was fixed transversely to the apex of a leaf, one inch in length, and which had already diverged considerably from its originally upright position. It continued to diverge (see A, Fig. 113), and to descend from 11.45 A.M. July 31st to 6.40 A.M. Aug. 1st. On August 1st it circumnutated about the same small space, and again descended at night. Next morning the pot was moved nearly one inch to the right, and a new tracing was begun (B). From this time, viz., 7 A.M. August 2nd to 8.20 A.M. on the 4th, the leaf manifestly circumnutated. It does not appear from the diagram that the leaves move periodically, for the descending course during the first two nights, was clearly due to epinastic growth, and at the close of our observations the leaf was not nearly so horizontal as it would ultimately become.

Fig. 113. Pinus pinaster: circumnutation of young leaf, traced from 11.45 A.M. July 31st to 8.20 A.M. Aug. 4th. At 7 A.M. Aug. 2nd the pot was moved an inch to one side, so that the tracing consists of two figures. Apex of leaf 14½ inches from the vertical glass, so movements much magnified.

Pinus austriaca.—Two leaves, 3 inches in length, but not quite fully grown, produced by a lateral shoot, on a young tree 3 feet in height, were observed during 29 h. (July 31st), in the same manner as the leaves of the previous species. Both these leaves certainly circumnutated, making within the above period two, or two and a half, small, irregular ellipses.

(26.) Cycas pectinata (Cycadeæ, Fam. 224).—A young leaf, 11½ inches in length, of which the leaflets had only recently become uncurled, was observed during 47 h. 30 m. The main petiole was secured to a stick at the base of the two terminal leaflets. To one of the latter, 3 3/4 inches in length, a filament was fixed; the leaflet was much bowed downward, but as the terminal part was upturned, the filament projected almost horizontally. The leaflet moved (see Fig. 114) largely and periodically, for it fell until about 7 P.M. and rose during the night, falling again next morning after 6.40 A.M. The descending lines are in a marked manner zigzag, and so probably would have been the ascending lines, if they had been traced throughout the night.

Fig. 114. Cycas pectinata: circumnutation of one of the terminal leaflets, traced from 8.30 A.M. June 22nd to 8 A.M. June 24th. Apex of leaflet 7 3/4 inches from the vertical glass, so tracing not greatly magnified, and here reduced to one-third of original scale; temp. 19°–21° C.

CIRCUMNUTATION OF LEAVES: MONOCOTYLEDONS.

(27.) Canna Warscewiczii (Cannaceae, Fam. 2).—The movements of a young leaf, 8 inches in length and 3½ in breadth, produced by a vigorous young plant, were observed during 45 h. 50 m., as shown in Fig. 115. The pot was slided about an inch to the right on the morning of the 11th, as a single figure would have been too complicated; but the two figures are continuous in time. The movement is periodical, as the leaf descended from the early morning until about 5 P.M., and ascended during the rest of the evening and part of the night. On the evening of the 11th it circumnutated on a small scale for some time about the same spot.

Fig. 115. Canna Warscewiczii: circumnutation of leaf, traced (A) from 11.30 A.M. June 10th to 6.40 A.M. 11th; and (B) from 6.40 A.M. 11th to 8.40 A.M. 12th. Apex of leaf 9 inches from the vertical glass.

(28.) Iris pseudo-acorus (Irideae, Fam. 10).—The movements of a young leaf, rising 13 inches above the water in which the plant grew, were traced as shown in the figure (Fig. 116), during 27 h. 30 m. It manifestly circumnutated, though only to a small extent. On the second morning, between 6.40 A.M. and 2 P.M. (at which latter hour the figure here given ends), the apex changed its course five times. During the next 8 h. 40 m. it zigzagged much, and descended as far as the lowest dot in the figure, making in its course two very small ellipses; but if these lines had been added to the diagram it would have been too complex.

Fig. 116. Iris pseudo-acorus: circumnutation of leaf, traced from 10.30 A.M. May 28th to 2 P.M. 29th. Tracing continued to 11 P.M., but not here copied. Apex of leaf 12 inches beneath the horizontal glass, so figure considerably magnified. Temp. 15°–16° C.

(29.) Crinum Capense (Amaryllideae, Fam. 11).—The leaves of this plant are remarkable for their great length and narrowness: one was measured and found to be 53 inches long and only 1.4 broad at the base. Whilst quite young they stand up almost vertically to the height of about a foot;

afterwards their tips begin to bend over, and subsequently hang vertically down, and thus continue to grow. A rather young leaf was selected, of which the dependent tapering point was as yet only 5½ inches in length, the upright basal part being 20 inches high, though this part would ultimately become shorter by being more bent over. A large bell-glass was placed over the plant, with a black dot on one side; and by bringing the dependent apex of the leaf into a line with this dot, the accompanying figure (Fig. 117) was traced on the other side of the bell, during 2½ days. During the first day (22nd) the tip travelled laterally far to the left, perhaps in consequence of the plant having been disturbed; and the last dot made at 10.30 P.M. on this day is alone here given. As we see in the figure, there can be no doubt that the apex of this leaf circumnutated.

Fig. 117. Crinum Capense: circumnutation of dependent tip of young leaf, traced on a bell-glass, from 10.30 P.M. May 22nd to 10.15 A.M. 25th. Figure not greatly magnified.

A glass filament with little triangles of paper was at the same time fixed obliquely across the tip of a still younger leaf, which stood vertically up and was as yet straight. Its movements were traced from 3 P.M. May 22nd to 10.15 A.M. 25th. The leaf was growing rapidly, so that the apex ascended greatly during this period; as it zigzagged much it was clearly circumnutating, and it apparently tended to form one ellipse each day. The lines traced during the night were much more vertical than those traced during the day; and this indicates that the tracing would have exhibited a nocturnal rise and a diurnal fall, if the leaf had not grown so quickly. The movement of this same leaf after an interval of six days (May 31st), by which time the tip had curved outwards into a horizontal position, and had thus made the first step towards becoming dependent, was traced orthogonically by the aid of a cube of wood (in the manner before explained); and it was thus ascertained that the actual distance travelled by the apex, and due to circumnutation, was 3 1/8 inches in the course of 20½ h. During the next 24 h. it travelled 2½ inches. The circumnutating movement, therefore, of this young leaf was strongly marked.

(30.) Pancratium littorale (Amaryllideae).—The movements, much magnified, of a leaf, 9 inches in length and inclined at about 45° above the horizon, were traced during two days. On the first day it changed its course completely, upwards and downwards and laterally, 9 times in 12 h.; and the figure traced apparently represented five ellipses. On the second day it was observed seldomer, and was therefore not seen to change its course so often, viz., only 6 times, but in the same complex manner as before. The movements were small in extent, but there could be no doubt about the circumnutation of the leaf.

(31.) Imatophyllum vel Clivia (sp.?) (Amaryllideae).—A long glass filament was fixed to a leaf, and the angle formed by it with the horizon was measured occasionally during three successive days. It fell each morning until between 3 and 4 P.M., and rose at night. The smallest angle at any time above the horizon was 48°, and the largest 50°; so that it rose only 2° at night; but as this was observed each day, and as similar observations were nightly made on another leaf on a distinct plant, there can be no doubt that the leaves move periodically, though to a very small extent. The position of the apex when it stood highest was .8 of an inch above its lowest point.

(32.) Pistia stratiotes (Aroideae, Fam. 30).—Hofmeister remarks that the leaves of this floating water-plant are more highly inclined at night than by day.[13] We therefore fastened a fine glass filament to the midrib of a moderately young leaf, and on Sept. 19th measured the angle which it formed with the horizon 14 times between 9 A.M. and 11.50 P.M. The temperature of the hot-house varied during the two days of observation between 18½° and 23½° C. At 9 A.M. the filament stood at 32° above the horizon; at 3.34 P.M. at 10° and at 11.50 P.M. at 55°; these two latter angles being the highest and the lowest observed during the day, showing a difference of 45°. The rising did not become strongly marked until between 5 and 6 P.M. On the next day the leaf stood at only 10° above the horizon at 8.25 A.M., and it remained at about 15° till past 3 P.M.; at 5.40 P.M. it was 23°, and at 9.30 P.M. 58°; so that the rise was more sudden this evening than on the previous one, and the difference in the angle amounted to 48°. The movement is obviously periodical, and as the leaf stood on the first night at 55°, and on the second night at 58° above the horizon, it appeared very steeply inclined. This case, as we shall see in a future chapter, ought perhaps to have been included under the head of sleeping plants.

[13] 'Die Lehre von der Pflanzenzelle,' 1867, p. 327.

(33.) Pontederia (sp.?) (from the highlands of St. Catharina, Brazil) (Pontederiaceae, Fam. 46).—A filament was fixed across the apex of a moderately young leaf, 7½ inches in height, and its movements were traced during 42½ h. (see Fig. 118). On the first evening, when the tracing was begun, and during the night, the leaf descended considerably. On the next morning it ascended in a strongly marked zigzag line, and descended again in the evening and during the night. The movement, therefore, seems to be periodic, but some doubt is thrown on this conclusion, because another leaf, 8 inches in height, appearing older and standing more highly inclined, behaved differently. During the first 12 h. it circumnutated over a small space, but during the night and the whole following day it ascended in the same general direction; the ascent being effected by repeated up and down well-pronounced oscillations.

Fig. 118. Pontederia (sp.?): circumnutation of leaf, traced from 4.50 P.M. July 2nd to 10.15 A.M. 4th. Apex of leaf 16½ inches from the vertical glass, so tracing greatly magnified. Temp. about 17° C., and therefore rather too low.

CRYPTOGAMS.

(34.) Nephrodium molle (Filices, Fam. 1).—A filament was fixed near the apex of a young frond of this Fern, 17 inches in height, which was not as yet fully uncurled; and its movements were traced during 24 h. We see in Fig. 119 that it plainly circumnutated. The movement was not greatly magnified as the frond was placed near to the vertical glass, and would probably have been greater and more rapid had the day been warmer. For the plant was brought out of a warm greenhouse and observed under a skylight, where the temperature was between 15° and 16° C. We have seen in Chap. I. that a frond of this Fern, as yet only slightly lobed and with a rachis only .23 inch in height, plainly circumnutated.[14]

[14] Mr. Loomis and Prof. Asa Gray have described ('Botanical Gazette,' 1880, pp. 27, 43), an extremely curious case of movement in the fronds, but only in the fruiting fronds, of Asplenium trichomanes. They move almost as rapidly as the little leaflets of Desmodium gyrans, alternately backwards and forwards through from 20 to 40 degrees, in a plane at right angles to that of the frond. The apex of the frond describes "a long and very narrow ellipse," so that it circumnutates. But the movement differs from ordinary circumnutation as it occurs only when the plant is exposed to the light; even artificial light "is sufficient to excite motion for a few minutes."

Fig. 119. Nephrodium molle: circumnutation of rachis, traced from 9.15 A.M. May 28th to 9 A.M. 29th. Figure here given two-thirds of original scale.

In the chapter on the Sleep of Plants the conspicuous circumnutation of Marsilea quadrifoliata (Marsileaceae, Fam. 4) will be described.

It has also been shown in Chap. I. that a very young Selaginella (Lycopodiaceæ, Fam. 6), only .4 inch in height, plainly circumnutated; we may therefore conclude that older plants, whilst growing, would do the same.

(35.) Lunularia vulgaris (Hepaticae, Fam. 11, Muscales).—The earth in an old flower-pot was coated with this plant, bearing gemmae. A highly inclined frond, which projected .3 inch above the soil and was .4 inch in breadth, was selected for observation. A glass hair of extreme tenuity, .75 inch in length, with its end whitened, was cemented with shellac to the frond at right angles to its breadth; and a white stick with a minute black spot was driven into the soil close behind the end of the hair. The white end could be accurately brought into a line with the black spot, and dots could thus be successively made on the vertical glass-plate in front. Any movement of the frond would of course be exhibited and increased by the long glass hair; and the black spot was placed so close behind the end of the hair, relatively to the distance of the glass-plate in front, that the movement of the end was magnified about 40 times. Nevertheless, we are convinced that our tracing gives a fairly faithful representation of the movements of the frond. In the intervals between each observation, the plant was covered by a small bell-glass. The frond, as already stated, was highly inclined, and the pot stood in front of a north-east window. During the five first days the frond moved downwards or became less inclined; and the long line which was traced was strongly zigzag, with loops occasionally formed or nearly formed; and this indicated circumnutation. Whether the sinking was due to epinastic growth, or apheliotropism, we do not know. As the sinking was slight on the fifth day, a new tracing was begun on the sixth day (Oct. 25th), and was continued for 47 h.; it is here given (Fig. 120). Another tracing was made on the next day (27th) and the frond was found to be still circumnutating, for during 14 h. 30 m. it changed its course completely (besides minor changes) 10 times. It was casually observed for two more days, and was seen to be continually moving.

Fig. 120. Lunularia vulgaris: circumnutation of a frond, traced from 9 A.M. Oct 25th to 8 A.M. 27th.

The lowest members of the vegetable series, the Thallogens, apparently circumnutate. If an Oscillaria be watched under the microscope, it may be seen to describe circles about every 40 seconds. After it has bent to one side, the tip first begins to bend back to the opposite side and then the whole filament curves over in the same direction. Hofmeister[15] has given a minute account of the curious, but less regular though constant, movements of Spirogyra: during 2½ h. the filament moved 4 times to the left and 3 times to the right, and he refers to a movement at right angles to the above. The tip moved at the rate of about 0.1 mm. in five minutes. He compares the movement with the nutation of the higher plants.[16] We shall hereafter see that heliotropic movements result from modified circumnutation, and as unicellular Moulds bend to the light we may infer that they also circumnutate.

[15] 'Ueber die Bewegungen der Faden der *Spirogyra princeps:* Jahreshefte des Vereins für vaterländische Naturkunde in Württemberg,' 1874, p. 211.

[16] Zukal also remarks (as quoted in 'Journal R. Microscop. Soc.,' 1880, vol. iii. p. 320) that the movements of Spirulina, a member of the Oscillatorieae, are closely analogous "to the well-known rotation of growing shoots and tendrils."

CONCLUDING REMARKS ON THE CIRCUMNUTATION OF LEAVES.

The circumnutating movements of young leaves in 33 genera, belonging to 25 families, widely distributed amongst ordinary and gymnospermous Dicotyledons and amongst Monocotyledons, together with several Cryptogams, have now been described. It would, therefore, not be rash to assume that the growing leaves of all plants circumnutate, as we have seen reason to conclude is the case with cotyledons. The seat of movement generally lies in the petiole, but sometimes both in the petiole and blade, or in the blade alone. The extent of the movement differed much in different plants; but the distance passed over was never great, except with Pistia, which ought perhaps to have been included amongst sleeping plants. The angular movement of the leaves was only occasionally measured; it commonly varied from only 2° (and probably even less in some instances) to about 10°; but it amounted to 23° in the common bean. The movement is chiefly in a vertical plane, but as the ascending and descending lines never coincided, there was always some lateral movement, and thus irregular ellipses were formed. The movement, therefore, deserves to be called one of circumnutation; for all circumnutating organs tend to describe ellipses,—that is, growth on one side is succeeded by growth on nearly but not quite the opposite side. The ellipses, or the zigzag lines representing drawn-out ellipses, are generally very narrow; yet with the Camellia, their minor axes were half as long, and with the Eucalyptus more than half as long as their major axes. In the case of Cissus, parts of the figure more nearly represented circles than ellipses. The amount of lateral movement is therefore sometimes considerable. Moreover, the longer axes of the successively formed ellipses (as with the Bean, Cissus, and Sea-kale), and in several instances the zigzag lines representing ellipses, were extended in very different directions during the same day or on the next day. The course followed was curvilinear or straight, or slightly or strongly zigzag, and little loops or triangles were often formed. A single large irregular ellipse may be described on one day, and two smaller ones by the same plant on the next day. With Drosera two, and with Lupinus, Eucalyptus and Pancratium, several were formed each day.

The oscillatory and jerking movements of the leaves of Dionaea, which resemble those of the hypocotyl of the cabbage, are highly remarkable, as seen under the microscope. They continue night and day for some months, and are displayed by young unexpanded leaves, and by old ones which have lost their sensibility to a touch, but which, after absorbing animal matter, close their lobes. We shall hereafter meet with the same kind of movement in the joints of certain Gramineæ, and it is probably common to many plants while circumnutating. It is, therefore, a strange fact that no such movement could be detected in the tentacles of Drosera rotundifolia, though a member of the same family with Dionaea; yet the tentacle which was observed was so sensitive, that it began to curl inwards in 23 seconds after being touched by a bit of raw meat.

One of the most interesting facts with respect to the circumnutation of leaves is the periodicity of their movements; for they often, or even generally, rise a little in the evening and early part of the night, and sink again on the following morning. Exactly the same phenomenon was observed in the case of cotyledons. The leaves in 16 genera out of the 33 which were observed behaved in this manner, as did probably 2 others. Nor must it be supposed that in the remaining 15 genera there was no periodicity in their movements; for 6 of them were observed during too short a period for any judgment to be formed on this head, and 3 were so young that their epinastic growth, which serves to bring them down into a horizontal position, overpowered every other kind of movement. In only one genus, Cannabis, did the leaves sink in the evening, and Kraus attributes this movement to the prepotency of their epinastic growth. That the periodicity is determined by the daily alternations of light and darkness there can hardly be a doubt, as will hereafter be shown. Insectivorous plants are very little affected, as far as their movements are concerned, by light; and hence probably it is that their leaves, at least in the cases of Sarracenia, Drosera, and Dionaea, do not move periodically. The upward movement in the evening is at first slow, and with different plants begins at very different hours;—with Glaucium as early as 11 A.M., commonly between 3 and 5 P.M., but sometimes as late as 7 P.M. It should be observed that none of the leaves described in this chapter (except, as we believe, those of Lupinus speciosus) possess a pulvinus; for the periodical movements of leaves thus provided have generally been amplified into so-called sleep-movements, with which we are not here concerned. The fact of leaves and cotyledons frequently, or even generally, rising a little in the evening and sinking in the morning, is of interest as giving the foundation from which the specialised sleep-movements of many leaves and cotyledons, not provided with a pulvinus, have been developed. the above periodicity should be kept in mind, by any one considering the problem of the horizontal position of leaves and cotyledons during the day, whilst illuminated from above.

CHAPTER V.
MODIFIED CIRCUMNUTATION: CLIMBING PLANTS; EPINASTIC AND HYPONASTIC MOVEMENTS.

Circumnutation modified through innate causes or through the action of external conditions—Innate causes—Climbing plants; similarity of their movements with those of ordinary plants; increased amplitude; occasional points of difference—Epinastic growth of young leaves—Hyponastic growth of the hypocotyls and epicotyls of seedlings—Hooked tips of climbing and other plants due to modified circumnutation—Ampelopsis tricuspidata—Smithia Pfundii—Straightening of the tip due to hyponasty—Epinastic growth and circumnutation of the flower-peduncles of Trifolium repens and Oxalis carnosa.

The radicles, hypocotyls and epicotyls of seedling plants, even before they emerge from the ground, and afterwards the cotyledons, are all continually circumnutating. So it is with the stems, stolons, flower-peduncles, and leaves of older plants. We may, therefore, infer with a considerable degree of safety that all the growing parts of all plants circumnutate. Although this movement, in its ordinary or unmodified state, appears in some cases to be of service to plants, either directly or indirectly—for instance, the circumnutation of the radicle in penetrating the ground, or that of the arched hypocotyl and epicotyl in breaking through the surface—yet circumnutation is so general, or rather so universal a phenomenon, that we cannot suppose it to have been gained for any special purpose. We must believe that it follows in some unknown way from the manner in which vegetable tissues grow.

We shall now consider the many cases in which circumnutation has been modified for various special purposes; that is, a movement already in progress is temporarily increased in some one direction, and temporarily diminished or quite arrested in other directions. These cases may be divided in two sub-classes; in one of which the modification depends on innate or constitutional causes, and is independent of external conditions, excepting in so far that the proper ones for growth must be present. In the second sub-class the modification depends to a large extent on external agencies, such as the daily alternations of light and darkness, or light alone, temperature, or the attraction of gravity. The first small sub-class will be considered in the present chapter, and the second sub-class in the remainder of this volume.

THE CIRCUMNUTATION OF CLIMBING PLANTS.

The simplest case of modified circumnutation is that offered by climbing plants, with the exception of those which climb by the aid of motionless hooks or of rootlets: for the modification consists chiefly in the greatly increased amplitude of the movement. This would follow either from greatly increased growth over a small length, or more probably from moderately increased growth spread over a considerable length of the moving organ, preceded by turgescence, and acting successively on all sides. The circumnutation of climbers is more regular than that of ordinary plants; but in almost every other respect there is a close similarity between their movements, namely, in their tendency to describe ellipses directed successively to all points of the compass—in their courses being often interrupted by zigzag lines, triangles, loops, or small ellipses—in the rate of movement, and in different species revolving once or several times within the same length of time. In the same internode, the movements cease first in the lower part and then slowly upwards. In both sets of cases the movement may be modified in a closely analogous manner by geotropism and by heliotropism; though few climbing plants are heliotropic. Other points of similarity might be pointed out.

That the movements of climbing plants consist of ordinary circumnutation, modified by being increased in amplitude, is well exhibited whilst the plants are very young; for at this early age they move like other seedlings, but as they grow older their movements gradually increase without undergoing any other change. That this power is innate, and is not excited by any external agencies, beyond those necessary for growth and vigour, is obvious. No one doubts that this power has been gained for the sake of enabling climbing plants to ascend to a height, and thus to reach the light. This is effected by two very different methods; first, by twining spirally round a support, but to do so their stems must be long and flexible; and, secondly, in the case of leaf-climbers and tendril-bearers, by bringing these organs into contact with a support, which is then seized by the aid of their sensitiveness. It may be here remarked that these latter movements have no relation, as far as we can judge, with circumnutation. In other cases the tips of tendrils, after having been brought into contact with a support, become developed into little discs which adhere firmly to it.

We have said that the circumnutation of climbing plants differs from that of ordinary plants chiefly by its greater amplitude. But most leaves circumnutate in an almost vertical plane, and therefore describe very narrow ellipses, whereas the many kinds of tendrils which consist of metamorphosed leaves, make much broader ellipses or nearly circular figures; and thus they have a far better chance of catching hold of a support on any side. The movements of climbing plants have also been modified in some few other special ways. Thus the circumnutating stems of Solnanum dulcamara can twine round a support only when this is as thin and flexible as a string or thread. The twining stems of several British plants cannot twine round a support when it is more than a few inches in thickness; whilst in tropical forests some can embrace thick trunks;[1] and this great difference in power depends on some unknown difference in their manner of circumnutation. The most remarkable special modification of this movement which we have observed is in the tendrils of Echinocystis lobata; these are usually inclined at about 45° above the horizon, but they stiffen and straighten themselves so as to stand upright in a part of their circular course, namely, when they approach and have to pass over the summit or the shoot from which they arise. If they had not possessed and exercised this curious power, they would infallibly have struck against the summit of the shoot and been arrested in their course. As soon as one of these tendrils with its three branches begins to stiffen itself and rise up vertically, the revolving motion becomes more rapid; and as soon as it has passed over the point of difficulty, its motion coinciding with that from its own weight, causes it to fall into its previously inclined position so quickly, that the apex can be seen travelling like the hand of a gigantic clock.

[1] 'The Movements and Habits of Climbing Plants,' p. 36.

A large number of ordinary leaves and leaflets and a few flower-peduncles are provided with pulvini; but this is not the case with a single tendril at present known. The cause of this difference probably lies in the fact, that the chief service of a pulvinus is to prolong the movement of the part thus provided after growth has ceased; and as tendrils or other climbing-organs are of use only whilst the plant is increasing in height or growing, a pulvinus which served to prolong their movements would be useless.

It was shown in the last chapter that the stolons or runners of certain plants circumnutate largely, and that this movement apparently aids them in finding a passage between the crowded stems of adjoining plants. If it could be proved that their movements had been modified and increased for this special purpose, they ought to have been included in the present chapter; but as the amplitude of their revolutions is not so conspicuously different from that of ordinary plants, as in the case of climbers, we have no evidence on this head. We encounter the same doubt in the case of some plants which bury their pods in the ground. This burying process is certainly favoured by the circumnutation of the flower-peduncle; but we do not know whether it has been increased for this special purpose.

EPINASTY—HYPONASTY.

The term epinasty is used by De Vries[2] to express greater longitudinal growth along the upper than along the lower side of a part, which is thus caused to bend downwards; and hyponasty is used for the reversed process, by which the part is made to bend upwards. These actions come into play so frequently that the use of the above two terms is highly convenient. The movements thus induced result from a modified form of circumnutation; for, as we shall immediately see, an organ under the influence of epinasty does not generally move in a straight line downwards, or under that of hyponasty upwards, but oscillates up and down with some lateral movement: it moves, however, in a preponderant manner in one direction. This shows that there is some growth on all sides of the part, but more on the upper side in the case of epinasty, and more on the lower side in that of hyponasty, than on the other sides. At the same time there may be in addition, as De Vries insists, increased growth on one side due to geotropism, and on another side due to heliotropism; and thus the effects of epinasty or of hyponasty may be either increased or lessened.

[2] 'Arbeiten des Bot. Inst., in Würzburg,' Heft ii. 1872, p. 223. De Vries has slightly modified (p. 252) the meaning of the above two terms as first used by Schimper, and they have been adopted in this sense by Sachs.

He who likes, may speak of ordinary circumnutation as being combined with epinasty, hyponasty, the effects of gravitation, light, etc.; but it seems to us, from reasons hereafter to be given, to be more correct to say that circumnutation is modified by these several agencies. We will therefore speak of circumnutation, which is always in progress, as modified by epinasty, hyponasty, geotropism, or other agencies, whether internal or external.

One of the commonest and simplest cases of epinasty is that offered by leaves, which at an early age are crowded together round the buds, and diverge as they grow older. Sachs first remarked that this was due to increased growth along the upper side of the petiole and blade; and De Vries has now shown in more detail that the movement is thus caused, aided slightly by the weight of the leaf, and resisted as he believes by apogeotropism, at least after the leaf has somewhat diverged. In our observations on the circumnutation of leaves, some were selected which were rather too young, so that they continued to diverge or sink downwards whilst their movements were being traced. This may be seen in the diagrams (Figs. 98 and 112, pp. 232 and 249) representing the circumnutation of the young leaves of Acanthus mollis and Pelargonium zonale. Similar cases were observed with Drosera. The movements of a young leaf, only 3/4 inch in length, of Petunia violacea were traced during four days, and offers a better instance (Fig. 111, p. 248) as it diverged during the whole of this time in a curiously zigzag line with some of the angles sharply acute, and during the latter days plainly circumnutated. Some young leaves of about the same age on a plant of this Petunia, which had been laid horizontally, and on another plant which was left upright, both being kept in complete darkness, diverged in the same manner for 48 h., and apparently were not affected by apogeotropism; though their stems were in a state of high tension, for when freed from the sticks to which they had been tied, they instantly curled upwards.

The leaves, whilst very young, on the leading shoots of the Carnation (Dianthus caryophyllus) are highly inclined or vertical; and if the plant is growing vigorously they diverge so quickly that they become almost horizontal in a day. But they move downwards in a rather oblique line and continue for some time afterwards to move in the same direction, in connection, we presume, with their spiral arrangement on the stem. The course pursued by a young leaf whilst thus obliquely descending was traced, and the line was distinctly yet not strongly zigzag; the larger angles formed by the successive lines amounting only to 135°, 154°, and 163°. The subsequent lateral movement (shown in Fig. 96, p. 231) was strongly zigzag with occasional circumnutations. The divergence and sinking of the young leaves of this plant seem to be very little affected by geotropism or heliotropism; for a plant, the leaves of which were growing rather slowly (as ascertained by measurement) was laid horizontally, and the opposite young leaves diverged from one another symmetrically in the usual manner, without any upturning in the direction of gravitation or towards the light.

The needle-like leaves of Pinus pinaster form a bundle whilst young; afterwards they slowly diverge, so that those on the upright shoots become horizontal. The movements of one such young leaf was traced during 4½ days, and the tracing here given (Fig. 121) shows that it descended at first in a nearly straight line, but afterwards zigzagged, making one or two little loops. The diverging and descending movements of a rather older leaf were also traced (see former Fig. 113, p. 251): it descended during the first day and night in a somewhat zigzag line; it then circumnutated round a small space and again descended. By this time the leaf had nearly assumed its final position, and now plainly circumnutated. As in the case of the Carnation, the leaves, whilst very young, do not seem to be much affected by geotropism or heliotropism, for those on a young plant laid horizontally, and those on another plant left upright, both kept in the dark, continued to diverge in the usual manner without bending to either side.

Fig. 121. Pinus pinaster: epinastic downward movement of a young leaf, produced by a young plant in a pot, traced on a vertical glass under a skylight, from 6.45 A.M. June 2nd to 10.40 P.M. 6th.

With Cobœa scandens, the young leaves, as they successively diverge from the leading shoot which is bent to one side, rise up so as to project vertically, and they retain this position for some time whilst the tendril is revolving. The diverging and ascending movements of the petiole of one such a leaf, were traced on a vertical glass under a skylight; and the course pursued was in most parts nearly straight, but there were two well-marked zigzags (one of them forming an angle of 112°), and this indicates circumnutation.

The still closed lobes of a young leaf of Dionaea projected at right angles to the petiole, and were in the act of slowly rising. A glass filament was attached to the under side of the midrib, and its movements were traced on a vertical glass. It circumnutated once in the evening, and on the next day rose, as already described (see Fig. 106, p. 240), by a number of acutely zigzag lines, closely approaching in character to ellipses. This movement no doubt was due to epinasty, aided by apogeotropism, for the closed lobes of a very young leaf on a plant which had been placed horizontally, moved into nearly the same line with the petiole, as if the plant had stood upright; but at the same time the lobes curved laterally upwards, and thus occupied an unnatural position, obliquely to the plane of the foliaceous petiole.

As the hypocotyls and epicotyls of some plants protrude from the seed-coats in an arched form, it is doubtful whether the arching of these parts, which is invariably present when they break through the ground, ought always to be attributed to epinasty; but when they are at first straight and afterwards become arched, as often happens, the arching is certainly due to epinasty. As long as the arch is surrounded by compact earth it must retain its form; but as soon as it rises above the surface, or even before this period if artificially freed from the surrounding pressure, it begins to straighten itself, and this no doubt is mainly due to hyponasty. The movement of the upper and lower half of the arch, and of the crown, was occasionally traced; and the course was more or less zigzag, showing modified circumnutation.

With not a few plants, especially climbers, the summit of the shoot is hooked, so that the apex points vertically downwards. In seven genera of twining plants[3] the hooking, or as it has been called by Sachs, the nutation of the tip, is mainly due to an exaggerated form of circumnutation. That is, the growth is so great along one side that it bends the shoot completely over to the opposite side, thus forming a hook; the longitudinal line or zone of growth then travels a little laterally round the shoot, and the hook points in a slightly different direction, and so onwards until the hook is completely reversed. Ultimately it comes back to the point whence it started. This was ascertained by painting narrow lines with Indian ink along the convex surface of several hooks, and the line was found slowly to become at first lateral, then to appear along the concave surface, and ultimately back again on the convex surface. In the case of Lonicera brachypoda the hooked terminal part of the revolving shoot straightens itself periodically, but is never reversed; that is, the periodically increased growth of the concave side of the hook is sufficient only to straighten it, and not to bend it over to the opposite side. The hooking of the tip is of service to twining plants by aiding them to catch hold of a support, and afterwards by enabling this part to embrace the support much more closely than it could otherwise have done at first, thus preventing it, as we often observed, from being blown away by a strong wind. Whether the advantage thus gained by twining plants accounts for their summits being so frequently hooked, we do not know, as this structure is not very rare with plants which do not climb, and with some climbers (for instance, Vitis, Ampelopsis, Cissus, etc.) to whom it does not afford any assistance in climbing.

[3] 'The Movements and Habits of Climbing Plants,' 2nd edit. p. 13.

With respect to those cases in which the tip remains always bent or hooked towards the same side, as in the genera just named, the most obvious explanation is that the bending is due to continued growth in excess along the convex side. Wiesner, however, maintains[4] that in all cases the hooking of the tip is the result of its plasticity and weight,—a conclusion which from what we have already seen with several climbing plants is certainly erroneous. Nevertheless, we fully admit that the weight of the part, as well as geotropism, etc., sometimes come into play.

[4] 'Sitzb. der k. Akad. der Wissensch.,' Vienna, Jan. 1880, p. 16.

Ampelopsis tricuspidata.—This plant climbs by the aid of adhesive tendrils, and the hooked tips of the shoots do not appear to be of any service to it. The hooking depends chiefly, as far as we could ascertain, on the tip being affected by epinasty and geotropism; the lower and older parts continually straightening themselves through hyponasty and apogeotropism. We believe that the weight of the apex is an unimportant element, because on horizontal or inclined shoots the hook is often extended horizontally or even faces upwards. Moreover shoots frequently form loops instead of hooks; and in this case the extreme part, instead of hanging vertically down as would follow if weight was the efficient cause, extends horizontally or even points upwards. A shoot, which terminated in a rather open hook, was fastened in a highly inclined downward position, so that the concave side faced upwards, and the result was that the apex at first curved upwards. This apparently was due to epinasty and not to apogeotropism, for the apex, soon after passing the perpendicular, curved so rapidly downwards that we could not doubt that the movement was at least aided by geotropism. In the course of a few hours the hook was thus converted into a loop with the apex of the shoot pointing straight downwards. The longer axis of the loop was at first horizontal, but afterwards became vertical. During this same time the basal part of the hook (and subsequently of the loop) curved itself slowly upwards; and this must have been wholly due to apogeotropism in opposition to hyponasty. The loop was then fastened upside down, so that its basal half would be simultaneously acted on by hyponasty (if present) and by apogeotropism; and now it curved itself so greatly upwards in the course of only 4 h. that there could hardly be a doubt that both forces were acting together. At the same time the loop became open and was thus reconverted into a hook, and this apparently was effected by the geotropic movement of the apex in opposition to epinasty. In the case of Ampelopsis hederacea, weight plays, as far as we could judge, a more important part in the hooking of the tip.

Fig. 122. Ampelopsis tricuspidata: hyponastic movement of hooked tip of leading shoot, traced from 8.10 A.M. July 13th to 8 A.M. 15th. Apex of shoot 5½ inches from the vertical glass. Plant illuminated through a skylight. Temp. 17½°–19° C. Diagram reduced to one-third of original scale.

Fig. 123. Smithia Pfundii: hyponastic movement of the curved summit of a stem, whilst straightening itself, traced from 9 A.M. July 10th to 3 P.M. 13th. Apex 9½ inches from the vertical glass. Diagram reduced to one-fifth of original scale. Plant illuminated through skylight; temp. 17½°–19° C.

In order to ascertain whether the shoots of A. tricuspidata in straightening themselves under the combined action of hyponasty and apogeotropism moved in a simple straight course, or whether they circumnutated, glass filaments were fixed to the crowns of four hooked tips standing in their natural position; and the movements of the filaments were traced on a vertical glass. All four tracings resembled each other in a general manner; but we will give only one (see Fig. 122, p. 273). The filament rose at first, which shows that the hook was straightening itself; it then zigzagged, moving a little to the left between 9.25 A.M. and 9 P.M. From this latter hour on the 13th to 10.50 A.M. on the following morning (14th) the hook continued to straighten itself, and then zigzagged a short distance to the right. But from 1 P.M. to 10.40 P.M. on the 14th the movement was reversed and the shoot became more hooked. During the night, after 10.40 P.M. to 8.15 A.M. on the 15th, the hook again opened or straightened itself. By this time the glass filament had become so highly inclined that its movements could no longer be traced with accuracy; and by 1.30 P.M. on this same day, the crown of the former arch or hook had become perfectly straight and vertical. There can therefore be no doubt that the straightening of the hooked shoot of this plant is effected by the circumnutation of the arched portion—that is, by growth alternating between the upper and lower surface, but preponderant on the lower surface, with some little lateral movement.

We were enabled to trace the movement of another straightening shoot for a longer period (owing to its slower growth and to its having been placed further from the vertical glass), namely, from the early morning on July 13th to late in the evening of the 16th. During the whole daytime of the 14th, the hook straightened itself very little, but zigzagged and plainly circumnutated about nearly the same spot. By the 16th it had become nearly straight, and the tracing was no longer accurate, yet it was manifest that there was still a considerable amount of movement both up and down and laterally; for the crown whilst continuing to straighten itself occasionally became for a short time more curved, causing the filament to descend twice during the day.

Smithia Pfundii.—The stiff terminal shoots of this Leguminous water-plant from Africa project so as to make a rectangle with the stem below; but this occurs only when the plants are growing vigorously, for when kept in a cool place, the summits of the stems become straight, as they likewise did at the close of the growing season. The direction of the rectangularly bent part is independent of the chief source of light. But from observing the effects of placing plants in the dark, in which case several shoots became in two or three days upright or nearly upright, and when brought back into the light again became rectangularly curved, we believe that the bending is in part due to apheliotropism, apparently somewhat opposed by apogeotropism. On the other hand, from observing the effects of tying a shoot downwards, so that the rectangle faced upwards, we are led to believe that the curvature is partly due to epinasty. As the rectangularly bent portion of an upright stem grows older, the lower part straightens itself; and this is effected through hyponasty. He who has read Sachs' recent Essay on the vertical and inclined positions of the parts of plants[5] will see how difficult a subject this is, and will feel no surprise at our expressing ourselves doubtfully in this and other such cases.

[5] 'Ueber Orthotrope und Plagiotrope Pflanzentheile;' 'Arbeiten des Bot. Inst., in Würzburg,' Heft ii. 1879, p. 226.

A plant, 20 inches in height, was secured to a stick close beneath the curved summit, which formed rather less than a rectangle with the stem below. The shoot pointed away from the observer; and a glass filament pointing towards the vertical glass on which the tracing was made, was fixed to the convex surface of the curved portion. Therefore the descending lines in the figure represent the straightening of the curved portion as it grew older. The tracing (Fig. 123, p. 274) was begun at 9 A.M.

on July 10th; the filament at first moved but little in a zigzag line, but at 2 P.M. it began rising and continued to do so till 9 P.M.; and this proves that the terminal portion was being more bent downwards. After 9 P.M. on the 10th an opposite movement commenced, and the curved portion began to straighten itself, and this continued till 11.10 A.M. on the 12th, but was interrupted by some small oscillations and zigzags, showing movement in different directions. After 11.10 A.M. on the 12th this part of the stem, still considerably curved, circumnutated in a conspicuous manner until nearly 3 P.M. on the 13th; but during all this time a downward movement of the filament prevailed, caused by the continued straightening of the stem. By the afternoon of the 13th, the summit, which had originally been deflected more than a right angle from the perpendicular, had grown so nearly straight that the tracing could no longer be continued on the vertical glass. There can therefore be no doubt that the straightening of the abruptly curved portion of the growing stem of this plant, which appears to be wholly due to hyponasty, is the result of modified circumnutation. We will only add that a filament was fixed in a different manner across the curved summit of another plant, and the same general kind of movement was observed.

Trifolium repens.—In many, but not in all the species of Trifolium, as the separate little flowers wither, the sub-peduncles bend downwards, so as to depend parallel to the upper part of the main peduncle. In Tr. subterraneum the main peduncle curves downwards for the sake of burying its capsules, and in this species the sub-peduncles of the separate flowers bend upwards, so as to occupy the same position relatively to the upper part of the main peduncle as in Tr. repens. This fact alone would render it probable that the movements of the sub-peduncles in Tr. repens were independent of geotropism. Nevertheless, to make sure, some flower-heads were tied to little sticks upside down and others in a horizontal position; their sub-peduncles, however, all quickly curved upwards through the action of heliotropism. We therefore protected some flower-heads, similarly secured to sticks, from the light, and although some of them rotted, many of their sub-peduncles turned very slowly from their reversed or from their horizontal positions, so as to stand in the normal manner parallel to the upper part of the main peduncle. These facts show that the movement is independent of geotropism or apheliotropism; it must there[fore] be attributed to epinasty, which however is checked, at least as long as the flowers are young, by heliotropism. Most of the above flowers were never fertilised owing to the exclusion of bees; they consequently withered very slowly, and the movements of the sub-peduncles were in like manner much retarded.

Fig. 124. Trifolium repens: circumnutating and epinastic movements of the sub-peduncle of a single flower, traced on a vertical glass under a skylight, in A from 11.30 A.M. Aug. 27th to 7 A.M. 30th; in B from 7 A.M. Aug. 30th to a little after 6 P.M. Sept. 8th.

To ascertain the nature of the movement of the sub-peduncle, whilst bending downwards, a filament was fixed across the summit of the calyx of a not fully expanded and almost upright flower, nearly in the centre of the head. The main peduncle was secured to a stick close beneath the head. In order to see the marks on the glass filament, a few flowers had to be cut away on the lower side of the head. The flower under observation at first diverged a little from its upright position, so as to occupy the open space caused by the removal of the adjoining flowers. This required two days, after which time a new tracing was begun (Fig. 124). In A we see the complex circumnutating course pursued from 11.30 A.M. Aug. 26th to 7 A.M. on the 30th. The pot was then moved a very little to the right, and the tracing (B) was continued without interruption from 7 A.M. Aug. 30th to after 6 P.M. Sept. 8th. It should be observed that on most of these days, only a single dot was made each morning at the same hour. Whenever the flower was observed carefully, as on Aug. 30th and Sept. 5th and 6th, it was found to be circumnutating over a small space. At last, on Sept. 7th, it began to bend downwards, and continued to do so until after 6 P.M. on the 8th, and indeed until the morning of the 9th, when its movements could no longer be traced on the vertical glass. It was carefully observed during the whole of the 8th, and by 10.30 P.M. it had descended to a point lower down by two-thirds of the length of the figure as here given; but from want of space the tracing has been copied in B, only to a little after 6 P.M. On the morning of the 9th the flower was withered, and the sub-peduncle now stood at an angle of 57° beneath the horizon. If the flower had been fertilised it would have withered much sooner, and have moved much more quickly. We thus see that the sub-peduncle oscillated up and down, or circumnutated, during its whole downward epinastic course.

The sub-peduncles of the fertilised and withered flowers of Oxalis carnosa likewise bend downwards through epinasty, as will be shown in a future chapter; and their downward course is strongly zigzag, indicating circumnutation.

The number of instances in which various organs move through epinasty or hyponasty, often in combination with other forces, for the most diversified purposes, seems to be inexhaustibly great; and from the several cases which have been here given, we may safely infer that such movements are due to modified circumnutation.

CHAPTER VI.
MODIFIED CIRCUMNUTATION: SLEEP OR NYCTITROPIC MOVEMENTS, THEIR USE: SLEEP OF COTYLEDONS.

Preliminary sketch of the sleep or nyctitropic movements of leaves—Presence of pulvini—The lessening of radiation the final cause of nyctitropic movements—Manner of trying experiments on leaves of Oxalis, Arachis, Cassia, Melilotus, Lotus and Marsilea and on the cotyledons of Mimosa—Concluding remarks on radiation from leaves—Small differences in the conditions make a great difference in the result—Description of the nyctitropic position and movements of the cotyledons of various plants—List of species—Concluding remarks—Independence of the nyctitropic movements of the leaves and cotyledons of the same species—Reasons for believing that the movements have been acquired for a special purpose.

The so-called sleep of leaves is so conspicuous a phenomenon that it was observed as early as the time of Pliny;[1] and since Linnæus published his famous Essay, 'Somnus Plantarum,' it has been the subject of several memoirs. Many flowers close at night, and these are likewise said to sleep; but we are not here concerned with their movements, for although effected by the same mechanism as in the case of young leaves, namely, unequal growth on the opposite sides (as first proved by Pfeffer), yet they differ essentially in being excited chiefly by changes of temperature instead of light; and in being effected, as far as we can judge, for a different purpose. Hardly any one supposes that there is any real analogy between the sleep of animals and that of plants,[2] whether of leaves or flowers. It seems therefore, advisable to give a distinct name to the so-called sleep-movements of plants. These have also generally been confounded, under the term "periodic," with the slight daily rise and fall of leaves, as described in the fourth chapter; and this makes it all the more desirable to give some distinct name to sleep-movements. Nyctitropism and nyctitropic, i.e. night-turning, may be applied both to leaves and flowers, and will be occasionally used by us; but it would be best to confine the term to leaves. The leaves of some few plants move either upwards or downwards when the sun shines intensely on them, and this movement has sometimes been called diurnal sleep; but we believe it to be of an essentially different nature from the nocturnal movement, and it will be briefly considered in a future chapter.

[1] Pfeffer has given a clear and interesting sketch of the history of this subject in his 'Die Periodischen Bewegungen der Blattorgane,' 1875, P. 163.

[2] Ch. Royer must, however, be excepted; see 'Annales des Sc. Nat.' (5th series), Bot. vol. ix. 1868, p. 378.

The sleep or nyctitropism of leaves is a large subject, and we think that the most convenient plan will be first to give a brief account of the position which leaves assume at night, and of the advantages apparently thus gained. Afterwards the more remarkable cases will be described in detail, with respect to cotyledons in the present chapter, and to leaves in the next chapter. Finally, it will be shown that these movements result from circumnutation, much modified and regulated by the alternations of day and night, or light and darkness; but that they are also to a certain extent inherited.

Leaves, when they go to sleep, move either upwards or downwards, or in the case of the leaflets of compound leaves, forwards, that is, towards the apex of the leaf, or backwards, that is, towards its base; or, again, they may rotate on their own axes without moving either upwards or downwards. But in almost every case the plane of the blade is so placed as to stand nearly or quite vertically at night. Therefore the apex, or the base, or either lateral edge, may be directed towards the zenith. Moreover, the upper surface of each leaf, and more especially of each leaflet, is often brought into close contact with that of the opposite one; and this is sometimes effected by singularly complicated movements. This fact suggests that the upper surface requires more protection than the lower one. For instance, the terminal leaflet in Trifolium, after turning up at night so as to stand vertically, often continues to bend over until the upper surface is directed downwards whilst the lower surface is fully exposed to the sky; and an arched roof is thus formed over the two lateral leaflets, which have their upper surfaces pressed closely together. Here we have the unusual case of one of the leaflets not standing vertically, or almost vertically, at night.

Considering that leaves in assuming their nyctitropic positions often move through an angle of 90°; that the movement is rapid in the evening; that in some cases, as we shall see in the next chapter, it is extraordinarily complicated; that with certain seedlings, old enough to bear true leaves, the cotyledons move vertically upwards at night, whilst at the same time the leaflets move vertically downwards; and that in the same genus the leaves or cotyledons of some species move upwards, whilst those of other species move downwards;—from these and other such facts, it is hardly possible to doubt that plants must derive some great advantage from such remarkable powers of movement.

The nyctitropic movements of leaves and cotyledons are effected in two ways,[3] firstly, by means of pulvini which become, as Pfeffer has shown, alternately more turgescent on opposite sides; and secondly, by increased growth along one side of the petiole or midrib, and then on the opposite side, as was first proved by Batalin.[4] But as it has been shown by De Vries[5] that in these latter cases increased growth is preceded by the increased turgescence of the cells, the difference between the above two means of movement is much diminished, and consists chiefly in the turgescence of the cells of a fully developed pulvinus, not being followed by growth. When the movements of leaves or cotyledons, furnished with a pulvinus and destitute of one, are compared, they are seen to be closely similar, and are apparently effected for the same purpose. Therefore, with our object in view, it does not appear advisable to separate the above two sets of cases into two distinct classes. There is, however, one important distinction between them, namely, that movements effected by growth on the alternate sides, are confined to young growing leaves, whilst those effected by means of a pulvinus last for a long time. We have already seen well-marked instances of this latter fact with cotyledons, and so it is with leaves, as has been observed by Pfeffer and by ourselves. The long endurance of the nyctitropic movements when effected by the aid of pulvini indicates, in addition to the evidence already advanced, the functional importance of such movements to the plant. There is another difference between the two sets of cases, namely, that there is never, or very rarely, any torsion of the leaves, excepting when a pulvinus is present;[6] but this statement applies only to periodic and nyctitropic movements as may be inferred from other cases given by Frank.[7] The fact that the leaves of many plants place themselves at night in widely different positions from what they hold during the day, but with the one point in common, that their upper surfaces avoid facing the zenith, often with the additional fact that they come into close contact with opposite leaves or leaflets, clearly indicates, as it seems to us, that the object gained is the protection of the upper surfaces from being chilled at night by radiation. There is nothing improbable in the upper surface needing protection more than the lower, as the two differ in function and structure. All gardeners know that plants suffer from radiation. It is this and not cold winds which the peasants of Southern Europe fear for their olives.[8] Seedlings are often protected from radiation by a very thin covering of straw; and fruit-trees on walls by a few fir-branches, or even by a fishing-net, suspended over them. There is a variety of the gooseberry,[9] the flowers of which from being produced before the leaves, are not protected by them from radiation, and consequently often fail to yield fruit. An excellent observer[10] has remarked that one variety of the cherry has the petals of its flowers much curled backwards, and after a severe frost all the stigmas were killed; whilst at the same time, in another variety with incurved petals, the stigmas were not in the least injured.

[3] This distinction was first pointed out (according to Pfeffer, 'Die Periodischen Bewegungen der Blattorgane,' 1875, p. 161) by Dassen in 1837.

[4] 'Flora,' 1873, p. 433.

[5] 'Bot. Zeitung,' 1879, Dec. 19th, p. 830.

[6] Pfeffer, 'Die Period. Beweg. der Blattorgane.' 1875, p. 159.

[7] 'Die Nat. Wagerechte Richtung von Pflanzentheilen,' 1870, p. 52

[8] Martins in 'Bull. Soc. Bot. de France,' tom. xix. 1872. Wells, in his famous 'Essay on Dew,' remarks that an exposed thermometer rises as soon as even a fleecy cloud, high in the sky, passes over the zenith.

[9] 'Loudon's Gardener's Mag.,' vol. iv. 1828, p. 112.

[10] Mr. Rivers in 'Gardener's Chron.,' 1866, p. 732

This view that the sleep of leaves saves them from being chilled at night by radiation, would no doubt have occurred to Linnæus, had the principle of radiation been then discovered; for he suggests in many parts of his 'Somnus Plantarum' that the position of the leaves at night protects the young stems and buds, and often the young inflorescence, against cold winds. We are far from doubting that an additional advantage may be thus gained; and we have observed with several plants, for instance, Desmodium gyrans, that whilst the blade of the leaf sinks vertically down at night, the petiole rises, so that the blade has to move through a greater angle in order to assume its vertical position than would otherwise have been necessary; but with the result that all the leaves on the same plant are crowded together as if for mutual protection.

We doubted at first whether radiation would affect in any important manner objects so thin as are many cotyledons and leaves, and more especially affect differently their upper and lower surfaces; for although the temperature of their upper surfaces would undoubtedly fall when freely exposed to a clear sky, yet we thought that they would so quickly acquire by conduction the temperature of the surrounding air, that it could hardly make any sensible difference to them, whether they stood horizontally and radiated into the open sky, or vertically and radiated chiefly in a lateral direction towards neighbouring plants and other objects. We endeavoured, therefore, to ascertain something on this head by preventing the leaves of several plants from going to sleep, and by exposing to a clear sky when the temperature was beneath the freezing-point, these, as well as the other leaves on the same plants which had already assumed their nocturnal vertical position. Our experiments show that leaves thus compelled to remain horizontal at night, suffered much more injury from frost than those which were allowed to assume their normal vertical position. It may, however, be said that conclusions drawn from such observations are not applicable to sleeping plants, the inhabitants of countries where frosts do not occur. But in every country, and at all seasons, leaves must be exposed to nocturnal chills through radiation, which might be in some degree injurious to them, and which they would escape by assuming a vertical position.

In our experiments, leaves were prevented from assuming their nyctitropic position, generally by being fastened with the finest entomological pins (which did not sensibly injure them) to thin sheets of cork supported on sticks. But in some instances they were fastened down by narrow strips of card, and in others by their petioles being passed through slits in the cork. The leaves were at first fastened close to the cork, for as this is a bad conductor, and as the leaves were not exposed for long periods, we thought that the cork, which had been kept in the house, would very slightly warm them; so that if they were injured by the frost in a greater degree than the free vertical leaves, the evidence would be so much the stronger that the horizontal position was injurious. But we found that when there was any slight difference in the result, which could be detected only occasionally, the leaves which had been fastened closely down suffered rather more than those fastened with very long and thin pins, so as to stand from ½ to 3/4 inch above the cork. This difference in the result, which is in itself curious as showing what a very slight difference in the conditions influences the amount of injury inflicted, may be attributed, as we believe, to the surrounding warmer air not circulating freely beneath the closely pinned leaves and thus slightly warming them. This conclusion is supported by some analogous facts hereafter to be given.

We will now describe in detail the experiments which were tried. These were troublesome from our not being able to predict how much cold the leaves of the several species could endure. Many plants had every leaf killed, both those which were secured in a horizontal position and those which were allowed to sleep—that is, to rise up or sink down vertically. Others again had not a single leaf in the least injured, and these had to be re-exposed either for a longer time or to a lower temperature.

Oxalis acetosella.—A very large pot, thickly covered with between 300 and 400 leaves, had been kept all winter in the greenhouse. Seven leaves were pinned horizontally open, and were exposed on March 16th for 2 h. to a clear sky, the temperature on the surrounding grass being –4° C. (24° to 25° F.). Next morning all seven leaves were found quite killed, so were many of the free ones which had previously gone to sleep, and about 100 of them, either dead or browned and injured were picked off. Some leaves showed that they had been slightly injured by not expanding during the whole of the next day, though they afterwards recovered. As all the leaves which were pinned open were killed, and only about a third or fourth of the others were either killed or injured, we had some little evidence that those which were prevented from assuming their vertically dependent position suffered most.

The following night (17th) was clear and almost equally cold (–3° to –4° C. on the grass), and the pot was again exposed, but this time for only 30 m. Eight leaves had been pinned out, and in the morning two of them were dead, whilst not a single other leaf on the many plants was even injured.

On the 23rd the pot was exposed for 1 h. 30 m., the temperature on the grass being only –2° C., and not one leaf was injured: the pinned open leaves, however, all stood from ½ to 3/4 of an inch above the cork.

On the 24th the pot was again placed on the ground and exposed to a clear sky for between 35 m. and 40 m. By a mistake the thermometer was left on an adjoining sun-dial 3 feet high, instead of being placed on the grass; it recorded 25° to 26° F. (–3.3° to –3.8° C.), but when looked at after 1 h. had fallen to 22° F. (–5.5° C.); so that the pot was perhaps exposed to rather a lower temperature than on the two first occasions. Eight leaves had been pinned out, some close to the cork and some above it, and on the following morning five of them (i.e. 63 per cent.) were found killed. By counting a portion of the leaves we estimated that about 250 had been allowed to go to sleep, and of these about 20 were killed (i.e. only 8 per cent.), and about 30 injured.

Considering these cases, there can be no doubt that the leaves of this Oxalis, when allowed to assume their normal vertically dependent position at night, suffer much less from frost than those (23 in number) which had their upper surfaces exposed to the zenith.

Oxalis carnosa.—A plant of this Chilian species was exposed for 30 m. to a clear sky, the thermometer on the grass standing at –2° C., with some of its leaves pinned open, and not one leaf on the whole bushy plant was in the least injured. On the 16th of March another plant was similarly exposed for 30 m., when the temperature on the grass was only a little lower, viz., –3° to –4° C. Six of the leaves had been pinned open, and next morning five of them were found much browned. The plant was a large one, and none of the free leaves, which were asleep and depended vertically, were browned, excepting four very young ones. But three other leaves, though not browned, were in a rather flaccid condition, and retained their nocturnal position during the whole of the following day. In this case it was obvious that the leaves which were exposed horizontally to the zenith suffered most. This same pot was afterwards exposed for 35–40 m. on a slightly colder night, and every leaf, both the pinned open and the free ones, was killed. It may be added that two pots of O. corniculata (var. Atro-purpurea) were exposed for 2 h. and 3 h. to a clear sky with the temp. on grass –2° C., and none of the leaves, whether free or pinned open, were at all injured.

Arachis hypogoea.—Some plants in a pot were exposed at night for 30 m. to a clear sky, the temperature on the surrounding grass being –2° C., and on two nights afterwards they were again exposed to the same temperature, but this time during 1 h. 30 m. On neither occasion was a single leaf, whether pinned open or free, injured; and this surprised us much, considering its native tropical African home. Two plants were next exposed (March 16th) for 30 m. to a clear sky, the temperature of the surrounding grass being now lower, viz., between –3° and –4° C., and all four pinned-open leaves were killed and blackened. These two plants bore 22 other and free leaves (excluding some very young bud-like ones) and only two of these were killed and three somewhat injured; that is, 23 per cent. were either killed or injured, whereas all four pinned-open leaves were utterly killed.

On another night two pots with several plants were exposed for between 35 m. and 40 m. to a clear sky, and perhaps to a rather lower temperature, for a thermometer on a dial, 3 feet high, close by stood at –3.3° to –3.8° C. In one pot three leaves were pinned open, and all were badly injured; of the 44 free leaves, 26 were injured, that is, 59 per cent. In the other pot 3 leaves were pinned open and all were killed; four other leaves were prevented from sleeping by narrow strips of stiff paper gummed across them, and all were killed; of 24 free leaves, 10 were killed, 2 much injured, and 12 unhurt; that is, 50 per cent. of the free leaves were either killed or much injured. Taking the two pots together, we may say that rather more than half of the free leaves, which were asleep, were either killed or injured, whilst all the ten horizontally extended leaves, which had been prevented from going to sleep, were either killed or much injured.

Cassia floribunda.—A bush was exposed at night for 40 m. to a clear sky, the temperature on the surrounding grass being –2° C., and not a leaf was injured.[11] It was again exposed on another night for 1 h., when the temperature of the grass was –4° C.; and now all the leaves on a large bush, whether pinned flat open or free, were killed, blackened, and shrivelled, with the exception of those on one small branch, low down, which was very slightly protected by the leaves on the branches above. Another tall bush, with four of its large compound leaves pinned out horizontally, was afterwards exposed (temp. of surrounding grass exactly the same, viz., –4° C.), but only for 30 m. On the following morning every single leaflet on these four leaves was dead, with both their upper and lower surfaces completely blackened. Of the many free leaves on the bush, only seven were blackened, and of these only a single one (which was a younger and more tender leaf than any of the pinned ones) had both surfaces of the leaflets blackened. The contrast in this latter respect was well shown by a free leaf, which stood between two pinned-open ones; for these latter had the lower surfaces of their leaflets as black as ink, whilst the intermediate free leaf, though badly injured, still retained a plain tinge of green on the lower surface of the leaflets. This bush exhibited in a striking manner the evil effects of the leaves not being allowed to assume at night their normal dependent position; for had they all been prevented from doing so, assuredly every single leaf on the bush would have been utterly killed by this exposure of only 30 m. The leaves whilst sinking downwards in the evening twist round, so that the upper surface is turned inwards, and is thus better protected than the outwardly turned lower surface. Nevertheless, it was always the upper surface which was more blackened than the lower, whenever any difference could be perceived between them; but whether this was due to the cells near the upper surface being more tender, or merely to their containing more chlorophyll, we do not know.

[11] Cassia laevigata was exposed to a clear sky for 35 m., and C. calliantha (a Guiana species) for 60 m., the temperature on the surrounding grass being –2° C., and neither was in the least injured. But when C. laevigata was exposed for 1 h., the temp. on the surrounding grass being between –3° and –4° C., every leaf was killed.

Melilotus officinalis.—A large pot with many plants, which had been kept during the winter in the greenhouse, was exposed during 5 h. at night to a slight frost and clear sky. Four leaves had been pinned out, and these died after a few days; but so did many of the free leaves. Therefore nothing certain could be inferred from this trial, though it indicated that the horizontally extended leaves suffered most. Another large pot with many plants was next exposed for 1 h., the temperature on the surrounding grass being lower, viz., -3° to –4° C. Ten leaves had been pinned out, and the result was striking, for on the following morning all these were found much injured or killed, and none of the many free leaves on the several plants were at all injured, with the doubtful exception of two or three very young ones.

Melilotus Italica.—Six leaves were pinned out horizontally, three with their upper and three with their lower surfaces turned to the zenith. The plants were exposed for 5 h. to a clear sky, the temperature on ground being about –1° C. Next morning the six pinned-open leaves seemed more injured even than the younger and more tender free ones on the same branches. The exposure, however, had been too long, for after an interval of some days many of the free leaves seemed in almost as bad a condition as the pinned-out ones. It was not possible to decide whether the leaves with their upper or those with their lower surfaces turned to the zenith had suffered most.

Melilotus suaveolens.—Some plants with 8 leaves pinned out were exposed to a clear sky during 2 h., the temperature on the surrounding grass being –2° C. Next morning 6 out of these 8 leaves were in a flaccid condition. There were about 150 free leaves on the plant, and none of these were injured, except 2 or 3 very young ones. But after two days, the plants having been brought back into the greenhouse, the 6 pinned-out leaves all recovered.

Melilotus Taurica.—Several plants were exposed for 5 h. during two nights to a clear sky and slight frost, accompanied by some wind; and 5 leaves which had been pinned out suffered more than those both above and below on the same branches which had gone to sleep. Another pot, which had likewise been kept in the greenhouse, was exposed for 35–40 m. to a clear sky, the temperature of the surrounding grass being between –3° and –4° C. Nine leaves had been pinned out, and all of these were killed. On the same plants there were 210 free leaves, which had been allowed to go to sleep, and of these about 80 were killed, i.e. only 38 per cent.

Melilotus Petitpierreana.—The plants were exposed to a clear sky for 35–40 m.: temperature on surrounding grass –3° to –4° C. Six leaves had been pinned out so as to stand about ½ inch above the cork, and four had been pinned close to it. These 10 leaves were all killed, but the closely pinned ones suffered most, as 4 of the 6 which stood above the cork still retained small patches of a green colour. A considerable number, but not nearly all, of the free leaves, were killed or much injured, whereas all the pinned out ones were killed.

Melilotus macrorrhiza.—The plants were exposed in the same manner as in the last case. Six leaves had been pinned out horizontally, and five of them were killed, that is, 83 percent. We estimated that there were 200 free leaves on the plants, and of these about 50 were killed and 20 badly injured, so that about 35 per cent of the free leaves were killed or injured.

Lotus aristata.—Six plants were exposed for nearly 5 h. to a clear sky; temperature on surrounding grass –1.5° C. Four leaves had been pinned out horizontally, and 2 of these suffered more than those above or below on the same branches, which had been allowed to go to sleep. It is rather a remarkable fact that some plants of Lotus Jacoboeus, an inhabitant of so hot a country as the Cape Verde Islands, were exposed one night to a clear sky, with the temperature of the surrounding grass –2° C., and on a second night for 30 m. with the temperature of the grass between –3° and –4° C., and not a single leaf, either the pinned-out or free ones, was in the least injured.

Marsilea quadrifoliata.—A large plant of this species—the only Cryptogamic plant known to sleep—with some leaves pinned open, was exposed for 1 h. 35 m. to a clear sky, the temperature on the surrounding ground being –2° C., and not a single leaf was injured. After an interval of some days the plant was again exposed for 1 h. to a clear sky, with the temperature on the surrounding ground lower, viz., –4° C. Six leaves had been pinned out horizontally, and all of them were utterly killed. The plant had emitted long trailing stems, and these had been wrapped round with a blanket, so as to protect them from the frozen ground and from radiation; but a very large number of leaves were left freely exposed, which had gone to sleep, and of these only 12 were killed. After another interval, the plant, with 9 leaves pinned out, was again exposed for 1 h., the temperature on the ground being again –4° C. Six of the leaves were killed, and one which did not at first appear injured afterwards became streaked with brown. The trailing branches, which rested on the frozen ground, had one-half or three-quarters of their leaves killed, but of the many other leaves on the plant, which alone could be fairly compared with the pinned-out ones, none appeared at first sight to have been killed, but on careful search 12 were found in this state. After another interval, the plant with 9 leaves pinned out, was exposed for 35–40 m. to a clear sky and to nearly the same, or perhaps a rather lower, temperature (for the thermometer by an accident had been left on a sun-dial close by), and 8 of these leaves were killed. Of the free leaves (those on the trailing branches not being considered), a good many were killed, but their number, compared with the uninjured ones, was small. Finally, taking the three trials together, 24 leaves, extended horizontally, were exposed to the zenith and to unobstructed radiation, and of these 20 were killed and 1 injured; whilst a relatively very small proportion of the leaves, which had been allowed to go to sleep with their leaflets vertically dependent, were killed or injured.

The cotyledons of several plants were prepared for trial, but the weather was mild and we succeeded only in a single instance in having seedlings of the proper age on nights which were clear and cold. The cotyledons of 6 seedlings of Mimosa pudica were fastened open on cork and were thus exposed for 1 h. 45 m. to a clear sky, with the temperature on the surrounding ground at 29° F.; of these, 3 were killed. Two other seedlings, after their cotyledons had risen up and had closed together, were bent over and fastened so that they stood horizontally, with the lower surface of one cotyledon fully exposed to the zenith, and both were killed. Therefore of the 8 seedlings thus tried 5, or more than half, were killed. Seven other seedlings with their cotyledons in their normal nocturnal position, viz., vertical and closed, were exposed at the same time, and of these only 2 were killed.[12] Hence it appears, as far as these few trials tell anything, that the vertical position at night of the cotyledons of Mimosa pudica protects them to a certain degree from the evil effects of radiation and cold.

[12] We were surprised that young seedlings of so tropical a plant as Mimosa pudica were able to resist, as well as they did, exposure for 1 hr. 45 m. to a clear sky, the temperature on the surrounding ground being 29° F. It may be added that seedlings of the Indian 'Cassia pubescens' were exposed for 1 h. 30 m. to a clear sky, with the temp. on the surrounding ground at –2° C., and they were not in the least injured.

Concluding Remarks on the Radiation from Leaves at Night.—We exposed on two occasions during the summer to a clear sky several pinned-open leaflets of Trifolium pratense, which naturally rise at night, and of Oxalis purpurea, which naturally sink at night (the plants growing out of doors), and looked at them early on several successive mornings, after they had assumed their diurnal positions. The difference in the amount of dew on the pinned-open leaflets and on those which had gone to sleep was generally conspicuous; the latter being sometimes absolutely dry, whilst the leaflets which had been horizontal were coated with large beads of dew. This shows how much cooler the leaflets fully exposed to the zenith must have become, than those which stood almost vertically, either upwards or downwards, during the night.

From the several cases above given, there can be no doubt that the position of the leaves at night affects their temperature through radiation to such a degree, that when exposed to a clear sky during a frost, it is a question of life and death. We may therefore admit as highly probable, seeing that their nocturnal position is so well adapted to lessen radiation, that the object gained by their often complicated sleep movements, is to lessen the degree to which they are chilled at night. It should be kept in mind that it is especially the upper surface which is thus protected, as it is never directed towards the zenith, and is often brought into close contact with the upper surface of an opposite leaf or leaflet.

We failed to obtain sufficient evidence, whether the better protection of the upper surface has been gained from its being more easily injured than the lower surface, or from its injury being a greater evil to the plant. That there is some difference in constitution between the two surfaces is shown by the following cases. Cassia floribunda was exposed to a clear sky on a sharp frosty night, and several leaflets which had assumed their nocturnal dependent position with their lower surfaces turned outwards so as to be exposed obliquely to the zenith, nevertheless had these lower surfaces less blackened than the upper surfaces which were turned inwards and were in close contact with those of the opposite leaflets. Again, a pot full of plants of Trifolium resupinatum, which had been kept in a warm room for three days, was turned out of doors (Sept. 21st) on a clear and almost frosty night. Next morning ten of the terminal leaflets were examined as opaque objects under the microscope. These leaflets, in going to sleep, either turn vertically upwards, or more commonly bend a little over the lateral leaflets, so that their lower surfaces are more exposed to the zenith than their upper surfaces. Nevertheless, six of these ten leaflets were distinctly yellower on the upper than on the lower and more exposed surface. In the remaining four, the result was not so plain, but certainly whatever difference there was leaned to the side of the upper surface having suffered most.

It has been stated that some of the leaflets experimented on were fastened close to the cork, and others at a height of from ½ to 3/4 of an inch above it; and that whenever, after exposure to a frost, any difference could be detected in their states, the closely pinned ones had suffered most. We attributed this difference to the air, not cooled by radiation, having been prevented from circulating freely beneath the closely pinned leaflets. That there was really a difference in the temperature of leaves treated in these two different methods, was plainly shown on one occasion; for after the exposure of a pot with plants of Melilotus dentata for 2 h. to a clear sky (the temperature on the surrounding grass being –2° C.), it was manifest that more dew had congealed into hoar-frost on the closely pinned leaflets, than on those which stood horizontally a little above the cork. Again, the tips of some few leaflets, which had been pinned close to the cork, projected a little beyond the edge, so that the air could circulate freely round them. This occurred with six leaflets of Oxalis acetosella, and their tips certainly suffered rather less then the rest of the same leaflets; for on the following morning they were still slightly green. The same result followed, even still more clearly, in two cases with leaflets of Melilotus officinalis which projected a little beyond the cork; and in two other cases some leaflets which were pinned close to the cork were injured, whilst other free leaflets on the same leaves, which had not space to rotate and assume their proper vertical position, were not at all injured.

Another analogous fact deserves notice: we observed on several occasions that a greater number of free leaves were injured on the branches which had been kept motionless by some of their leaves having been pinned to the corks, than on the other branches. This was conspicuously the case with those of Melilotus Petitpierreana, but the injured leaves in this instance were not actually counted. With Arachis hypogaea, a young plant with 7 stems bore 22 free leaves, and of these 5 were injured by the frost, all of which were on two stems, bearing four leaves pinned to the cork-supports. With Oxalis carnosa, 7 free leaves were injured, and every one of them belonged to a cluster of leaves, some of which had been pinned to the cork. We could account for these cases only by supposing that the branches which were quite free had been slightly waved about by the wind, and that their leaves had thus been a little warmed by the surrounding warmer air. If we hold our hands motionless before a hot fire, and then wave them about, we immediately feel relief; and this is evidently an analogous, though reversed, case. These several facts—in relation to leaves pinned close to or a little above the cork-supports—to their tips projecting beyond it—and to the leaves on branches kept motionless—seem to us curious, as showing how a difference, apparently trifling, may determine the greater or less injury of the leaves. We may even infer as probable that the less or greater destruction during a frost of the leaves on a plant which does not sleep, may often depend on the greater or less degree of flexibility of their petioles and of the branches which bear them.

NYCTITROPIC OR SLEEP MOVEMENTS OF COTYLEDONS.

We now come to the descriptive part of our work, and will begin with cotyledons, passing on to leaves in the next chapter. We have met with only two brief notices of cotyledons sleeping. Hofmeister,[13] after stating that the cotyledons of all the observed seedlings of the Caryophylleae (Alsineae and Sileneae) bend upwards at night (but to what angle he does not state), remarks that those of Stellaria media rise up so as to touch one another; they may therefore safely be said to sleep. Secondly, according to Ramey,[14] the cotyledons of Mimosa pudica and of Clianthus Dampieri rise up almost vertically at night and approach each other closely. It has been shown in a previous chapter that the cotyledons of a large number of plants bend a little upwards at night, and we here have to meet the difficult question at what inclination may they be said to sleep? According to the view which we maintain, no movement deserves to be called nyctitropic, unless it has been acquired for the sake of lessening radiation; but this could be discovered only by a long series of experiments, showing that the leaves of each species suffered from this cause, if prevented from sleeping. We must therefore take an arbitrary limit. If a cotyledon or leaf is inclined at 60° above or beneath the horizon, it exposes to the zenith about one-half of its area; consequently the intensity of its radiation will be lessened by about half, compared with what it would have been if the cotyledon or leaf had remained horizontal. This degree of diminution certainly would make a great difference to a plant having a tender constitution. We will therefore speak of a cotyledon and hereafter of a leaf as sleeping, only when it rises at night to an angle of about 60°, or to a still higher angle, above the horizon, or sinks beneath it to the same amount. Not but that a lesser diminution of radiation may be advantageous to a plant, as in the case of Datura stramonium, the cotyledons of which rose from 31° at noon to 55° at night above the horizon. The Swedish turnip may profit by the area of its leaves being reduced at night by about 30 per cent., as estimated by Mr. A. S. Wilson; though in this case the angle through which the leaves rose was not observed. On the other hand, when the angular rise of cotyledons or of leaves is small, such as less than 30°, the diminution of radiation is so slight that it probably is of no significance to the plant in relation to radiation. For instance, the cotyledons of Geranium Ibericum rose at night to 27° above the horizon, and this would lessen radiation by only 11 per cent.: those of Linum Berendieri rose to 33°, and this would lessen radiation by 16 per cent.

[13] 'Die Lehre von der Pflanzenzelle,' 1867, p. 327.

[14] 'Adansonia,' March 10th, 1869.

There are, however, some other sources of doubt with respect to the sleep of cotyledons. In certain cases, the cotyledons whilst young diverge during the day to only a very moderate extent, so that a small rise at night, which we know occurs with the cotyledons of many plants, would necessarily cause them to assume a vertical or nearly vertical position at night; and in this case it would be rash to infer that the movement was effected for any special purpose. On this account we hesitated long whether we should introduce several Cucurbitaceous plants into the following list; but from reasons, presently to be given, we thought that they had better be at least temporarily included. This same source of doubt applies in some few other cases; for at the commencement of our observations we did not always attend sufficiently to whether the cotyledons stood nearly horizontally in the middle of the day. With several seedlings, the cotyledons assume a highly inclined position at night during so short a period of their life, that a doubt naturally arises whether this can be of any service to the plant. Nevertheless, in most of the cases given in the following list, the cotyledons may be as certainly said to sleep as may the leaves of any plant. In two cases, namely with the cabbage and radish, the cotyledons of which rise almost vertically during the few first nights of their life, it was ascertained by placing young seedlings in the klinostat, that the upward movement was not due to apogeotropism.

The names of the plants, the cotyledons of which stand at night at an angle of at least 60° with the horizon, are arranged in the appended list on the same system as previously followed. The numbers of the Families, and with the Leguminosae the numbers of the Tribes, have been added to show how widely the plants in question are distributed throughout the dicotyledonous series. A few remarks will have to be made about many of the plants in the list. In doing so, it will be convenient not to follow strictly any systematic order, but to treat of the Oxalidæ and the Leguminosae at the close; for in these two Families the cotyledons are generally provided with a pulvinus, and their movements endure for a much longer time than those of the other plants in the list.

List of Seedling Plants, the cotyledons of which rise or sink at night to an angle of at least 60° above or beneath the horizon.

Brassica oleracea. Cruciferae (Fam. 14). — napus (as we are informed by Prof. Pfeffer). Raphanus sativus. Cruciferae. Githago segetum. Caryophylleae (Fam. 26). Stellaria media (according to Hofmeister, as quoted). Caryophylleae. Anoda Wrightii. Malvaceae (Fam. 36). Gossypium (var. Nankin cotton). Malvaceae. Oxalis rosea. Oxalidæ (Fam. 41). — floribunda. — articulata. — Valdiviana. — sensitiva. Geranium rotundifolium. Geraniaceae (Fam. 47). Trifolium subterraneum. Leguminosae (Fam. 75, Tribe 3). — strictum. — leucanthemum. Lotus ornithopopoides. Leguminosae (Tribe 4). — peregrinus. — Jacobæus. Clianthus Dampieri. Leguminosae (Tribe 5)—according to M. Ramey. Smithia sensitiva. Leguminosae (Tribe 6). Haematoxylon Campechianum. Leguminosae (Tribe 13)—according to Mr. R. I. Lynch. Cassia mimosoides. Leguminosae (Tribe 14). — glauca. — florida. — corymbosa. — pubescens. — tora. — neglecta. — 3 other Brazilian unnamed species. Bauhinia (sp.?. Leguminosae (Tribe 15). Neptunia oleracea. Leguminosae (Tribe 20). Mimosa pudica. Leguminosae (Tribe 21). — albida. Cucurbita ovifera. Cucurbitaceæ (Fam. 106). — aurantia. Lagenaria vulgaris. Cucurbitaceæ. Cucumis dudaim. Cucurbitaceæ. Apium petroselinum. Umbelliferae (Fam. 113). — graveolens. Lactuca scariola. Compositæ (Fam. 122). Helianthus annuus (?). Compositæ. Ipomœa caerulea. Convolvulaceae (Fam. 151). — purpurea. — bona-nox. — coccinea. Solanum lycopersicum. Solaneae (Fam. 157.) Mimulus, (sp. ?) Scrophularineae (Fam. 159)—from information given us by Prof. Pfeffer. Mirabilis jalapa. Nyctagineae (Fam. 177). Mirabilis longiflora. Beta vulgaris. Polygoneae (Fam. 179). Amaranthus caudatus. Amaranthaceae (Fam. 180). Cannabis sativa (?). Cannabineae (Fam. 195).

Brassica oleracea (Cruciferae).—It was shown in the first chapter that the cotyledons of the common cabbage rise in the evening and stand vertically up at night with their petioles in contact. But as the two cotyledons are of unequal height, they frequently interfere a little with each other's movements, the shorter one often not standing quite vertically. They awake early in the morning; thus at 6.45 A.M. on Nov. 27th, whilst if was still dark, the cotyledons, which had been vertical and in contact on the previous evening, were reflexed, and thus presented a very different appearance. It should be borne in mind that seedlings in germinating at the proper season, would not be subjected to darkness at this hour in the morning. The above amount of movement of the cotyledons is only temporary, lasting with plants kept in a warm greenhouse from four to six days; how long it would last with seedlings growing out of doors we do not know.

Raphanus sativus.—In the middle of the day the blades of the cotyledons of 10 seedlings stood at right angles to their hypocotyls, with their petioles a little divergent; at night the blades stood vertically, with their bases in contact and with their petioles parallel. Next morning, at 6.45 A.M., whilst it was still dark, the blades were horizontal. On the following night they were much raised, but hardly stood sufficiently vertical to be said to be asleep, and so it was in a still less degree on the third night. Therefore the cotyledons of this plant (kept in the greenhouse) go to sleep for even a shorter time than those of the cabbage. Similar observations were made, but only during a single day and night, on 13 other seedlings likewise raised in the greenhouse, with the same result.

The petioles of the cotyledons of 11 young seedlings of Sinapis nigra were slightly divergent at noon, and the blades stood at right angles to the hypocotyls; at night the petioles were in close contact, and the blades considerably raised, with their bases in contact, but only a few stood sufficiently upright to be called asleep. On the following morning, the petioles diverged before it was light. The hypocotyl is slightly sensitive, so that if rubbed with a needle it bends towards the rubbed side. In the case of Lepidium sativum, the petioles of the cotyledons of young seedlings diverge during the day and converge so as to touch each other during the night, by which means the bases of the tripartite blades are brought into contact; but the blades are so little raised that they cannot be said to sleep. The cotyledons of several other cruciferous plants were observed, but they did not rise sufficiently during the night to be said to sleep.

Githago segetum (Caryophylleae).—On the first day after the cotyledons had burst through the seed-coats, they stood at noon at an angle of 75° above the horizon; at night they moved upwards, each through an angle of 15° so as to stand quite vertical and in contact with one another. On the second day they stood at noon at 59° above the horizon, and again at night were completely closed, each having risen 31°. On the fourth day the cotyledons did not quite close at night. The first and succeeding pairs of young true leaves behaved in exactly the same manner. We think that the movement in this case may be called nyctitropic, though the angle passed through was small. The cotyledons are very sensitive to light and will not expand if exposed to an extremely dim one.

Anoda Wrightii (Malvaceae).—The cotyledons whilst moderately young, and only from .2 to .3 inch in diameter, sink in the evening from their mid-day horizontal position to about 35° beneath the horizon. But when the same seedlings were older and had produced small true leaves, the almost orbicular cotyledons, now .55 inch in diameter, moved vertically downwards at night. This fact made us suspect that their sinking might be due merely to their weight; but they were not in the least flaccid, and when lifted up sprang back through elasticity into their former dependent position. A pot with some old seedlings was turned upside down in the afternoon, before the nocturnal fall had commenced, and at night they assumed in opposition to their own weight (and to any geotropic action) an upwardly directed vertical position. When pots were thus reversed, after the evening fall had already commenced, the sinking movement appeared to be somewhat disturbed; but all their movements were occasionally variable without any apparent cause. This latter fact, as well as that of the young cotyledons not sinking nearly so much as the older ones, deserves notice. Although the movement of the cotyledons endured for a long time, no pulvinus was exteriorly visible; but their growth continued for a long time. The cotyledons appear to be only slightly heliotropic, though the hypocotyl is strongly so.

Gossypium arboreum (?) (var. Nankin cotton) (Malvaceae).—The cotyledons behave in nearly the same manner as those of the Anoda. On June 15th the cotyledons of two seedlings were .65 inch in length (measured along the midrib) and stood horizontally at noon; at 10 P.M. they occupied the same position and had not fallen at all. On June 23rd, the cotyledons of one of these seedlings were 1.1 inch in length, and by 10 P.M. they had fallen from a horizontal position to 62° beneath the horizon. The cotyledons of the other seedling were 1.3 inch in length, and a minute truc leaf had been formed; they had fallen at 10 P.M. to 70° beneath the horizon. On June 25th, the true leaf of this latter seedling was .9 inch in length, and the cotyledons occupied nearly the same position at night. By July 9th the cotyledons appeared very old and showed signs of withering; but they stood at noon almost horizontally, and at 10 P.M. hung down vertically.

Gossypium herbaceum.—It is remarkable that the cotyledons of this species behave differently from those of the last. They were observed during 6 weeks from their first development until they had grown to a very large size (still appearing fresh and green), viz. 2½ inches in breadth. At this age a true leaf had been formed, which with its petiole was 2 inches long. During the whole of these 6 weeks the cotyledons did not sink at night; yet when old their weight was considerable and they were borne by much elongated petioles. Seedlings raised from some seed sent us from Naples, behaved in the same manner; as did those of a kind cultivated in Alabama and of the Sea-island cotton. To what species these three latter forms belong we do not know. We could not make out in the case of the Naples cotton, that the position of the cotyledons at night was influenced by the soil being more or less dry; care being taken that they were not rendered flaccid by being too dry. The weight of the large cotyledons of the Alabama and Sea-island kinds caused them to hang somewhat downwards, when the pots in which they grew were left for a time upside down. It should, however, be observed that these three kinds were raised in the middle of the winter, which sometimes greatly interferes with the proper nyctitropic movements of leaves and cotyledons.

Cucurbitaceæ.—The cotyledons of Cucurbita aurantia and ovifera, and of Lagenaria vulgaris, stand from the 1st to the 3rd day of their life at about 60° above the horizon, and at night rise up so as to become vertical and in close contact with one another. With Cucumis dudaim they stood at noon at 45° above the horizon, and closed at night. The tips of the cotyledons of all these species are, however, reflexed, so that this part is fully exposed to the zenith at night; and this fact is opposed to the belief that the movement is of the same nature as that of sleeping plants. After the first two or three days the cotyledons diverge more during the day and cease to close at night. Those of Trichosanthes anguina are somewhat thick and fleshy, and did not rise at night; and they could perhaps hardly be expected to do so. On the other hand, those of *Acanthosicyos horrida*[15] present nothing in their appearance opposed to their moving at night in the same manner as the preceding species; yet they did not rise up in any plain manner. This fact leads to the belief that the nocturnal movements of the above-named species has been acquired for some special purpose, which may be to protect the young plumule from radiation, by the close contact of the whole basal portion of the two cotyledons.

[15] This plant, from Dammara Land in S. Africa, is remarkable from being the one known member of the Family which is not a climber; it has been described in 'Transact. Linn. Soc.,' xxvii. p. 30.

Geranium rotundifolium (Geraniaceae).—A single seedling came up accidentally in a pot, and its cotyledons were observed to bend perpendicularly downwards during several successive nights, having been horizontal at noon. It grew into a fine plant but died before flowering: it was sent to Kew and pronounced to be certainly a Geranium, and in all probability the above-named species. This case is remarkable because the cotyledons of G. cinereum, Endressii, Ibericum, Richardsoni, and subcaulescens were observed during some weeks in the winter, and they did not sink, whilst those of G. Ibericum rose 27° at night.

Apium petroselinum (Umbelliferae).—A seedling had its cotyledons (Nov. 22nd) almost fully expanded during the day; by 8.30 P.M. they had risen considerably, and at 10.30 P.M. were almost closed, their tips being only 8/100 of an inch apart. On the following morning (23rd) the tips were 58/100 of an inch apart, or more than seven times as much. On the next night the cotyledons occupied nearly the same position as before. On the morning of the 24th they stood horizontally, and at night were 60° above the horizon; and so it was on the night of the 25th. But four days afterwards (on the 29th), when the seedlings were a week old, the cotyledons had ceased to rise at night to any plain degree.

Apium graveolens.—The cotyledons at noon were horizontal, and at 10 P.M. stood at an angle of 61° above the horizon.

Lactuca scariola (Compositæ).—The cotyledons whilst young stood sub-horizontally during the day, and at night rose so as to be almost vertical, and some were quite vertical and closed; but this movement ceased when they had grown old and large, after an interval of 11 days.

Helianthus annuus (Compositæ).—This case is rather doubtful; the cotyledons rise at night, and on one occasion they stood at 73° above the horizon, so that they might then be said to have been asleep.

Ipomœa caerulea vel Pharbitis nil (Convolvulaceae).—The cotyledons behave in nearly the same manner as those of the Anoda and Nankin cotton, and like them grow to a large size. Whilst young and small, so that their blades were from .5 to .6 of an inch in length, measured along the middle to the base of the central notch, they remained horizontal both during the middle of the day and at night. As they increased in size they began to sink more and more in the evening and early night; and when they had grown to a length (measured in the above manner) of from 1 to 1.25 inch, they sank between 55° and 70° beneath the horizon. They acted, however, in this manner only when they had been well illuminated during the day. Nevertheless, the cotyledons have little or no power of bending towards a lateral light, although the hypocotyl is strongly heliotropic. They are not provided with a pulvinus, but continue to grow for a long time.

Ipomœa purpurea (vel Pharbitis hispida).—The cotyledons behave in all respects like those of I. caerulea. A seedling with cotyledons .75 inch in length (measured as before) and 1.65 inch in breadth, having a small true leaf developed, was placed at 5.30 P.M. on a klinostat in a darkened box, so that neither weight nor geotropism could act on them. At 10 P.M. one cotyledon stood at 77° and the other at 82° beneath the horizon. Before being placed in the klinostat they stood at 15° and 29° beneath the horizon. The nocturnal position depends chiefly on the curvature of the petiole close to the blade, but the whole petiole becomes slightly curved downwards. It deserves notice that seedlings of this and the last-named species were raised at the end of February and another lot in the middle of March, and the cotyledons in neither case exhibited any nyctitropic movement.

Ipomœa bona-nox.—The cotyledons after a few days grow to an enormous size, those on a young seedling being 3 1/4 inches in breadth. They were extended horizontally at noon, and at 10 P.M. stood at 63° beneath the horizon. five days afterwards they were 4½ inches in breadth, and at night one stood at 64° and the other 48° beneath the horizon. Though the blades are thin, yet from their great size and from the petioles being long, we imagined that their depression at night might be determined by their weight; but when the pot was laid horizontally, they became curved towards the hypocotyl, which movement could not have been in the least aided by their weight, at the same time they were somewhat twisted upwards through apogeotropism. Nevertheless, the weight of the cotyledons is so far influential, that when on another night the pot was turned upside down, they were unable to rise and thus to assume their proper nocturnal position.

Ipomœa coccinea.—The cotyledons whilst young do not sink at night, but when grown a little older, but still only .4 inch in length (measured as before) and .82 in breadth, they became greatly depressed. In one case they were horizontal at noon, and at 10 P.M. one of them stood at 64° and the other at 47° beneath the horizon. The blades are thin, and the petioles, which become much curved down at night, are short, so that here weight can hardly have produced any effect. With all the above species of Ipomœa, when the two cotyledons on the same seedling were unequally depressed at night, this seemed to depend on the position which they had held during the day with reference to the light.

Solanum lycopersicum (Solaneae).—The cotyledons rise so much at night as to come nearly in contact. Those of 'S. palinacanthum' were horizontal at noon, and by 10 P.M. had risen only 27° 30 minutes; but on the following morning before it was light they stood at 59° above the horizon, and in the afternoon of the same day were again horizontal. The behaviour of the cotyledons of this latter species seems, therefore, to be anomalous.

Mirabilis jalapa and longiflora (Nyctagineae).—The cotyledons, which are of unequal size, stand horizontally during the middle of the day, and at night rise up vertically and come into close contact with one another. But this movement with M. longiflora lasted for only the three first nights.

Beta vulgaris (Polygoneae).—A large number of seedlings were observed on three occasions. During the day the cotyledons sometimes stood sub-horizontally, but more commonly at an angle of about 50° above the horizon, and for the first two or three nights they rose up vertically so as to be completely closed. During the succeeding one or two nights they rose only a little, and afterwards hardly at all.

Amaranthus caudatus (Amaranthaceae).—At noon the cotyledons of many seedlings, which had just germinated, stood at about 45° above the horizon, and at 10.15 P.M. some were nearly and the others quite closed. On the following morning they were again well expanded or open.

Cannabis sativa (Cannabineae).—We are very doubtful whether this plant ought to be here included. The cotyledons of a large number of seedlings, after being well illuminated during the day, were curved downwards at night, so that the tips of some pointed directly to the ground, but the basal part did not appear to be at all depressed. On the following morning they were again flat and horizontal. the cotyledons of many other seedlings were at the same time not in any way affected. Therefore this case seems very different from that of ordinary sleep, and probably comes under the head of epinasty, as is the case with the leaves of this plant according to Kraus. The cotyledons are heliotropic, and so is the hypocotyl in a still stronger degree.

Oxalis.—We now come to cotyledons provided with a pulvinus, all of which are remarkable from the continuance of the nocturnal movements during several days or even weeks, and apparently after growth has ceased. The cotyledons of O. rosea, floribunda and articulata sink vertically down at night and clasp the upper part of the hypocotyl. Those of O. Valdiviana and sensitiva, on the contrary, rise vertically up, so that their upper surfaces come into close contact; and after the young leaves are developed these are clasped by the cotyledons. As in the daytime they stand horizontally, or are even a little deflected beneath the horizon, they move in the evening through an angle of at least 90°. Their complicated circumnutating movements during the day have been described in the first chapter. The experiment was a superfluous one, but pots with seedlings of O. rosea and floribunda were turned upside down, as soon as the cotyledons began to show any signs of sleep, and this made no difference in their movements.

Leguminosae.—It may be seen in our list that the cotyledons of several species in nine genera, widely distributed throughout the Family, sleep at night; and this probably is the case with many others. The cotyledons of all these species are provided with a pulvinus; and the movement in all is continued during many days or weeks. In Cassia the cotyledons of the ten species in the list rise up vertically at night and come into close contact with one another. We observed that those of C. florida opened in the morning rather later than those of C. glauca and pubescens. The movement is exactly the same in C. mimosoides as in the other species, though its subsequently developed leaves sleep in a different manner. The cotyledons of an eleventh species, namely, C. nodosa, are thick and fleshy, and do not rise up at night. The circumnutation of the cotyledons during the day of C. tora has been described in the first chapter. Although the cotyledons of Smithia sensitiva rose from a horizontal position in the middle of the day to a vertical one at night, those of S. Pfundii, which are thick and fleshy, did not sleep. When Mimosa pudica and albida have been kept at a sufficiently high temperature during the day, the cotyledons come into close contact at night; otherwise they merely rise up almost vertically. The circumnutation of those of M. pudica has been described. The cotyledons of a Bauhinia from St. Catharina in Brazil stood during the day at an angle of about 50° above the horizon, and at night rose to 77°; but it is probable that they would have closed completely, if the seedlings had been kept in a warmer place.

Lotus.—In three species of Lotus the cotyledons were observed to sleep. Those of L. Jacoboeus present the singular case of not rising at night in any conspicuous manner for the first 5 or 6 days of their life, and the pulvinus is not well developed at this period. Afterwards the sleeping movement is well displayed, though to a variable degree, and is long continued. We shall hereafter meet with a nearly parallel case with the leaves of Sida rhombifolia. The cotyledons of L. Gebelii are only slightly raised at night, and differ much in this respect from the three species in our list.

Trifolium.—The germination of 21 species was observed. In most of them the cotyledons rise hardly at all, or only slightly, at night; but those of T. glomeratum, striatum and incarnactum rose from 45° to 55° above the horizon. With T. subterraneum, leucanthemum and strictum, they stood up vertically; and with T. strictum the rising movement is accompanied, as we shall see, by another movement, which makes us believe that the rising is truly nyctitropic. We did not carefully examine the cotyledons of all the species for a pulvinus, but this organ was distinctly present in those of T. subterraneum and strictum; whilst there was no trace of a pulvinus in some species, for instance, in T. resupinatum, the cotyledons of which do not rise at night.

Trifolium subterraneum.—The blades of the cotyledons on the first day after germination (Nov. 21st) were not fully expanded, being inclined at about 35° above the horizon; at night they rose to about 75°. Two days afterwards the blades at noon were horizontal, with the petioles highly inclined upwards; and it is remarkable that the nocturnal movement is almost wholly confined to the blades, being effected by the pulvinus at their bases; whilst the petioles retain day and night nearly the same inclination. On this night (Nov. 23rd), and for some few succeeding nights, the blades rose from a horizontal into a vertical position, and then became bowed inwards at about an average angle of 10°; so that they had passed through an angle of 100°. Their tips now almost touched one another, their bases being slightly divergent. The two blades thus formed a highly inclined roof over the axis of the seedling. This movement is the same as that of the terminal leaflet of the tripartite leaves of many species of Trifolium. After an interval of 8 days (Nov. 29th) the blades were horizontal during the day, and vertical at night, and now they were no longer bowed inwards. They continued to move in the same manner for the following two months, by which time they had increased greatly in size, their petioles being no less than .8 of an inch in length, and two true leaves had by this time been developed.

Trifolium strictum.—On the first day after germination the cotyledons, which are provided with a pulvinus, stood at noon horizontally, and at night rose to only about 45° above the horizon. Four days afterwards the seedlings were again observed at night, and now the blades stood vertically and were in contact, excepting the tips, which were much deflexed, so that they faced the zenith. At this age the petioles are curved upwards, and at night, when the bases of the blades are in contact, the two petioles together form a vertical ring surrounding the plumule. The cotyledons continued to act in nearly the same manner for 8 or 10 days from the period of germination; but the petioles had by this time become straight and had increased much in length. After from 12 to 14 days the first simple true leaf was formed, and during the ensuing fortnight a remarkable movement was repeatedly observed. At I. (Fig. 125) we have a sketch, made in the middle of the day, of a seedling about a fortnight old. The two cotyledons, of which Rc is the right and Lc the left one, stand directly opposite one another, and the first true leaf (F) projects at right angles to them. At night (see II. and III.) the right cotyledon (Rc) is greatly raised, but is not otherwise changed in position. The left cotyledon (Lc) is likewise raised, but it is also twisted so that its blade, instead of exactly facing the opposite one, now stands at nearly right angles to it. This nocturnal twisting movement is effected not by means of the pulvinus, but by the twisting of the whole length of the petiole, as could be seen by the curved line of its upper concave surface. At the same time the true leaf (F) rises up, so as to stand vertically, or it even passes the vertical and is inclined a little inwards. It also twists a little, by which means the upper surface of its blade fronts, and almost comes into contact with, the upper surface of the twisted left cotyledon. This seems to be the object gained by these singular movements. Altogether 20 seedlings were examined on successive nights, and in 19 of them it was the left cotyledon alone which became twisted, with the true leaf always so twisted that its upper surface approached closely and fronted that of the left cotyledon. In only one instance was the right cotyledon twisted, with the true leaf twisted towards it; but this seedling was in an abnormal condition, as the left cotyledon did not rise up properly at night. This whole case is remarkable, as with the cotyledons of no other plant have we seen any nocturnal movement except vertically upwards or downwards. It is the more remarkable, because we shall meet with an analogous case in the leaves of the allied genus Melilotus, in which the terminal leaflet rotates at night so as to present one edge to the zenith and at the same time bends to one side, so that its upper surface comes into contact with that of one of the two now vertical lateral leaflets.

Fig. 125. Trifolium strictum: diurnal and nocturnal positions of the two cotyledons and of the first leaf. I. Seedling viewed obliquely from above, during the day: Rc, right cotyledon; Lc, left cotyledon; F, first true leaf. II. A rather younger seedling, viewed at night: Rc, right cotyledon raised, but its position not otherwise changed; Lc, left cotyledon raised and laterally twisted; F, first leaf raised and twisted so as to face the left twisted cotyledon. III. Same seedling viewed at night from the opposite side. The back of the first leaf, F, is here shown instead of the front, as in II.

Concluding Remarks on the Nyctitropic Movements of Cotyledons.—The sleep of cotyledons (though this is a subject which has been little attended to), seems to be a more common phenomenon than that of leaves. We observed the position of the cotyledons during the day and night in 153 genera, widely distributed throughout the dicotyledonous series, but otherwise selected almost by hazard; and one or more species in 26 of these genera placed their cotyledons at night so as to stand vertically or almost vertically, having generally moved through an angle of at least 60°. If we lay on one side the Leguminosae, the cotyledons of which are particularly liable to sleep, 140 genera remain; and out of these, the cotyledons of at least one species in 19 genera slept. Now if we were to select by hazard 140 genera, excluding the Leguminosae, and observed their leaves at night, assuredly not nearly so many as 19 would be found to include sleeping species. We here refer exclusively to the plants observed by ourselves.

In our entire list of seedlings, there are 30 genera, belonging to 16 Families, the cotyledons of which in some of the species rise or sink in the evening or early night, so as to stand at least 60° above or beneath the horizon. In a large majority of the genera, namely, 24, the movement is a rising one; so that the same direction prevails in these nyctitropic movements as in the lesser periodic ones described in the second chapter. The cotyledons move downwards during the early part of the night in only 6 of the genera; and in one of them, Cannabis, the curving down of the tip is probably due to epinasty, as Kraus believes to be the case with the leaves. The downward movement to the amount of 90° is very decided in Oxalis Valdiviana and sensitiva, and in Geranium rotundifolium. It is a remarkable fact that with Anoda Wrightii, one species of Gossypium and at least 3 species of Ipomœa, the cotyledons whilst young and light sink at night very little or not at all; although this movement becomes well pronounced as soon as they have grown large and heavy. Although the downward movement cannot be attributed to the weight of the cotyledons in the several cases which were investigated, namely, in those of the Anoda, Ipomœa purpurea and bona-nox, nor in that of I. coccinea, yet bearing in mind that cotyledons are continually circumnutating, a slight cause might at first have determined whether the great nocturnal movement should be upwards or downwards. We may therefore suspect that in some aboriginal member of the groups in question, the weight of the cotyledons first determined the downward direction. The fact of the cotyledons of these species not sinking down much whilst they are young and tender, seems opposed to the belief that the greater movement when they are grown older, has been acquired for the sake of protecting them from radiation at night; but then we should remember that there are many plants, the leaves of which sleep, whilst the cotyledons do not; and if in some cases the leaves are protected from cold at night whilst the cotyledons are not protected, so in other cases it may be of more importance to the species that the nearly full-grown cotyledons should be better protected than the young ones.

In all the species of Oxalis observed by us, the cotyledons are provided with pulvini; but this organ has become more or less rudimentary in O. corniculata, and the amount of upward movement of its cotyledons at night is very variable, but is never enough to be called sleep. We omitted to ascertain whether the cotyledons of Geranium rotundifolium possess pulvini. In the Leguminosae all the cotyledons which sleep, as far as we have seen, are provided with pulvini. But with Lotus Jacobæus, these are not fully developed during the first few days of the life of the seedling, and the cotyledons do not then rise much at night. With Trifolium strictum the blades of the cotyledons rise at night by the aid of their pulvini; whilst the petiole of one cotyledon twists half-round at the same time, independently of its pulvinus.

As a general rule, cotyledons which are provided with pulvini continue to rise or sink at night during a much longer period than those destitute of this organ. In this latter case the movement no doubt depends on alternately greater growth on the upper and lower side of the petiole, or of the blade, or of both, preceded probably by the increased turgescence of the growing cells. Such movements generally last for a very short period—for instance, with Brassica and Githago for 4 or 5 nights, with Beta for 2 or 3, and with Raphanus for only a single night. There are, however, some strong exceptions to this rule, as the cotyledons of Gossypium, Anoda and Ipomœa do not possess pulvini, yet continue to move and to grow for a long time. We thought at first that when the movement lasted for only 2 or 3 nights, it could hardly be of any service to the plant, and hardly deserved to be called sleep; but as many quickly-growing leaves sleep for only a few nights, and as cotyledons are rapidly developed and soon complete their growth, this doubt now seems to us not well-founded, more especially as these movements are in many instances so strongly pronounced. We may here mention another point of similarity between sleeping leaves and cotyledons, namely, that some of the latter (for instance, those of Cassia and Githago) are easily affected by the absence of light; and they then either close, or if closed do not open; whereas others (as with the cotyledons of Oxalis) are very little affected by light. In the next chapter it will be shown that the nyctitropic movements both of cotyledons and leaves consist of a modified form of circumnutation.

As in the Leguminosae and Oxalidæ, the leaves and the cotyledons of the same species generally sleep, the idea at first naturally occurred to us, that the sleep of the cotyledons was merely an early development of a habit proper to a more advanced stage of life. But no such explanation can be admitted, although there seems to be some connection, as might have been expected, between the two sets of cases. For the leaves of many plants sleep, whilst their cotyledons do not do so—of which fact Desmodium gyrans offers a good instance, as likewise do three species of Nicotiana observed by us; also Sida rhombifolia, Abutilon Darwinii, and Chenopodium album. On the other hand, the cotyledons of some plants sleep and not the leaves, as with the species of Beta, Brassica, Geranium, Apium, Solanum, and Mirabilis, named in our list. Still more striking is the fact that, in the same genus, the leaves of several or of all the species may sleep, but the cotyledons of only some of them, as occurs with Trifolium, Lotus, Gossypium, and partially with Oxalis. Again, when both the cotyledons and the leaves of the same plant sleep, their movements may be of a widely dissimilar nature: thus with Cassia the cotyledons rise vertically up at night, whilst their leaves sink down and twist round so as to turn their lower surfaces outwards. With seedlings of Oxalis Valdiviana, having 2 or 3 well-developed leaves, it was a curious spectacle to behold at night each leaflet folded inwards and hanging perpendicularly downwards, whilst at the same time and on the same plant the cotyledons stood vertically upwards.

These several facts, showing the independence of the nocturnal movements of the leaves and cotyledons on the same plant, and on plants belonging to the same genus, lead to the belief that the cotyledons have acquired their power of movement for some special purpose. Other facts lead to the same conclusion, such as the presence of pulvini, by the aid of which the nocturnal movement is continued during some weeks. In Oxalis the cotyledons of some species move vertically upwards, and of others vertically downwards at night; but this great difference within the same natural genus is not so surprising as it may at first appear, seeing that the cotyledons of all the species are continually oscillating up and down during the day, so that a small cause might determine whether they should rise or sink at night. Again, the peculiar nocturnal movement of the left-hand cotyledon of Trifolium strictum, in combination with that of the first true leaf. Lastly, the wide distribution in the dicotyledonous series of plants with cotyledons which sleep. Reflecting on these several facts, our conclusion seems justified, that the nyctitropic movements of cotyledons, by which the blade is made to stand either vertically or almost vertically upwards or downwards at night, has been acquired, at least in most cases, for some special purpose; nor can we doubt that this purpose is the protection of the upper surface of the blade, and perhaps of the central bud or plumule, from radiation at night.

CHAPTER VII. MODIFIED CIRCUMNUTATION: NYCTITROPIC OR SLEEP MOVEMENTS OF LEAVES.

Conditions necessary for these movements—List of Genera and Families, which include sleeping plants—Description of the movements in the several Genera—Oxalis: leaflets folded at night—Averrhoa: rapid movements of the leaflets—Porlieria: leaflets close when plant kept very dry—Tropaeolum: leaves do not sleep unless well illuminated during day—Lupinus: various modes of sleeping—Melilotus: singular movements of terminal leaflet—Trifolium—Desmodium: rudimentary lateral leaflets, movements of, not developed on young plants, state of their pulvini—Cassia: complex movements of the leaflets—Bauhinia: leaves folded at night—Mimosa pudica: compounded movements of leaves, effect of darkness—Mimosa albida, reduced leaflets of—Schrankia: downward movement of the pinnae—Marsilea: the only cryptogam known to sleep—Concluding remarks and summary—Nyctitropism consists of modified circumnutation, regulated by the alternations of light and darkness—Shape of first true leaves.

We now come to the nyctitropic or sleep movements of leaves. It should be remembered that we confine this term to leaves which place their blades at night either in a vertical position or not more than 30° from the vertical,—that is, at least 60° above or beneath the horizon. In some few cases this is effected by the rotation of the blade, the petiole not being either raised or lowered to any considerable extent. The limit of 30° from the vertical is obviously an arbitrary one, and has been selected for reasons previously assigned, namely, that when the blade approaches the perpendicular as nearly as this, only half as much of the surface is exposed at night to the zenith and to free radiation as when the blade is horizontal. Nevertheless, in a few instances, leaves which seem to be prevented by their structure from moving to so great an extent as 60° above or beneath the horizon, have been included amongst sleeping plants.

It should be premised that the nyctitropic movements of leaves are easily affected by the conditions to which the plants have been subjected. If the ground is kept too dry, the movements are much delayed or fail: according to Dassen,[1] even if the air is very dry the leaves of Impatiens and Malva are rendered motionless. Carl Kraus has also lately insisted[2] on the great influence which the quantity of water absorbed has on the periodic movements of leaves; and he believes that this cause chiefly determines the variable amount of sinking of the leaves of Polygonum convolvulus at night; and if so, their movements are not in our sense strictly nyctitropic. Plants in order to sleep must have been exposed to a proper temperature: Erythrina crista-galli, out of doors and nailed against a wall, seemed in fairly good health, but the leaflets did not sleep, whilst those on another plant kept in a warm greenhouse were all vertically dependent at night. In a kitchen-garden the leaflets of Phaseolus vulgaris did not sleep during the early part of the summer. Ch. Royer says,[3] referring I suppose to the native plants in France, that they do not sleep when the temperature is below 5° C. or 41° F. In the case of several sleeping plants, viz., species of Tropaeolum, Lupinus, Ipomœa, Abutilon, Siegesbeckia, and probably other genera, it is indispensable that the leaves should be well illuminated during the day in order that they may assume at night a vertical position; and it was probably owing to this cause that seedlings of Chenopodium album and Siegesbeckia orientalis, raised by us during the middle of the winter, though kept at a proper temperature, did not sleep. Lastly, violent agitation by a strong wind, during a few minutes, of the leaves of Maranta arundinacea (which previously had not been disturbed in the hot-house), prevented their sleeping during the two next nights.

[1] Dassen,'Tijdschrift vor. Naturlijke Gesch. en Physiologie,' 1837, vol. iv. p. 106. See also Ch. Royer on the importance of a proper state of turgescence of the cells, in 'Annal. des Sc. Nat. Bot.' (5th series), ix. 1868, p. 345.

[2] 'Beiträge zur Kentniss der Bewegungen,' etc., in 'Flora,' 1879, pp. 42, 43, 67, etc.

[3] 'Annal. des Sc. Nat. Bot.' (5th Series), ix. 1868, p. 366.

We will now give our observations on sleeping plants, made in the manner described in the Introduction. The stem of the plant was always secured (when not stated to the contrary) close to the base of the leaf, the movements of which were being observed, so as to prevent the stem from circumnutating. As the tracings were made on a vertical glass in front of the plant, it was obviously impossible to trace its course as soon as the leaf became in the evening greatly inclined either upwards or downwards; it must therefore be understood that the broken lines in the diagrams, which represent the evening and nocturnal courses, ought always to be prolonged to a much greater distance, either upwards or downwards, than appears in them. The conclusions which may be deduced from our observations will be given near the end of this chapter.

In the following list all the genera which include sleeping plants are given, as far as known to us. The same arrangement is followed as in former cases, and the number of the Family is appended. This list possesses some interest, as it shows that the habit of sleeping is common to some few plants throughout the whole vascular series. The greater number of the genera in the list have been observed by ourselves with more or less care; but several are given on the authority of others (whose names are appended in the list), and about these we have nothing more to say. No doubt the list is very imperfect, and several genera might have been added from the 'Somnus Plantarum' by Linnæus; but we could not judge in some of his cases, whether the blades occupied at night a nearly vertical position. He refers to some plants as sleeping, for instance, Lathyrus odoratus and Vicia faba, in which we could observe no movement deserving to be called sleep, and as no one can doubt the accuracy of Linnæus, we are left in doubt.

[List of Genera, including species the leaves of which sleep.

CLASS I. DICOTYLEDONS.

Sub-class I. ANGIOSPERMS.

Genus Family.

Githago Caryophylleae (26). Stellaria (Batalin). " Portulaca (Ch.Royer). Portulaceae (27). Sida Malvaceae (36). Abutilon. " Malva (Linnæus and Pfeffer). " Hibiscus (Linnæus). " Anoda. " Gossypium. " Ayenia (Linnæus). Sterculaceae (37). Triumfetta (Linnæus). Tiliaceae (38). Linum (Batalin). Lineae (39). Oxalis. Oxalidæ (41). Averrhoa. " Porlieria. Zygophylleæ (45). Guiacum. " Impatiens (Linnæus, Pfeffer, Batalin). Balsamineae (48). Tropaeolum. Tropaeoleae (49). Crotolaria (Thiselton Dyer). Leguminosae (75) Tribe II. Lupinus. " " Cytisus. " " Trigonella. " Tr. III. Medicago. " Melilotus. " " Trifolium. " " Securigera. " Tr. IV. Lotus. " " Psoralea. " Tr. V. Amorpha (Cuchartre). " " Daelea. " " Indigofera. " " Tephrosia. " " Wistaria. " " Robinia. " " Sphaerophysa. " " Colutea. " " Astragalus. " " Glycyrrhiza. " " Coronilla. " Tr. VI. Hedysarum. " " Genus Family. Onobrychis. Leguminosae (75) Tr. VI. Smithia. " " Arachis. " " Desmodium. " " Urania. " " Vicia. " Tr. VII. Centrosema. " Tr. VIII. Amphicarpæa. " " Glycine. " " Erythrina. " " Apios. " " Phaseolus. " " Sophora. " Tr. X. Caesalpinia. " Tr. XIII. Haematoxylon. " " Gleditschia (Duchartre). " " Poinciana. " " Cassia. " Tr. XIV. Bauhinia. " Tr. XV. Tamarindus. " Tr. XVI. Adenanthera. " Tr. XX. Prosopis. " " Neptunia. " " Mimosa. " " Schrankia. " " Acacia. " Tr. XXII. Albizzia. " Tr. XXIII. Melaleuca (Bouché). Myrtaceae (94). Genus Family. Aenothera (Linnæus). Omagrarieae (100). Passiflora. Passifloracea (105). Siegesbeckia. Compositæ (122). Ipomœa. Convolvulacea (151). Nicotiana. Solaneae (157). Mirabilis.

Nyctagineae (177). Polygonum (Batalin). Polygoneae (179). Amaranthus. Amaranthaceae (180). Chenopodium. Chenopodieae (181). Pimelia (Bouché). Thymeteae (188). Euphorbia. Euphorbiaceae (202) Phyllanthus (Pfeffer). "

Sub-class II. GYMNOSPERMS. Aies (Chatin).

CLASS II. MONOCOTYLEDONS.

Thalia. Cannaceae (21). Maranta. " Colocasia. Aroideae (30). Strephium. Gramineæ (55).

CLASS III. ACOTYLEDONS.

Marsilea. Marsileaceae (4).

Githago segetum (Caryophylleae).—The first leaves produced by young seedlings, rise up and close together at night. On a rather older seedling, two young leaves stood at noon at 55° above the horizon, and at night at 86°, so each had risen 31°. The angle, however, was less in some cases. Similar observations were occasionally made on young leaves (for the older ones moved very little) produced by nearly full-grown plants. Batalin says ('Flora,' Oct. 1st, 1873, p. 437) that the young leaves of Stellaria close up so completely at night that they form together great buds.

Sida (Malvaceae).—the nyctitropic movements of the leaves in this genus are remarkable in some respects. Batalin informs us (see also 'Flora,' Oct. 1st, 1873, p. 437) that those of S. napaea fall at night, but to what angle he cannot remember. The leaves of S. rhombifolia and retusa, on the other hand, rise up vertically, and are pressed against the stem. We have therefore here within the same genus, directly opposite movements. Again, the leaves of S. rhombifolia are furnished with a pulvinus, formed of a mass of small cells destitute of chlorophyll, and with their longer axes perpendicular to the axis of the petiole. As measured along this latter line, these cells are only 1/5th of the length of those of the petiole; but instead of being abruptly separated from them (as is usual with the pulvinus in most plants), they graduate into the larger cells of the petiole. On the other hand, S. napaea, according to Batalin, does not possess a pulvinus; and he informs us that a gradation may be traced in the several species of the genus between these two states of the petiole. Sida rhombifolia presents another peculiarity, of which we have seen no other instance with leaves that sleep: for those on very young plants, though they rise somewhat in the evening, do not go to sleep, as we observed on several occasions; whilst those on rather older plants sleep in a conspicuous manner. For instance a leaf (.85 of an inch in length) on a very young seedling 2 inches high, stood at noon 9° above the horizon, and at 10 P.M. at 28°, so it had risen only 19°; another leaf (1.4 inch in length) on a seedling of the same height, stood at the same two periods at 7° and 32°, and therefore had risen 25°. These leaves, which moved so little, had a fairly well-developed pulvinus. After an interval of some weeks, when the same seedlings were 2½ and 3 inches in height, some of the young leaves stood up at night quite vertically, and others were highly inclined; and so it was with bushes which were fully grown and were flowering.

Fig. 126. Sida rhombifolia: circumnutation and nyctitropic (or sleep) movements of a leaf on a young plant, 9½ inches high; filament fixed to midrib of nearly full-grown leaf, 2 3/8 inches in length; movement traced under a sky-light. Apex of leaf 5 5/8 inches from the vertical glass, so diagram not greatly enlarged.

The movement of a leaf was traced from 9.15 A.M. on May 28th to 8.30 A.M. on the 30th. The temperature was too low (15°–16° C.), and the illumination hardly sufficient; consequently the leaves did not become quite so highly inclined at night, as they had done previously and as they did subsequently in the hot-house: but the movements did not appear otherwise disturbed. On the first day the leaf sank till 5.15 P.M.; it then rose rapidly and greatly till 10.5 P.M., and only a little higher during the rest of the night (Fig. 126). Early on the next day (29th) it fell in a slightly zigzag line rapidly until 9 A.M., by which time it had reached nearly the same place as on the previous morning. During the remainder of the day it fell slowly, and zigzagged laterally. The evening rise began after 4 P.M. in the same manner as before, and on the second morning it again fell rapidly. The ascending and descending lines do not coincide, as may be seen in the diagram. On the 30th a new tracing was made (not here given) on a rather enlarged scale, as the apex of the leaf now stood 9 inches from the vertical glass. In order to observe more carefully the course pursued at the time when the diurnal fall changes into the nocturnal rise, dots were made every half-hour between 4 P.M. and 10.30 P.M. This rendered the lateral zigzagging movement during the evening more conspicuous than in the diagram given, but it was of the same nature as there shown. The impression forced on our minds was that the leaf was expending superfluous movement, so that the great nocturnal rise might not occur at too early an hour.

Abutilon Darwinii (Malvaceae).—The leaves on some very young plants stood almost horizontally during the day, and hung down vertically at night. Very fine plants kept in a large hall, lighted only from the roof, did not sleep at night for in order to do so the leaves must be well illuminated during the day. The cotyledons do not sleep. Linnæus says that the leaves of his Sida abutilon sink perpendicularly down at night, though the petioles rise. Prof. Pfeffer informs us that the leaves of a Malva, allied to M.

sylvestris, rise greatly at night; and this genus, as well as that of Hibiscus, are included by Linnæus in his list of sleeping plants.

Anoda Wrightii (Malvaceae).—The leaves, produced by very young plants, when grown to a moderate size, sink at night either almost vertically down or to an angle of about 45° beneath the horizon; for there is a considerable degree of variability in the amount of sinking at night, which depends in part on the degree to which they have been illuminated during the day. But the leaves, whilst quite young, do not sink down at night, and this is a very unusual circumstance. The summit of the petiole, where it joins the blade, is developed into a pulvinus, and this is present in very young leaves which do not sleep; though it is not so well defined as in older leaves.

Gossypium (var. Nankin cotton, Malvaceae).—Some young leaves, between 1 and 2 inches in length, borne by two seedlings 6 and 7½ inches in height, stood horizontally, or were raised a little above the horizon at noon on July 8th and 9th; but by 10 P.M. they had sunk down to between 68° and 90° beneath the horizon. When the same plants had grown to double the above height, their leaves stood at night almost or quite vertically dependent. The leaves on some large plants of G. maritimum and Brazilense, which were kept in a very badly lighted hot-house, only occasionally sank much downwards at night, and hardly enough to be called sleep.

Oxalis (Oxalidæ).—In most of the species in this large genus the three leaflets sink vertically down at night; but as their sub-petioles are short the blades could not assume this position from the want of space, unless they were in some manner rendered narrower; and this is effected by their becoming more or less folded (Fig. 127). The angle formed by the two halves of the same leaflet was found to vary in different individuals of several species between 92° and 150°; in three of the best folded leaflets of O. fragrans it was 76°, 74°, and 54°. The angle is often different in the three leaflets of the same leaf. As the leaflets sink down at night and become folded, their lower surfaces are brought near together (see B), or even into close contact; and from this circumstance it might be thought that the object of the folding was the protection of their lower surfaces. If this had been the case, it would have formed a strongly marked exception to the rule, that when there is any difference in the degree of protection from radiation of the two surfaces of the leaves, it is always the upper surface which is the best protected. But that the folding of the leaflets, and consequent mutual approximation of their lower surfaces, serves merely to allow them to sink down vertically, may be inferred from the fact that when the leaflets do not radiate from the summit of a common petiole, or, again, when there is plenty of room from the sub-petioles not being very short, the leaflets sink down without becoming folded. This occurs with the leaflets of O. sensitiva, Plumierii, and bupleurifolia.

Fig. 127. Oxalis acetosella: A, leaf seen from vertically above; B, diagram of leaf asleep, also seen from vertically above.

There is no use in giving a long list of the many species which sleep in the above described manner. This holds good with species having rather fleshy leaves, like those of O. carnosa, or large leaves like those of O. Ortegesii, or four leaflets like those of O. variabilis. There are, however, some species which show no signs of sleep, viz., O. pentaphylla, enneaphylla, hirta, and rubella. We will now describe the nature of the movements in some of the species.

Oxalis acetosella.—The movement of a leaflet, together with that of the main petiole, are shown in the following diagram (Fig. 128), traced between 11 A.M. on October 4th and 7.45 A.M. on the 5th. After 5.30 P.M. on the 4th the leaflet sank rapidly, and at 7 P.M. depended vertically. for some time before it assumed this latter position, its movements could, of course, no longer be traced on the vertical glass, and the broken line in the diagram ought to be extended much further down in this and all other cases. By 6.45 A.M. on the following morning it had risen considerably, and continued to rise for the next hour; but, judging from other observations, it would soon have begun to fall again. Between 11 A.M. and 5.30 P.M. the leaflet moved at least four times up and four times down before the great nocturnal fall commenced; it reached its highest point at noon. Similar observations were made on two other leaflets, with nearly the same results. Sachs and Pfeffer have also described briefly[4] the autonomous movements of the leaves of this plant.

[4] Sachs in 'Flora,' 1863, p. 470, etc; Pfeffer, 'Die Period. Bewegungen,' etc., 1875, p. 53.

Fig 128. Oxalis acetosella: circumnutation and nyctitropic movements of a nearly full-grown leaf, with filament attached to the midrib of one of the leaflets; traced on vertical glass during 20 h. 45m.

On another occasion the petiole of a leaf was secured to a little stick close beneath the leaflets, and a filament tipped with a bead of sealing-wax was affixed to the mid-rib of one of them, and a mark was placed close behind. At 7 P.M., when the leaflets were asleep, the filament depended vertically down, and the movements of the bead were then traced till 10.40 P.M., as shown in the following diagram (Fig. 129). We here see that the leaflet moved a little from side to side, as well as a little up and down, whilst asleep.

Oxalis Valdiviana.—The leaves resemble those of the last species, and the movements of two leaflets (the main petioles of both having been secured) were traced during two days; but the tracings are not given, as they resembled that of O. acetosella, with the exception that the up and down oscillations were not so frequent during the day, and there was more lateral movement, so that broader ellipses were described. The leaves awoke early in the morning, for by 6.45 A.M. on June 12th and 13th they had not only risen to their full height, but had already begun to fall, that is, they were circumnutating. We have seen in the last chapter that the cotyledons, instead of sinking, rise up vertically at night.

Fig 129. Oxalis acetosella: circumnutation of leaflet when asleep; traced on vertical glass during 3 h. 40 m.

Oxalis Ortegesii.—The large leaves of this plant sleep like those of the previous species. The main petioles are long, and that of a young leaf rose 20° between noon and 10 P.M., whilst the petiole of an older leaf rose only 13°. Owing to this rising of the petioles, and the vertical sinking of the large leaflets, the leaves become crowded together at night, and the whole plant then exposes a much smaller surface to radiation than during the day.

Oxalis Plumierii.—In this species the three leaflets do not surround the summit of the petiole, but the terminal leaflet projects in the line of the petiole, with a lateral leaflet on each side. They all sleep by bending vertically downwards, but do not become at all folded. The petiole is rather long, and, one having been secured to a stick, the movement of the terminal leaflet was traced during 45 h. on a vertical glass. It moved in a very simple manner, sinking rapidly after 5 P.M., and rising rapidly early next morning. During the middle of the day it moved slowly and a little laterally. Consequently the ascending and descending lines did not coincide, and a single great ellipse was formed each day. There was no other evidence of circumnutation, and this fact is of interest, as we shall hereafter see.

Oxalis sensitiva.—The leaflets, as in the last species, bend vertically down at night, without becoming folded. The much elongated main petiole rises considerably in the evening, but in some very young plants the rise did not commence until late at night. We have seen that the cotyledons, instead of sinking like the leaflets, rise up vertically at night.

Oxalis bupleurifolia.—This species is rendered remarkable by the petioles being foliaceous, like the phyllodes of many Acacias. The leaflets are small, of a paler green and more tender consistence than the foliaceous petioles. The leaflet which was observed was .55 inch in length, and was borne by a petiole 2 inches long and .3 inch broad. It may be suspected that the leaflets are on the road to abortion or obliteration, as has actually occurred with those of another Brazilian species, O. rusciformis. Nevertheless, in the present species the nyctitropic movements are perfectly performed. The foliaceous petiole was first observed during 48 h., and found to be in continued circumnutation, as shown in the accompanying figure (Fig. 130). It rose during the day and early part of the night, and fell during the remainder of the night and early morning; but the movement was not sufficient to be called sleep. The ascending and descending lines did not coincide, so that an ellipse was formed each day. There was but little zigzagging; if the filament had been fixed longitudinally, we should probably have seen that there was more lateral movement than appears in the diagram.

Fig. 130. Oxalis bupleurifolia: circumnutation of foliaceous petiole, filament fixed obliquely across end of petiole; movements traced on vertical glass from 9 A.M. June 26th to 8.50 A.M. 28th. Apex of leaflet 4½ inches from the glass, so movement not much magnified. Plant 9 inches high, illuminated from above. Temp. 23½°–24½° C.

A terminal leaflet on another leaf was next observed (the petiole being secured), and its movements are shown in Fig. 131. During the day the leaflets are extended horizontally, and at night depend vertically; and as the petiole rises during the day the leaflets have to bend down in the evening more than 90°, so as to assume at night their vertical position. On the first day the leaflet simply moved up and down; on the second day it plainly circumnutated between 8 A.M. and 4.30 P.M., after which hour the great evening fall commenced.

Fig. 131. Oxalis bupleurifolia: circumnutation and nyctitropic movement of terminal leaflet, with filament affixed along the midrib; traced on a vertical glass from 9 A.M. on June 26th to 8.45 A.M. 28th. Conditions the same as in the last case.

Averrhoa bilimbi (Oxalidæ).—It has long been known,[5] firstly, that the leaflets in this genus sleep; secondly, that they move spontaneously during the day; and thirdly, that they are sensitive to a touch; but in none of these respects do they differ essentially from the species of Oxalis. They differ, however, as Mr. R. I. Lynch[6] has lately shown, in their spontaneous movements being strongly marked. In the case of A. bilimbi, it is a wonderful spectacle to behold on a warm sunny day the leaflets one after the other sinking rapidly downwards, and again ascending slowly. Their movements rival those of Desmodium gyrans. At night the leaflets hang vertically down; and now they are motionless, but this may be due to the opposite ones being pressed together (Fig. 132). The main petiole is in constant movement during the day, but no careful observations were made on it. The following diagrams are graphic representations of the variations in the angle, which a given leaflet makes with the vertical. The observations were made as follows. The plant growing in a pot was kept in a high temperature, the petiole of the leaf to be observed pointing straight at the observer, being separated from him by a vertical pane of glass. The petiole was secured so that the basal joint, or pulvinus, of one of the lateral leaflets was at the centre of a graduated arc placed close behind the leaflet. A fine glass filament was fixed to the leaf, so as to project like a continuation of the midrib. This filament acted as an index; and as the leaf rose and fell, rotating about its basal joint, its angular movement could be recorded by reading off at short intervals of time the position of the glass filament on the graduated arc. In order to avoid errors of parallax, all readings were made by looking through a small ring painted on the vertical glass, in a line with the joint of the leaflet and the centre of the graduated arc. In the following diagrams the ordinates represent the angles which the leaflet made with the vertical at successive instants.[7] It follows that a fall in the curve represents an actual dropping of the leaf, and that the zero line represents a vertically dependent position. Fig. 133 represents the nature of the movements which occur in the evening, as soon as the leaflets begin to assume their nocturnal position. At 4.55 P.M. the leaflet formed an angle of 85° with the vertical, or was only 5° below the horizontal; but in order that the diagram might get into our page, the leaflet is represented falling from 75° instead of 85°. Shortly after 6 P.M. it hung vertically down, and had attained its nocturnal position. Between 6.10 and 6.35 P.M. it performed a number of minute oscillations of about 2° each, occupying periods of 4 or 5 m. The complete state of rest of the leaflet which ultimately followed is not shown in the diagram. It is manifest that each oscillation consists of a gradual rise, followed by a sudden fall. Each time the leaflet fell, it approached nearer to the nocturnal position than it did on the previous fall. The amplitude of the oscillations diminished, while the periods of oscillation became shorter.

[5] Dr. Bruce, 'Philosophical Trans.,' 1785, p. 356.

[6] 'Journal Linn. Soc.,' vol. xvi. 1877, p. 231.

[7] In all the diagrams 1 mm. in the horizontal direction represents one minute of time. Each mm. in the vertical direction represents one degree of angular movement. In Figs. 133 and 134 the temperature is represented (along the ordinates) in the scale of 1 mm. to each 0.1 degree C. In Fig. 135 each mm. equals 0.2° F.

Fig. 132. Averrhoa bilimbi: leaf asleep; drawing reduced.

Fig. 133. Averrhoa bilimbi: angular movements of a leaflet during its evening descent, when going to sleep. Temp. 78°–81° F.

In bright sunshine the leaflets assume a highly inclined dependent position. A leaflet in diffused light was observed rising for 25 m. A blind was then pulled up so that the plant was brightly illuminated (BR in Fig. 134), and within a minute it began to fall, and ultimately fell 47°, as shown in the diagram. This descent was performed by six descending steps, precisely similar to those by which the nocturnal fall is effected. The plant was then again shaded (SH), and a long slow rise occurred until another series of falls commenced at BR', when the sun was again admitted. In this experiment cool air was allowed to enter by the windows being opened at the same time that the blinds were pulled up, so that in spite of the sun shining on the plant the temperature was not raised.

The effect of an increase of temperature in diffused light is shown in Fig. 135. The temperature began to rise at 11.35 A.M. (in consequence of the fire being lighted), but by 12.42 a marked fall had occurred. It may be seen in the diagram that when the temperature was highest there were rapid oscillations of small amplitude, the mean position of the leaflet being at the time nearer the vertical. When the temperature began to fall, the oscillations became slower and larger, and the mean position of the leaf again approached the horizontal. The rate of oscillation was sometimes quicker than is represented in the above diagram. Thus, when the temperature was between 31° and 32° C., 14 oscillations of a few degrees occurred in 19 m. On the other hand, an oscillation may be much slower; thus a leaflet was observed (temperature 25° C.) to rise during 40 m. before it fell and completed its oscillation.

Fig. 134. Averrhoa bilimbi: angular movements of leaflet during a change from bright illumination to shade; temperature (broken line) remaining nearly the same.

Fig. 135. Averrhoa bilimbi: angular movement of leaflet during a change of temperature; light remaining the same. The broken line shows the change of temperature.

Fig. 136. Porlieria hygrometrica: circumnutation and nyctitropic movements of petiole of leaf, traced from 9.35 A.M. July 7th to about midnight on the 8th. Apex of leaf 7½ inches from the vertical glass. Temp. 19½°–20½° C.

Porlieria hygrometrica (Zygophylleæ).—The leaves of this plant (Chilian form) are from 1 to 1½ inch in length, and bear as many as 16 or 17 small leaflets on each side, which do not stand opposite one another. They are articulated to the petiole, and the petiole to the branch by a pulvinus. We must premise that apparently two forms are confounded under the same name: the leaves on a bush from Chili, which was sent to us from Kew, bore many leaflets, whilst those on plants in the Botanic Garden at

Würzburg bore only 8 or 9 pairs; and the whole character of the bushes appeared somewhat different. We shall also see that they differ in a remarkable physiological peculiarity. On the Chilian plant the petioles of the younger leaves on upright branches, stood horizontally during the day, and at night sank down vertically so as to depend parallel and close to the branch beneath. The petioles of rather older leaves did not become at night vertically depressed, but only highly inclined. In one instance we found a branch which had grown perpendicularly downwards, and the petioles on it moved in the same direction relatively to the branch as just stated, and therefore moved upwards. On horizontal branches the younger petioles likewise move at night in the same direction as before, that is, towards the branch, and are consequently then extended horizontally; but it is remarkable that the older petioles on the same branch, though moving a little in the same direction, also bend downwards; they thus occupy a somewhat different position, relatively to the centre of the earth and to the branch, from that of the petioles on the upright branches. With respect to the leaflets, they move at night towards the apex of the petiole until their midribs stand nearly parallel to it; and they then lie neatly imbricated one over the other. Thus half of the upper surface of each leaflet is in close contact with half of the lower surface of the one next in advance; and all the leaflets, excepting the basal ones, have the whole of their upper surfaces and half of their lower surfaces well protected. Those on the opposite sides of the same petiole do not come into close contact at night, as occurs with the leaflets of so many Leguminosae but are separated by an open furrow; nor could they exactly coincide, as they stand alternately with respect to one another.

The circumnutation of the petiole of a leaf 3/4 of an inch in length, on an upright branch, was observed during 36h., and is shown in the preceding diagram (Fig. 136). On the first morning, the leaf fell a little and then rose until 1 P.M., and this was probably due to its being now illuminated through a skylight from above; it then circumnutated on a very small scale round the same spot until about 4 P.M., when the great evening fall commenced. During the latter part of the night or very early on the next morning the leaf rose again. On the second day it fell during the morning till 1 P.M., and this no doubt is its normal habit. From 1 to 4 P.M. it rose in a zigzag line, and soon afterwards the great evening fall commenced. It thus completed a double oscillation during the 24 h.

The specific name given to this plant by Ruiz and Pavon, indicates that in its native arid home it is affected in some manner by the dryness or dampness of the atmosphere.[8] In the Botanic Garden at Würzburg, there was a plant in a pot out of doors which was daily watered, and another in the open ground which was never watered. After some hot and dry weather there was a great difference in the state of the leaflets on these two plants; those on the unwatered plant in the open ground remaining half, or even quite, closed during the day. But twigs cut from this bush, with their ends standing in water, or wholly immersed in it, or kept in damp air under a bell-glass, opened their leaves though exposed to a blazing sun; whilst those on the plant in the ground remained closed. The leaves on this same plant, after some heavy rain, remained open for two days; they then became half closed during two days, and after an additional day were quite closed. This plant was now copiously watered, and on the following morning the leaflets were fully expanded. The other plant growing in a pot, after having been exposed to heavy rain, was placed before a window in the Laboratory, with its leaflets open, and they remained so during the daytime for 48 h.; but after an additional day were half closed. The plant was then watered, and the leaflets on the two following days remained open. On the third day they were again half closed, but on being again watered remained open during the two next days. From these several facts we may conclude that the plant soon feels the want of water; and that as soon as this occurs, it partially or quite closes its leaflets, which in their then imbricated condition expose a small surface to evaporation. It is therefore probable that this sleep-like movement, which occurs only when the ground is dry, is an adaptation against the loss of moisture.

[8] 'Systema Veg. Florae Peruvianae et Chilensis,' tom. i. p. 95, 1798. We cannot understand the account given by the authors of the behaviour of this plant in its native home. There is much about its power of foretelling changes in the weather; and it appears as if the brightness of the sky largely determined the opening and closing of the leaflets.

A bush about 4 feet in height, a native of Chili, which was thickly covered with leaves, behaved very differently, for during the day it never closed its leaflets. On July 6th the earth in the small pot in which it grew appeared extremely dry, and it was given a very little water. After 21 and 22 days (on the 27th and 28th), during the whole of which time the plant did not receive a drop of water, the leaves began to droop, but they showed no signs of closing during the day. It appeared almost incredible that any plant, except a fleshy one, could have kept alive in soil so dry, which resembled the dust on a road. On the 29th, when the bush was shaken, some leaves fell off, and the remaining ones were unable to sleep at night. It was therefore moderately watered, as well as syringed, late in the evening. On the next morning (30th) the bush looked as fresh as ever, and at night the leaves went to sleep. It may be added that a small branch while growing on the bush was enclosed, by means of a curtain of bladder, during 13 days in a large bottle half full of quicklime, so that the air within must have been intensely dry; yet the leaves on this branch did not suffer in the least, and did not close at all during the hottest days. Another trial was made with the same bush on August 2nd and 6th (the soil appearing at this latter date extremely dry), for it was exposed out of doors during the whole day to the wind, but the leaflets showed no signs of closing. The Chilian form therefore differs widely from the one at Würzburg, in not closing its leaflets when suffering from the want of water; and it can live for a surprisingly long time without water.

Tropaeolum majus (?) (cultivated var.) (Tropaeoleae).—Several plants in pots stood in the greenhouse, and the blades of the leaves which faced the front-lights were during the day highly inclined and at night vertical; whilst the leaves on the back of the pots, though of course illuminated through the roof, did not become vertical at night. We thought, at first, that this difference in their positions was in some manner due to heliotropism, for the leaves are highly heliotropic. The true explanation, however, is that unless they are well illuminated during at least a part of the day they do not sleep at night; and a little difference in the degree of illumination determines whether or not they shall become vertical at night. We have observed no other so well-marked a case as this, of the influence of previous illumination on nyctitropic movements. The leaves present also another peculiarity in their habit of rising or awaking in the morning, being more strongly fixed or inherited than that of sinking or sleeping at night. The movements are caused by the bending of an upper part of the petiole, between ½ and 1 inch in length; but the part close to the blade, for about 1/4 of an inch in length, does not bend and always remains at right angles to the blade. The bending portion does not present any external or internal difference in structure from the rest of the petiole. We will now give the experiments on which the above conclusions are founded.

A large pot with several plants was brought on the morning of Sept. 3rd out of the greenhouse and placed before a north-east window, in the same position as before with respect to the light, as far as that was possible. On the front of the plants, 24 leaves were marked with thread, some of which had their blades horizontal, but the greater number were inclined at about 45°, beneath the horizon; at night all these, without exception, became vertical. Early on the following morning (4th) they reassumed their former positions, and at night again became vertical. On the 5th the shutters were opened at 6.15 A.M., and by 8.18 A.M., after the leaves had been illuminated for 2 h. 3 m. and had acquired their diurnal position, they were placed in a dark cupboard. They were looked at twice during the day and thrice in the evening, the last time at 10.30 P.M., and not one had become vertical. At 8 A.M. on the following morning (6th) they still retained the same diurnal position, and were now replaced before the north-east window. At night all the leaves which had faced the light had their petioles curved and their blades vertical; whereas none of the leaves on the back of the plants, although they had been moderately illuminated by the diffused light of the room, were vertical. They were now at night placed in the same dark cupboard; at 9 A.M. on the next morning (7th) all those which had been asleep had reassumed their diurnal position. The pot was then placed for 3 h. in the sunshine, so as to stimulate the plants; at noon they were placed before the same north-east window, and at night the leaves slept in the usual manner and awoke on the following morning. At noon on this day (8th) the plants, after having been left before the north-east window for 5 h. 45 m. and thus illuminated (though not brightly, as the sky was cloudy during the whole time), were replaced in the dark cupboard, and at 3 P.M. the position of the leaves was very little, if at all, altered, so that they are not quickly affected by darkness; but by 10.15 P.M. all the leaves which had faced the north-east sky during the 5 h. 45 m. of illumination stood vertical, whereas those on the back of the plant retained their diurnal position. On the following morning (9th) the leaves awoke as on the two former occasions in the dark, and they were kept in the dark during the whole day; at night a very few of them became vertical, and this was the one instance in which we observed any inherited tendency or habit in this plant to sleep at the proper time. That it was real sleep was shown by these same leaves reassuming their diurnal position on the following morning (10th) whilst still kept in the dark.

The pot was then (9.45 A.M. 10th) replaced, after having been kept for 36 h. in darkness, before the north-east window; and at night the blades of all the leaves (excepting a few on the back of the plants) became conspicuously vertical.

At 6.45 A.M. (11th) after the plants had been illuminated on the same side as before during only 25 m., the pot was turned round, so that the leaves which had faced the light now faced the interior of the room, and not one of these went to sleep at night; whilst some, but not many, of those which had formerly stood facing the back of the room and which had never before been well illuminated or gone to sleep, now assumed a vertical position at night. On the next day (12th) the plant was turned round into its original position, so that the same leaves faced the light as formerly, and these now went to sleep in the usual manner. We will only add that with some young seedlings kept in the greenhouse, the blades of the first pair of true leaves (the cotyledons being hypogean) stood during the day almost horizontally and at night almost vertically.

A few observations were subsequently made on the circumnutation of three leaves, whilst facing a north-east window; but the tracings are not given, as the leaves moved somewhat towards the light. It was, however, manifest that they rose and fell more than once during the daytime, the ascending and descending lines being in parts extremely zigzag. The nocturnal fall commenced about 7 P.M., and the leaves had risen considerably by 6.45 A.M. on the following morning.

Leguminosae.—This Family includes many more genera with sleeping species than all the other families put together. The number of the tribes to which each genus belongs, according to Bentham and Hooker's arrangement, has been added.

Crotolaria (sp.?) (Tribe 2).—This plant is monophyllous, and we are informed by Mr. T. Thiselton Dyer that the leaves rise up vertically at night and press against the stem.

Lupinus (Tribe 2).—The palmate or digitate leaves of the species in this large genus sleep in three different manners. One of the simplest, is that all the leaflets become steeply inclined downwards at night, having been during the day extended horizontally. This is shown in the accompanying figures (Fig. 137), of a leaf of L. pilosus, as seen during the day from vertically above, and of another leaf asleep with the leaflets inclined downwards. As in this position they are crowded together, and as they do not become folded like those in the genus Oxalis, they cannot occupy a vertically dependent position; but they are often inclined at an angle of 50° beneath the horizon. In this species, whilst the leaflets are sinking, the petioles rise up, in two instances when the angles were measured to the extent of 23°. The leaflets of L. sub-carnosus and arboreus, which were horizontal during the day, sank down at night in nearly the same manner; the former to an angle of 38° and the latter of 36°, beneath the horizon; but their petioles did not move in any plainly perceptible degree. It is, however, quite possible, as we shall presently see, that if a large number of plants of the three foregoing and of the following species were to be observed at all seasons, some of the leaves would be found to sleep in a different manner.

Fig. 137. Lupinus pilosus: A, leaf seen from vertically above in daytime; B, leaf asleep, seen laterally at night.

In the two following species the leaflets, instead of moving downwards, rise at night. With L. Hartwegii some stood at noon at a mean angle of 36° above the horizon, and at night at 51°, thus forming together a hollow cone with moderately steep sides. The petiole of one leaf rose 14° and of a second 11° at night. With L. luteus a leaflet rose from 47° at noon to 65° above the horizon at night, and another on a distinct leaf rose from 45° to 69°. The petioles, however, sink at night to a small extent, viz., in three instances by 2°, 6°, and 9° 30 seconds. Owing to this movement of the petioles, the outer and longer leaflets have to bend up a little more than the shorter and inner ones, in order that all should stand symmetrically at night.

We shall presently see that some leaves on the same individual plants of L. luteus sleep in a very different manner.

We now come to a remarkable position of the leaves when asleep, which is common to several species of Lupines. On the same leaf the shorter leaflets, which generally face the centre of the plant, sink at night, whilst the longer ones on the opposite side rise; the intermediate and lateral ones merely twisting on their own axes. But there is some variability with respect to which leaflets rise or fall. As might have been expected from such diverse and complicated movements, the base of each leaflet is developed (at least in the case of L. luteus) into a pulvinus. The result is that all the leaflets on the same leaf stand at night more or less highly inclined, or even quite vertically, forming in this latter case a vertical star. This occurs with the leaves of a species purchased under the name of L. pubescens; and in the accompanying figures we see at A (Fig. 138) the leaves in their diurnal position; and at B the same plant at night with the two upper leaves having their leaflets almost vertical. At C another leaf, viewed laterally, is shown with the leaflets quite vertical. It is chiefly or exclusively the youngest leaves which form at night vertical stars. But there is much variability in the position of the leaves at night on the same plant; some remaining with their leaflets almost horizontal, others forming more or less highly inclined or vertical stars, and some with all their leaflets sloping downwards, as in our first class of cases. It is also a remarkable fact, that although all the plants produced from the same lot of seeds were identical in appearance, yet some individuals at night had the leaflets of all their leaves arranged so as to form more or less highly inclined stars; others had them all sloping downwards and never forming a star; and others, again, retained them either in a horizontal position or raised them a little.

Fig. 138. Lupinus pubescens: A, leaf viewed laterally during the day; B, same leaf at night; C, another leaf with the leaflet forming a vertical star at night. Figures reduced.

We have as yet referred only to the different positions of the leaflets of L. pubescens at night; but the petioles likewise differ in their movements. That of a young leaf which formed a highly inclined star at night, stood at noon at 42° above the horizon, and during the night at 72°, so had risen 30°. The petiole of another leaf, the leaflets of which occupied a similar position at night, rose only 6°. On the other hand, the petiole of a leaf with all its leaflets sloping down at night, fell at this time 4°. The petioles of two rather older leaves were subsequently observed; both of which stood during the day at exactly the same angle, viz., 50° above the horizon, and one of these rose 7°–8°, and the other fell 3°–4° at night. We meet with cases like that of L. pubescens with some other species. On a single plant of L. mutabilis some leaves, which stood horizontally during the day, formed highly inclined stars at night, and the petiole of one rose 7°. Other leaves which likewise stood horizontally during the day, had at night all their leaflets sloping downwards at 46° beneath the horizon, but their petioles had hardly moved. Again, L. luteus offered a still more remarkable case, for on two leaves, the leaflets which stood at noon at about 45° above the horizon, rose at night to 65° and 69°, so that they formed a hollow cone with steep sides. Four leaves on the same plant, which had their leaflets horizontal at noon, formed vertical stars at night; and three other leaves equally horizontal at noon, had all their leaflets sloping downwards at night. So that the leaves on this one plant assumed at night three different positions. Though we cannot account for this fact, we can see that such a stock might readily give birth to species having widely different nyctitropic habits.

Little more need be said about the sleep of the species of Lupinus; several, namely, L. polyphyllus, nanus, Menziesii, speciosus, and albifrons, though observed out of doors and in the greenhouse, did not change the position of their leaves sufficiently at night to be said to sleep. From observations made on two sleeping species, it appears that, as with Tropaeolum majus, the leaves must be well illuminated during the day in order to sleep at night. For several plants, kept all day in a sitting-room with north-east windows, did not sleep at night; but when the pots were placed on the following day out of doors, and were brought in at night, they slept in the usual manner. the trial was repeated on the following day and night with the same result.

Some observations were made on the circumnutation of the leaves of L. luteus and arboreus. It will suffice to say that the leaflets of the latter exhibited a double oscillation in the course of 24 h.; for they fell from the early morning until 10.15 A.M., then rose and zigzagged greatly till 4 P.M., after which hour the great nocturnal fall commenced. By 8 A.M. on the following morning the leaflets had risen to their proper height. We have seen in the fourth chapter, that the leaves of Lupinus speciosus, which do not sleep, circumnutate to an extraordinary extent, making many ellipses in the course of the day.

Cytisus (Tribe 2), Trigonella and Medicago (Tribe 3).—Only a few observations were made on these three genera. The petioles on a young plant, about a foot in height, of Cytisus fragrans rose at night, on one occasion 23° and on another 33°. The three leaflets also bend upwards, and at the same time approach each other, so that the base of the central leaflet overlaps the bases of the two lateral leaflets. They bend up so much that they press against the stem; and on looking down on one of these young plants from vertically above, the lower surfaces of the leaflets are visible; and thus their upper surfaces, in accordance with the general rule, are best protected from radiation. Whilst the leaves on these young plants were thus behaving, those on an old bush in full flower did not sleep at night.

Fig. 139. Medicago marina: A, leaves during the day; B, leaves asleep at night.

Trigonella Cretica resembles a Melilotus in its sleep, which will be immediately described. According to M. Royer,[9] the leaves of Medicago maculata rise up at night, and "se renversent un peu de manière à presenter obliquement au ciel leur face inférieure." A drawing is here given (Fig. 139) of the leaves of M. marina awake and asleep; and this would almost serve for Cytisus fragrans in the same two states.

[9] 'Annales des Sc. Nat. Bot.' (5th series), ix. 1868, p. 368.

Melilotus (Tribe 3).—The species in this genus sleep in a remarkable manner. The three leaflets of each leaf twist through an angle of 90°, so that their blades stand vertically at night with one lateral edge presented to the zenith (Fig. 140). We shall best understand the other and more complicated movements, if we imagine ourselves always to hold the leaf with the tip of the terminal leaflet pointed to the north. The leaflets in becoming vertical at night could of course twist so that their upper surfaces should face to either side; but the two lateral leaflets always twist so that this surface tends to face the north, but as they move at the same time towards the terminal leaflet, the upper surface of the one faces about N.N.W., and that of the other N.N.E. The terminal leaflet behaves differently, for it twists to either side, the upper surface facing sometimes east and sometimes west, but rather more commonly west than east. The terminal leaflet also moves in another and more remarkable manner, for whilst its blade is twisting and becoming vertical, the whole leaflet bends to one side, and invariably to the side towards which the upper surface is directed; so that if this surface faces the west the whole leaflet bends to the west, until it comes into contact with the upper and vertical surface of the western lateral leaflet. Thus the upper surface of the terminal and of one of the two lateral leaflets is well protected.

The fact of the terminal leaflet twisting indifferently to either side and afterwards bending to the same side, seemed to us so remarkable, that we endeavoured to discover the cause. We imagined that at the commencement of the movement it might be determined by one of the two halves of the leaflet being a little heavier than the other. Therefore bits of wood were gummed on one side of several leaflets, but this produced no effect; and they continued to twist in the same direction as they had previously done. In order to discover whether the same leaflet twisted permanently in the same direction, black threads were tied to 20 leaves, the terminal leaflets of which twisted so that their upper surfaces faced west, and 14 white threads to leaflets which twisted to the east. These were observed occasionally during 14 days, and they all continued, with a single exception, to twist and bend in the same direction; for one leaflet, which had originally faced east, was observed after 9 days to face west. The seat of both the twisting and bending movement is in the pulvinus of the sub-petioles.

Fig. 140. Melilotus officinalis: A, leaf during the daytime. B, another leaf asleep. C, a leaf asleep as viewed from vertically above; but in this case the terminal leaflet did not happen to be in such close contact with the lateral one, as is usual.

We believe that the leaflets, especially the two lateral ones, in performing the above described complicated movements generally bend a little downwards; but we are not sure of this, for, as far as the main petiole is concerned, its nocturnal movement is largely determined by the position which the leaf happens to occupy during the day. Thus one main petiole was observed to rise at night 59°, whilst three others rose only 7° and 9°. The petioles and sub-petioles are continually circumnutating during the whole 24 h., as we shall presently see.

The leaves of the following 15 species, M. officinalis, suaveolens, parviflora, alba, infesta, dentata, gracilis, sulcata, elegans, coerulea, petitpierreana, macrorrhiza, Italica, secundiflora, and Taurica, sleep in nearly the same manner as just described; but the bending to one side of the terminal leaflet is apt to fail unless the plants are growing vigorously. With M. petitpierreana and secundiflora the terminal leaflet was rarely seen to bend to one side. In young plants of M. Italica it bent in the usual manner, but with old plants in full flower, growing in the same pot and observed at the same hour, viz., 8.30 P.M., none of the terminal leaflets on several scores of leaves had bent to one side, though they stood vertically; nor had the two lateral leaflets, though standing vertically, moved towards the terminal one. At 10.30 P.M., and again one hour after midnight, the terminal leaflets had become very slightly bent to one side, and the lateral leaflets had moved a very little towards the terminal one, so that the position of the leaflets even at this late hour was far from the ordinary one. Again, with M. Taurica the terminal leaflets were never seen to bend towards either of the two lateral leaflets, though these, whilst becoming vertical, had bent towards the terminal one. The sub-petiole of the terminal leaflet in this species is of unusual length, and if the leaflet had bent to one side, its upper surface could have come into contact only with the apex of either lateral leaflet; and this, perhaps, is the meaning of the loss of the lateral movement.

The cotyledons do not sleep at night. the first leaf consists of a single orbicular leaflet, which twists at night so that the blade stands vertically. It is a remarkable fact that with M. Taurica, and in a somewhat less degree with M. macrorrhiza and petitpierreana, all the many small and young leaves produced during the early spring from shoots on some cut-down plants in the greenhouse, slept in a totally different manner from the normal one; for the three leaflets, instead of twisting on their own axes so as to present their lateral edges to the zenith, turned upwards and stood vertically with their apices pointing to the zenith. They thus assumed nearly the same position as in the allied genus Trifolium; and on the same principle that embryological characters reveal the lines of descent in the animal kingdom, so the movements of the small leaves in the above three species of Melilotus, perhaps indicate that this genus is descended from a form which was closely allied to and slept like a Trifolium. Moreover, there is one species, M. messanensis, the leaves of which, on full-grown plants between 2 and 3 feet in height, sleep like the foregoing small leaves and like those of a Trifolium. We were so much surprised at this latter case that, until the flowers and fruit were examined, we thought that the seeds of some Trifolium had been sown by mistake instead of those of a Melilotus. It appears therefore probable that M. messanensis has either retained or recovered a primordial habit.

The circumnutation of a leaf of M. officinalis was traced, the stem being left free; and the apex of the terminal leaflet described three laterally extended ellipses, between 8 A.M. and 4 P.M.; after the latter hour the nocturnal twisting movement commenced. It was afterwards ascertained that the above movement was compounded of the circumnutation of the stem on a small scale, of the main petiole which moved most, and of the sub-petiole of the terminal leaflet. The main petiole of a leaf having been secured to a stick, close to the base of the sub-petiole of the terminal leaflet, the latter described two small ellipses between 10.30 A.M., and 2 P.M. At 7.15 P.M., after this same leaflet (as well as another) had twisted themselves into their vertical nocturnal position, they began to rise slowly, and continued to do so until 10.35 P.M., after which hour they were no longer observed.

As M. messanensis sleeps in an anomalous manner, unlike that of any other species in the genus, the circumnutation of a terminal leaflet, with the stem secured, was traced during two days. On each morning the leaflet fell, until about noon, and then began to rise very slowly; but on the first day the rising movement was interrupted between 1 and 3 P.M. by the formation of a laterally extended ellipse, and on the second day, at the same time, by two smaller ellipses. The rising movement then recommenced, and became rapid late in the evening, when the leaflet was beginning to go to sleep. The awaking or sinking movement had already commenced by 6.45 A.M. on both mornings.

Trifolium (Tribe 3).—The nyctitropic movements of 11 species were observed, and were found to be closely similar. If we select a leaf of T. repens having an upright petiole, and with the three leaflets expanded horizontally, the two lateral leaflets will be seen in the evening to twist and approach each other, until their upper surfaces come into contact. At the same time they bend downwards in a plane at right angles to that of their former position, until their midribs form an angle of about 45° with the upper part of the petiole. This peculiar change of position requires a considerable amount of torsion in the pulvinus. The terminal leaflet merely rises up without any twisting and bends over until it rests on and forms a roof over the edges of the now vertical and united lateral leaflets. Thus the terminal leaflet always passes through an angle of at least 90°, generally of 130° or 140°, and not rarely—as was often observed with T. subterraneum—of 180°. In this latter case the terminal leaflet stands at night horizontally (as in Fig. 141), with its lower surface fully exposed to the zenith. Besides the difference in the angles, at which the terminal leaflets stand at night in the individuals of the same species, the degree to which the lateral leaflets approach each other often likewise differs.

Fig. 141. Trifolium repens: A, leaf during the day; B, leaf asleep at night.

We have seen that the cotyledons of some species and not of others rise up vertically at night. The first true leaf is generally unifoliate and orbicular; it always rises, and either stands vertically at night or more commonly bends a little over so as to expose the lower surface obliquely to the zenith, in the same manner as does the terminal leaflet of the mature leaf. But it does not twist itself like the corresponding first simple leaf of Melilotus. With T. Pannonicum the first true leaf was generally unifoliate, but sometimes trifoliate, or again partially lobed and in an intermediate condition.

Circumnutation.—Sachs described in 1863[10] the spontaneous up and down movements of the leaflets of T. incarnatum, when kept in darkness. Pfeffer made many observations on the similar movements in T. pratense.[11] He states that the terminal leaflet of this species, observed at different times, passed through angles of from 30° to 120° in the course of from 1½ to 4 h. We observed the movements of T. subterraneum, resupinatum, and repens.

[10] 'Flora,' 1863, p. 497.

[11] 'Die Period. Bewegungen,' 1875, pp. 35, 52.

Trifolium subterraneum.—A petiole was secured close to the base of the three leaflets, and the movement of the terminal leaflet was traced during 26½ h., as shown in the figure on the next page.

Between 6.45 A.M. and 6 P.M. the apex moved 3 times up and 3 times down, completing 3 ellipses in 11 h. 15 m. The ascending and descending lines stand nearer to one another than is usual with most plants, yet there was some lateral motion. At 6 P.M. the great nocturnal rise commenced, and on the next morning the sinking of the leaflet was continued until 8.30 A.M., after which hour it circumnutated in the manner just described. In the figure the great nocturnal rise and the morning fall are greatly abbreviated, from the want of space, and are merely represented by a short curved line. The leaflet stood horizontally when at a point a little beneath the middle of the diagram; so that during the daytime it oscillated almost equally above and beneath a horizontal position. At 8.30 A.M. it stood 48° beneath the horizon, and by 11.30 A.M. it had risen 50° above the horizon; so that it passed through 98° in 3 h. By the aid of the tracing we ascertained that the distance travelled in the 3 h. by the apex of this leaflet was 1.03 inch. If we look at the figure, and prolong upwards in our mind's eye the short curved broken line, which represents the nocturnal course, we see that the latter movement is merely an exaggeration or prolongation of one of the diurnal ellipses. The same leaflet had been observed on the previous day, and the course then pursued was almost identically the same as that here described.

Fig. 142. Trifolium subterraneum: circumnutation and nyctitropic movement of terminal leaflet (.68 inch in length), traced from 6.45 A.M. July 4th to 9.15 A.M. 5th. Apex of leaf 3 7/8 inches from the vertical glass, and movement, as here shown, magnified 5 1/4 times, reduced to one-half of original scale. Plant illuminated from above; temp. 16°–17° C.

Trifolium resupinatum.—A plant left entirely free was placed before a north-east window, in such a position that a terminal leaflet projected at right angles to the source of the light, the sky being uniformly clouded all day. The movements of this leaflet were traced during two days, and on both were closely similar. Those executed on the second day are shown in Fig. 143. The obliquity of the several lines is due partly to the manner in which the leaflet was viewed, and partly to its having moved a little towards the light. From 7.50 A.M. to 8.40 A.M. the leaflet fell, that is, the awakening movement was continued. It then rose and moved a little laterally towards the light. At 12.30 it retrograded, and at 2.30 resumed its original course, having thus completed a small ellipse during the middle of the day. In the evening it rose rapidly, and by 8 A.M. on the following morning had returned to exactly the same spot as on the previous morning. The line representing the nocturnal course ought to be extended much higher up, and is here abbreviated into a short, curved, broken line. The terminal leaflet, therefore, of this species described during the daytime only a single additional ellipse, instead of two additional ones, as in the case of T. subterraneum. But we should remember that it was shown in the fourth chapter that the stem circumnutates, as no doubt does the main petiole and the sub-petioles; so that the movement represented in Fig. 143 is a compounded one. We tried to observe the movements of a leaf kept during the day in darkness, but it began to go to sleep after 2 h. 15 m., and this was well pronounced after 4 h. 30 m.

Fig 143. Trifolium resupinatum: circumnutation and nyctitropic movements of the terminal leaflet during 24 hours.

Trifolium repens.—A stem was secured close to the base of a moderately old leaf, and the movement of the terminal leaflet was observed during two days. This case is interesting solely from the simplicity of the movements, in contrast with those of the two preceding species. On the first day the leaflet fell between 8 A.M. and 3 P.M., and on the second between 7 A.M. and 1 P.M. On both days the descending course was somewhat zigzag, and this evidently represents the circumnutating movement of the two previous species during the middle of the day. After 1 P.M., Oct. 1st (Fig. 144), the leaflet began to rise, but the movement was slow on both days, both before and after this hour, until 4 P.M. The rapid evening and nocturnal rise then commenced. Thus in this species the course during 24 h. consists of a single great ellipse; in T. resupinatum of two ellipses, one of which includes the nocturnal movement and is much elongated; and in T. subterraneum of three ellipses, of which the nocturnal one is likewise of great length.

Securigera coronilla (Tribe 4).—The leaflets, which stand opposite one another and are numerous, rise up at night, come into close contact, and bend backwards at a moderate angle towards the base of the petiole.

Fig. 144. Trifolium repens: circumnutation and nyctitropic movements of a nearly full-grown terminal leaflet, traced on a vertical glass from 7 A.M. Sept. 30th to 8 A.M. Oct. 1st. Nocturnal course, represented by curved broken line, much abbreviated.

Lotus (Tribe 4).—The nyctitropic movements of 10 species in this genus were observed, and found to be the same. The main petiole rises a little at night, and the three leaflets rise till they become vertical, and at the same time approach each other. This was conspicuous with L. Jacoboeus, in which the leaflets are almost linear. In most of the species the leaflets rise so much as to press against the stem, and not rarely they become inclined a little inwards with their lower surfaces exposed obliquely to the zenith. This was clearly the case with L. major, as its petioles are unusually long, and the leaflets are thus enabled to bend further inwards. The young leaves on the summits of the stems close up at night so much, as often to resemble large buds. The stipule-like leaflets, which are often of large size, rise up like the other leaflets, and press against the stem (Fig. 145). All the leaflets of L. Gebelii, and probably of the other species, are provided at their bases with distinct pulvini, of a yellowish colour, and formed of very small cells. The circumnutation of a terminal leaflet of L. peregrinus (with the stem secured) was traced during two days, but the movement was so simple that it is not worth while to give the diagram. The leaflet fell slowly from the early morning till about 1 P.M. It then rose gradually at first, but rapidly late in the evening. It occasionally stood still for about 20 m. during the day, and sometimes zigzagged a little. The movement of one of the basal, stipule-like leaflets was likewise traced in the same manner and at the same time, and its course was closely similar to that of the terminal leaflet.

Fig. 145. Lotus Creticus: A, stem with leaves awake during the day; B, with leaves asleep at night. SS, stipule-like leaflets.

In Tribe 5 of Bentham and Hooker, the sleep-movements of species in 12 genera have been observed by ourselves and others, but only in Robinia with any care. Psoralea acaulis raises its three leaflets at night; whilst Amorpha fruticosa,[12] Dalea alopecuroides, and Indigofera tinctoria depress them. Ducharte[13] states that Tephrosia caribaea is the sole example of "folioles couchées le long du pétiole et vers la base;" but a similar movement occurs, as we have already seen, and shall again see in other cases. Wistaria Sinensis, according to Royer,[14] "abaisse les folioles qui par une disposition bizarre sont inclinées dans la même feuille, les supérieures vers le sommet, les inférieures vers la base du petiole commun;" but the leaflets on a young plant observed by us in the greenhouse merely sank vertically downwards at night. The leaflets are raised in Sphaerophysa salsola, Colutea arborea, and Astragalus uliginosus, but are depressed, according to Linnæus, in Glycyrrhiza. The leaflets of Robinia pseudo-acacia likewise sink vertically down at night, but the petioles rise a little, viz., in one case 3°, and in another 4°. The circumnutating movements of a terminal leaflet on a rather old leaf were traced during two days, and were simple. The leaflet fell slowly, in a slightly zigzag line, from 8 A.M. to 5 P.M., and then more rapidly; by 7 A.M. on the following morning it had risen to its diurnal position. There was only one peculiarity in the movement, namely, that on both days there was a distinct though small oscillation up and down between 8.30 and 10 A.M., and this would probably have been more strongly pronounced if the leaf had been younger.

[12] Ducharte, 'Eléments de Botanique', 1867, p. 349.

[13] Ibid., p. 347.

[14] 'Ann. des Sciences Nats. Bot.' (5th series), ix. 1868.

Coronilla rosea (Tribe 6).—the leaves bear 9 or 10 pairs of opposite leaflets, which during the day stand horizontally, with their midribs at right angles to the petiole. At night they rise up so that the opposite leaflets come nearly into contact, and those on the younger leaves into close contact. At the same time they bend back towards the base of the petiole, until their midribs form with it angles of from 40° to 50° in a vertical plane, as here figured (Fig. 146). The leaflets, however, sometimes bend so much back that their midribs become parallel to and lie on the petiole. They thus occupy a reversed position to what they do in several Leguminosae, for instance, in Mimosa pudica; but, from standing further apart, they do not overlap one another nearly so much as in this latter plant. The main petiole is curved slightly downwards during the day, but straightens itself at night. In three cases it rose from 3° above the horizon at noon, to 9° at 10 P.M.; from 11° to 33°; and from 5° to 33°—the amount of angular movement in this latter case amounting to 28°. In several other species of Coronilla the leaflets showed only feeble movements of a similar kind.

Fig. 146. Coronilla rosea: leaf asleep.

Hedysarum coronarium (Tribe 6).—The small lateral leaflets on plants growing out of doors rose up vertically at night, but the large terminal one became only moderately inclined. The petioles apparently did not rise at all.

Smithia Pfundii (Tribe 6).—The leaflets rise up vertically, and the main petiole also rises considerably.

Arachis hypogoea (Tribe 6).—The shape of a leaf, with its two pairs of leaflets, is shown at A (Fig. 147); and a leaf asleep, traced from a photograph (made by the aid of aluminium light), is given at B. The two terminal leaflets twist round at night until their blades stand vertically, and approach each other until they meet, at the same time moving a little upwards and backwards. The two lateral leaflets meet each other in this same manner, but move to a greater extent forwards, that is, in a contrary direction to the two terminal leaflets, which they partially embrace. Thus all four leaflets form together a single packet, with their edges directed to the zenith, and with their lower surfaces turned outwards. On a plant which was not growing vigorously the closed leaflets seemed too heavy for the petioles to support them in a vertical position, so that each night the main petiole became twisted, and all the packets were extended horizontally, with the lower surfaces of the leaflets on one side directed to the zenith in a most anomalous manner. This fact is mentioned solely as a caution, as it surprised us greatly, until we discovered that it was an anomaly. The petioles are inclined upwards during the day, but sink at night, so as to stand at about right angles with the stem. The amount of sinking was measured only on one occasion, and found to be 39°. A petiole was secured to a stick at the base of the two terminal leaflets, and the circumnutating movement of one of these leaflets was traced from 6.40 A.M. to 10.40 P.M., the plant being illuminated from above. The temperature was 17°–17½° C., and therefore rather too low. During the 16 h. the leaflet moved thrice up and thrice down, and as the ascending and descending lines did not coincide, three ellipses were formed.

Fig. 147. Arachis hypogoea: A, leaf during the day, seen from vertically above; B, leaf asleep, seen laterally, copied from a photograph. Figures much reduced.

Fig. 148. Desmodium gyrans: leaf seen from above, reduced to one-half natural size. The minute stipules unusually large.

Desmodium gyrans (Tribe 6).—A large and full-grown leaf of this plant, so famous for the spontaneous movements of the two little lateral leaflets, is here represented (Fig. 148). The large terminal leaflet sleeps by sinking vertically down, whilst the petiole rises up. The cotyledons do not sleep, but the first-formed leaf sleeps equally well as the older ones. The appearance presented by a sleeping branch and one in the day-time, copied from two photographs, are shown at A and B (Fig. 149), and we see how at night the leaves are crowded together, as if for mutual protection, by the rising of the petioles. The petioles of the younger leaves near the summits of the shoots rise up at night, so as to stand vertical and parallel to the stem; whilst those on the sides were found in four cases to have risen respectively 46½°, 36°, 20°, and 19.5° above the inclined positions which they had occupied during the day. For instance, in the first of these four cases the petiole stood in the day at 23°, and at night at 69½° above the horizon. In the evening the rising of the petioles is almost completed before the leaflets sink perpendicularly downwards.

Circumnutation.—The circumnutating movements of four young shoots were observed during 5 h. 15 m.; and in this time each completed an oval figure of small size. The main petiole also circumnutates rapidly, for in the course of 31 m. (temp. 91° F.) it changed its course by as much as a rectangle six times, describing a figure which apparently represented two ellipses. The movement of the terminal leaflet by means of its sub-petiole or pulvinus is quite as rapid, or even more so, than that of the main petiole, and has much greater amplitude. Pfeffer has seen[15] these leaflets move through an angle of 8° in the course of from 10 to 30 seconds.

[15] 'Die Period. Beweg.,' p. 35.

Fig. 149. Desmodium gyrans: A, stem during the day; B, stem with leaves asleep. Figures reduced.

A fine, nearly full-grown leaf on a young plant, 8 inches in height, with the stem secured to a stick at the base of the leaf, was observed from 8.30 A.M. June 22nd to 8 A.M. June 24th. In the diagram given on the next page (Fig. 150), the two curved broken lines at the base, which represent the nocturnal courses, ought to be prolonged far downwards. On the first day the leaflet moved thrice down and thrice up, and to a considerable distance laterally; the course was also remarkably crooked. The dots were generally made every hour; if they had been made every few minutes all the lines would have been zigzag to an extraordinary degree, with here and there a loop formed. We may infer that this would have been the case, because five dots were made in the course of 31 m. (between 12.34 and 1.5 P.M.), and we see in the upper part of the diagram how crooked the course here is; if only the first and last dots had been joined we should have had a straight line. Exactly the same fact may be seen in the lines representing the course between 2.24 P.M. and 3 P.M., when six intermediate dots were made; and again at 4.46 and 4.50. But the result was widely different after 6 P.M.,—that is, after the great nocturnal descent had commenced; for though nine dots were then made in the course of 32 m., when these were joined (see Figure) the line thus formed was almost straight. The leaflets, therefore, begin to descend in the afternoon by zigzag lines, but as soon as the descent becomes rapid their whole energy is expended in thus moving, and their course becomes rectilinear. After the leaflets are completely asleep they move very little or not at all.

Fig. 150. Desmodium gyrans: circumnutation and nyctitropic movement of leaf (3 3/4 inches in length, petiole included) during 48 h. Filament affixed to midrib of terminal leaflet; its apex 6 inches from the vertical glass. Diagram reduced to one-third of original scale. Plant illuminated from above. Temp. 19°–20° C.

Had the above plant been subjected to a higher temperature than 67°–70° F., the movements of the terminal leaflet would probably have been even more rapid and wider in extent than those shown in the diagram; for a plant was kept for some time in the hot-house at from 92°–93° F., and in the course of 35 m. the apex of a leaflet twice descended and once ascended, travelling over a space of 1.2 inch in a vertical direction and of .82 inch in a horizontal direction. Whilst thus moving the leaflet also rotated on its own axis (and this was a point to which no attention had been before paid), for the plane of the blade differed by 41° after an interval of only a few minutes. Occasionally the leaflet stood still for a short time. There was no jerking movement, which is so characteristic of the little lateral leaflets. A sudden and considerable fall of temperature causes the terminal leaflet to sink downwards; thus a cut-off leaf was immersed in water at 95° F., which was slowly raised to 103° F., and afterwards allowed to sink to 70° F., and the sub-petiole of the terminal leaflet then curved downwards. The water was afterwards raised to 120° F., and the sub-petiole straightened itself. Similar experiments with leaves in water were twice repeated, with nearly the same result. It should be added, that water raised to even 122° F. does not soon kill a leaf. A plant was placed in darkness at 8.37 A.M., and at 2 P.M. (i.e. after 5 h. 23 m.), though the leaflets had sunk considerably, they had by no means acquired their nocturnal vertically dependent position. Pfeffer, on the other hand, says[16] that this occurred with him in from 3/4 h. to 2 h.; perhaps the difference in our results may be due to the plant on which we experimented being a very young and vigorous seedling.

[16] 'Die Period. Beweg.,' p. 39.

The Movements of the little Lateral Leaflets .—These have been so often described, that we will endeavour to be as brief as possible in giving a few new facts and conclusions. The leaflets sometimes quickly change their position by as much as nearly 180°; and their sub-petioles can then be seen to become greatly curved. They rotate on their own axes, so that their upper surfaces are directed to all points of the compass. The figure described by the apex is an irregular oval or ellipse. They sometimes remain stationary for a period. In these several respects there is no difference, except in rapidity and extent, between their movements and the lesser ones performed by the large terminal leaflet whilst making its great oscillations. The movements of the little leaflets are much influenced, as is well known, by temperature. This was clearly shown by immersing leaves with motionless leaflets in cold water, which was slowly raised to 103° F., and the leaflets then moved quickly, describing about a dozen little irregular circles in 40 m. By this time the water had become much cooler, and the movements became slower or almost ceased; it was then raised to 100° F., and the leaflets again began to move quickly. On another occasion a tuft of fine leaves was immersed in water at 53° F., and the leaflets were of course motionless. The water was raised to 99°, and the leaflets soon began to move; it was raised to 105°, and the movements became much more rapid; each little circle or oval being completed in from 1 m. 30 s. to 1 m. 45 s. There was, however, no jerking, and this fact may perhaps be attributed to the resistance of the water.

Sachs states that the leaflets do not move until the surrounding air is as high as 71°–72° F., and this agrees with our experience on full-grown, or nearly full-grown, plants. But the leaflets of young seedlings exhibit a jerking movement at much lower temperatures. A seedling was kept (April 16th) in a room for half the day where the temperature was steady at 64° F., and the one leaflet which it bore was continually jerking, but not so rapidly as in the hot-house. The pot was taken in the evening into a bed-room where the temperature remained at 62° during nearly the whole night; at 10 and 11 P.M. and at 1 A.M. the leaflet was still jerking rapidly; at 3.30 A.M. it was not seen to jerk, but was observed during only a short time. It was, however, now inclined at a much lower angle than that occupied at 1 A.M. At 6.30 A.M. (temp. 61° F.) its inclination was still less than before, and again less at 6.45 A.M.; by 7.40 A.M. it had risen, and at 8.30 A.M. was again seen to jerk. This leaflet, therefore, was moving during the whole night, and the movement was by jerks up to 1 A.M. (and possibly later) and again at 8.30 A.M., though the temperature was only 61° to 62° F. We must therefore conclude that the lateral leaflets produced by young plants differ somewhat in constitution from those on older plants.

In the large genus Desmodium by far the greater number of the species are trifoliate; but some are unifoliate, and even the same plant may bear uni- and trifoliate leaves. In most of the species the lateral leaflets are only a little smaller than the terminal one. Therefore the lateral leaflets of D. gyrans (see Fig. 148) must be considered as almost rudimentary. They are also rudimentary in function, if this expression may be used; for they certainly do not sleep like the full-sized terminal leaflets. It is, however, possible that the sinking down of the leaflets between 1 A.M. and 6.45 A.M., as above described, may represent sleep. It is well known that the leaflets go on jerking during the early part of the night; but my gardener observed (Oct. 13th) a plant in the hot-house between 5 and 5.30 A.M., the temperature having been kept up to 82° F., and found that all the leaflets were inclined, but he saw no jerking movement until 6.55 A.M., by which time the terminal leaflet had risen and was awake. Two days afterwards (Oct. 15th) the same plant was observed by him at 4.47 A.M. (temp. 77° F.), and he found that the large terminal leaflets were awake, though not quite horizontal; and the only cause which we could assign for this anomalous wakefulness was that the plant had been kept for experimental purposes during the previous day at an unusually high temperature; the little lateral leaflets were also jerking at this hour, but whether there was any connection between this latter fact and the sub-horizontal position of the terminal leaflets we do not know. Anyhow, it is certain that the lateral leaflets do not sleep like the terminal leaflets; and in so far they may be said to be in a functionally rudimentary condition. They are in a similar condition in relation to irritability; for if a plant be shaken or syringed, the terminal leaflets sink down to about 45° beneath the horizon; but we could never detect any effect thus produced on the lateral leaflets; yet we are not prepared to assert positively that rubbing or pricking the pulvinus produces no effect.

As in the case of most rudimentary organs, the leaflets are variable in size; they often depart from their normal position and do not stand opposite one another; and one of the two is frequently absent. This absence appeared in some, but not in all the cases, to be due to the leaflet having become completely confluent with the main petiole, as might be inferred from the presence of a slight ridge along its upper margin, and from the course of the vessels. In one instance there was a vestige of the leaflet, in the shape of a minute point, at the further end of the ridge. The frequent, sudden and complete disappearance of one or both of the rudimentary leaflets is a rather singular fact; but it is a much more surprising one that the leaves which are first developed on seedling plants are not provided with them. Thus, on one seedling the seventh leaf above the cotyledons was the first which bore any lateral leaflets, and then only a single one. On another seedling, the eleventh leaf first bore a leaflet; of the nine succeeding leaves five bore a single lateral leaflet, and four bore none at all; at last a leaf, the twenty-first above the cotyledons, was provided with two rudimentary lateral leaflets. From a widespread analogy in the animal kingdom, it might have been expected that these rudimentary leaflets would have been better developed and more regularly present on very young than on older plants. But bearing in mind, firstly, that long-lost characters sometimes reappear late in life, and secondly, that the species of Desmodium are generally trifoliate, but that some are unifoliate, the suspicion arises that D. gyrans is descended from a unifoliate species, and that this was descended from a trifoliate one; for in this case both the absence of the little lateral leaflets on very young seedlings, and their subsequent appearance, may be attributed to reversion to more or less distant progenitors.[17]

[17] Desmodium vespertilionis is closely allied to D. gyrans, and it seems only occasionally to bear rudimentary lateral leaflets. Duchartre, 'Eléments de Botanique,' 1867, p. 353.

No one supposes that the rapid movements of the lateral leaflets of 'D. gyrans' are of any use to the plant; and why they should behave in this manner is quite unknown. We imagined that their power of movement might stand in some relation with their rudimentary condition, and therefore observed the almost rudimentary leaflets of Mimosa albida vel sensitiva (of which a drawing will hereafter be given, Fig. 159); but they exhibited no extraordinary movements, and at night they went to sleep like the full-sized leaflets. There is, however, this remarkable difference in the two cases; in Desmodium the pulvinus of the rudimentary leaflets has not been reduced in length, in correspondence with the reduction of the blade, to the same extent as has occurred in the Mimosa; and it is on the length and degree of curvature of the pulvinus that the amount of movement of the blade depends. Thus the average length of the pulvinus in the large terminal leaflets of Desmodium is 3 mm., whilst that of the rudimentary leaflets is 2.86 mm.; so that they differ only a little in length. But in diameter they differ much, that of the pulvinus of the little leaflets being only 0.3 mm. to 0.4 mm.; whilst that of the terminal leaflets is 1.33 mm. If we now turn to the Mimosa, we find that the average length of the pulvinus of the almost rudimentary leaflets is only 0.466 mm., or rather more than a quarter of the length of the pulvinus of the full-sized leaflets, namely, 1.66 mm. In this small reduction in length of the pulvinus of the rudimentary leaflets of Desmodium, we apparently have the proximate cause of their great and rapid circumnutating movement, in contrast with that of the almost rudimentary leaflets of the Mimosa. The small size and weight of the blade, and the little resistance opposed by the air to its movement, no doubt also come into play; for we have seen that these leaflets if immersed in water, when the resistance would be much greater, were prevented from jerking forwards. Why, during the reduction of the lateral leaflets of Desmodium, or during their reappearance—if they owe their origin to reversion—the pulvinus should have been so much less affected than the blade, whilst with the Mimosa the pulvinus has been greatly reduced, we do not know. Nevertheless, it deserves notice that the reduction of the leaflets in these two genera has apparently been effected by a different process and for a different end; for with the Mimosa the reduction of the inner and basal leaflets was necessary from the want of space; but no such necessity exists with Desmodium, and the reduction of its lateral leaflets seems to have been due to the principle of compensation, in consequence of the great size of the terminal leaflet. Uraria (Tribe 6) and Centrosema (Tribe 8).—The leaflets of Uraria lagopus and the leaves of a Centrosema from Brazil both sink vertically down at night. In the latter plant the petiole at the same time rose 16½°.

Amphicarpoea monoica (Tribe 8).—The leaflets sink down vertically at night, and the petioles likewise fall considerably. A petiole, which was carefully observed, stood during the day 25° above the horizon and at night 32° below it; it therefore fell 57°. A filament was fixed transversely across the terminal leaflet of a fine young leaf (2 1/4 inches in length including the petiole), and the movement of the whole leaf was traced on a vertical glass. This was a bad plan in some respects, because the rotation of the leaflet, independently of its rising or falling, raised and depressed the filament; but it was the best plan for our special purpose of observing whether the leaf moved much after it had gone to sleep. The plant had twined closely round a thin stick, so that the circumnutation of the stem was prevented. The movement of the leaf was traced during 48 h., from 9 A.M. July 10th to 9 A.M. July 12th. In the figure given (Fig. 151) we see how complicated its course was on both days: during the second day it changed its course greatly 13 times. The leaflets began to go to sleep a little after 6 P.M., and by 7.15 P.M. hung vertically down and were completely asleep; but on both nights they continued to move from 7.15 P.M. to 10.40 and 10.50 P.M., quite as much as during the day; and this was the point which we wished to ascertain. We see in the figure that the great sinking movement late in the evening does not differ essentially from the circumnutation during the day.

Fig. 151. Amphicarpoea monoica: circumnutation and nyctitropic movement of leaf during 48 h.; its apex 9 inches from the vertical glass. Figure reduced to one-third of original scale. Plant illuminated from above; temp 17½°–18½° C.

Glycine hispida (Tribe 8).—The three leaflets sink vertically down at night.

Erythrina (Tribe 8).—Five species were observed, and the leaflets of all sank vertically down at night; with E. caffra and with a second unnamed species, the petioles at the same time rose slightly. The movements of the terminal leaflet of E. crista-galli (with the main petiole secured to a stick) were traced from 6.40 A.M. June 8th, to 8 A.M. on the 10th. In order to observe the nyctitropic movements of this plant, it is necessary that it should have grown in a warm greenhouse, for out of doors in our climate it does not sleep. We see in the tracing (Fig. 152) that the leaflet oscillated twice up and down between early morning and noon; it then fell greatly, afterwards rising till 3 P.M. At this latter hour the great nocturnal fall commenced. On the second day (of which the tracing is not given) there was exactly the same double oscillation before noon, but only a very small one in the afternoon. On the third morning the leaflet moved laterally, which was due to its beginning to assume an oblique position, as seems invariably to occur with the leaflets of this species as they grow old. On both nights after the leaflets were asleep and hung vertically down, they continued to move a little both up and down, and from side to side.

Erythrina caffra.—A filament was fixed transversely across a terminal leaflet, as we wished to observe its movements when asleep. The plant was placed in the morning of June 10th under a skylight, where the light was not bright; and we do not know whether it was owing to this cause or to the plant having been disturbed, but the leaflet hung vertically down all day; nevertheless it circumnutated in this position, describing a figure which represented two irregular ellipses. On the next day it circumnutated in a greater degree, describing four irregular ellipses, and by 3 P.M. had risen into a horizontal position. By 7.15 P.M. it was asleep and vertically dependent, but continued to circumnutate as long as observed, until 11 P.M.

Fig. 152. Erythrina crista-galli: circumnutation and nyctitropic movement of terminal leaflet, 3 3/4 inches in length, traced during 25 h.; apex of leaf 3½ inches from the vertical glass. Figure reduced to one-half of original scale. Plant illuminated from above; temp. 17½°–18½° C.

Erythrina corallodendron.—The movements of a terminal leaflet were traced. During the second day it oscillated four times up and four times down between 8 A.M. and 4 P.M., after which hour the great nocturnal fall commenced. On the third day the movement was equally great in amplitude, but was remarkably simple, for the leaflet rose in an almost perfectly straight line from 6.50 A.M. to 3 P.M., and then sank down in an equally straight line until vertically dependent and asleep.

Apios tuberosa (Tribe 8).—The leaflets sink vertically down at night.

Phaseolus vulgaris (Tribe 8).—The leaflets likewise sink vertically down at night. In the greenhouse the petiole of a young leaf rose 16°, and that of an older leaf 10° at night. With plants growing out of doors the leaflets apparently do not sleep until somewhat late in the season, for on the nights of July 11th and 12th none of them were asleep; whereas on the night of August 15th the same plants had most of their leaflets vertically dependent and asleep. With Ph. caracalla and Hernandesii, the primary unifoliate leaves and the leaflets of the secondary trifoliate leaves sink vertically down at night. This holds good with the secondary trifoliate leaves of Ph. Roxburghii, but it is remarkable that the primary unifoliate leaves which are much elongated, rise at night from about 20° to about 60° above the horizon. With older seedlings, however, having the secondary leaves just developed, the primary leaves stand in the middle of the day horizontally, or are deflected a little beneath the horizon. In one such case the primary leaves rose from 26° beneath the horizon at noon, to 20° above it at 10 P.M.; whilst at this same hour the leaflets of the secondary leaves were vertically dependent. Here, then, we have the extraordinary case of the primary and secondary leaves on the same plant moving at the same time in opposite directions.

We have now seen that the leaflets in the six genera of Phaseoleae observed by us (with the exception of the primary leaves of Phaseolus Roxburghii) all sleep in the same manner, namely, by sinking vertically down. The movements of the petioles were observed in only three of these genera. They rose in Centrosema and Phaseolus, and sunk in Amphicarpæa.

Sophora chrysophylla (Tribe 10).—The leaflets rise at night, and are at the same time directed towards the apex of the leaf, as in Mimosa pudica.

Caesalpinia, Hoematoxylon, Gleditschia, Poinciana.—The leaflets of two species of Caesalpinia (Tribe 13) rose at night. With Haematoxylon Campechianum (Tribe 13) the leaflets move forwards at night, so that their midribs stand parallel to the petiole, and their now vertical lower surfaces are turned outwards (Fig. 153). The petiole sinks a little. In Gleditschia, if we understand correctly Duchartre's description, and in Poinciana Gilliesii (both belonging to Tribe 13), the leaves behave in the same manner.

Fig. 153. Haematoxylon Campechianum: A, branch during daytime; B, branch with leaves asleep, reduced to two-thirds of natural scale.

Cassia (Tribe 14).—The nyctitropic movements of the leaves in many species in this genus are closely alike, and are highly complex. They were first briefly described by Linnæus, and since by Duchartre. Our observations were made chiefly on C. floribunda[18] and corymbosa, but several other species were casually observed. The horizontally extended leaflets sink down vertically at night; but not simply, as in so many other genera, for each leaflet rotates on its own axis, so that its lower surface faces outwards. The upper surfaces of the opposite leaflets are thus brought into contact with one another beneath the petiole, and are well protected (Fig. 154). The rotation and other movements are effected by means of a well-developed pulvinus at the base of each leaflet, as could be plainly seen when a straight narrow black line had been painted along it during the day. The two terminal leaflets in the daytime include rather less than a right angle; but their divergence increases greatly whilst they sink downwards and rotate, so that they stand laterally at night, as may be seen in the figure. Moreover, they move somewhat backwards, so as to point towards the base of the petiole. In one instance we found that the midrib of a terminal leaflet formed at night an angle of 36°, with a line dropped perpendicularly from the end of the petiole. The second pair of leaflets likewise moves a little backwards, but less than the terminal pair; and the third pair moves vertically downwards, or even a little forwards. Thus all the leaflets, in those species which bear only 3 or 4 pairs, tend to form a single packet, with their upper surfaces in contact, and their lower surfaces turned outwards. Lastly, the main petiole rises at night, but with leaves of different ages to very different degrees, namely some rose through an angle of only 12°, and others as much as 41°.

[18] I am informed by Mr. Dyer that Mr. Bentham believes that C. floribunda (a common greenhouse bush) is a hybrid raised in France, and that it comes very near to C. laevigata. It is no doubt the same as the form described by Lindley ('Bot. Reg.,' Tab. 1422) as C. Herbertiana.

Fig. 154. Cassia corymbosa: A, plant during day; B, same plant at night. Both figures copied from photographs.

Cassia calliantha.—The leaves bear a large number of leaflets, which move at night in nearly the same manner as just described; but the petioles apparently do not rise, and one which was carefully observed certainly fell 3°. Cassia pubescens.—The chief difference in the nyctitropic movements of this species, compared with those of the former species, consists in the leaflets not rotating nearly so much; therefore their lower surfaces face but little outwards at night. The petioles, which during the day are inclined only a little above the horizon, rise at night in a remarkable manner, and stand nearly or quite vertically. This, together with the dependent position of the leaflets, makes the whole plant wonderfully compact at night. In the two foregoing figures, copied from photographs, the same plant is represented awake and asleep (Fig. 155), and we see how different is its appearance.

Fig. 155. Cassia pubescens: A, upper part of plant during the day; B, same plant at night. Figures reduced from photographs.

Cassia mimosoides.—At night the numerous leaflets on each leaf rotate on their axes, and their tips move towards the apex of the leaf; they thus become imbricated with their lower surfaces directed upwards, and with their midribs almost parallel to the petiole. Consequently, this species differs from all the others seen by us, with the exception of the following one, in the leaflets not sinking down at night. A petiole, the movement of which was measured, rose 8° at night.

Cassia Barclayana.—The leaflets of this Australian species are numerous, very narrow, and almost linear. At night they rise up a little, and also move towards the apex of the leaf. For instance, two opposite leaflets which diverged from one another during the day at an angle of 104°, diverted at night only 72°; so that each had risen 16° above its diurnal position. The petiole of a young leaf rose at night 34°, and that of an older leaf 19°. Owing to the slight movement of the leaflets and the considerable movement of the petiole, the bush presents a different appearance at night to what it does by day; yet the leaves can hardly be said to sleep.

The circumnutating movements of the leaves of C. floribunda, calliantha, and pubescens were observed, each during three or four days; they were essentially alike, those of the last-named species being the simplest. The petiole of C. floribunda was secured to a stick at the base of the two terminal leaflets, and a filament was fixed along the midrib of one of them. Its movements were traced from 1 P.M. on August 13th to 8.30 A.M. 17th; but those during the last 2 h. are alone given in Fig. 156. From 8 A.M. on each day (by which hour the leaf had assumed its diurnal position) to 2 or 3 P.M., it either zigzagged or circumnutated over nearly the same small space; at between 2 and 3 P.M. the great evening fall commenced. The lines representing this fall and the early morning rise are oblique, owing to the peculiar manner in which the leaflets sleep, as already described. After the leaflet was asleep at 6 P.M., and whilst the glass filament hung perpendicularly down, the movement of its apex was traced until 10.30 P.M.; and during this whole time it swayed from side to side, completing more than one ellipse.

Fig 156. Cassia floribunda: circumnutation and nyctitropic movement of a terminal leaflet (1 5/6 inch in length) traced from 8.30 A.M. to same hour on following morning. Apex of leaflet 5½ inches from the vertical glass. Main petiole 3 3/4 inches long. Temp. 16°–17½° C. Figure reduced to one-half of the original scale.

Bauhinia (Tribe 15).—The nyctitropic movements of four species were alike, and were highly peculiar. A plant raised from seed sent us from South Brazil by Fritz Müller, was more especially observed. The leaves are large and deeply notched at their ends. At night the two halves rise up and close completely together, like the opposite leaflets of many Leguminosae. With very young plants the petioles rise considerably at the same time; one, which was inclined at noon 45° above the horizon, at night stood at 75°; it thus rose 30°; another rose 34°. Whilst the two halves of the leaf are closing, the midrib at first sinks vertically downwards and afterwards bends backwards, so as to pass close along one side of its own upwardly inclined petiole; the midrib being thus directed towards the stem or axis of the plant. The angle which the midrib formed with the horizon was measured in one case at different hours: at noon it stood horizontally; late in the evening it depended vertically; then rose to the opposite side, and at 10.15 P.M. stood at only 27° beneath the horizon, being directed towards the stem. It had thus travelled through 153°. Owing to this movement—to the leaves being folded—and to the petioles rising, the whole plant is as much more compact at night than during the day, as a fastigiate Lombardy poplar is compared with any other species of poplar. It is remarkable that when our plants had grown a little older, viz., to a height of 2 or 3 feet, the petioles did not rise at night, and the midribs of the folded leaves were no longer bent back along one side of the petiole. We have noticed in some other genera that the petioles of very young plants rise much more at night than do those of older plants.

Tamarindus Indica (Tribe 16).—The leaflets approach or meet each other at night, and are all directed towards the apex of the leaf. They thus become imbricated with their midribs parallel to the petiole. The movement is closely similar to that of Haematoxylon (see Fig. 153), but more striking from the greater number of the leaflets.

Adenanthera, Prosopis, and Neptunia (Tribe 20).—With Adenanthera pavonia the leaflets turn edgeways and sink at night. In Prosopis they turn upwards. With Neptunia oleracea the leaflets on the opposite sides of the same pinna come into contact at night and are directed forwards. The pinnae themselves move downwards, and at the same time backwards or towards the stem of the plant. The main petiole rises.

Mimosa pudica (Tribe 20).—This plant has been the subject of innumerable observations; but there are some points in relation to our subject which have not been sufficiently attended to. At night, as is well known, the opposite leaflets come into contact and point towards the apex of the leaf; they thus become neatly imbricated with their upper surfaces protected. The four pinnae also approach each other closely, and the whole leaf is thus rendered very compact. The main petiole sinks downwards during the day till late in the evening, and rises until very early in the morning. The stem is continually circumnutating at a rapid rate, though not to a wide extent. Some very young plants, kept in darkness, were observed during two days, and although subjected to a rather low temperature of 57°–59° F., the stem of one described four small ellipses in the course of 12 h. We shall immediately see that the main petiole is likewise continually circumnutating, as is each separate pinna and each separate leaflet. Therefore, if the movement of the apex of any one leaflet were to be traced, the course described would be compounded of the movements of four separate parts.

A filament had been fixed on the previous evening, longitudinally to the main petiole of a nearly full-grown, highly-sensitive leaf (four inches in length), the stem having been secured to a stick at its base; and a tracing was made on a vertical glass in the hot-house under a high temperature. In the figure given (Fig. 157), the first dot was made at 8.30 A.M. August 2nd, and the last at 7 P.M. on the 3rd. During 12 h. on the first day the petiole moved thrice downwards and twice upwards. Within the same length of time on the second day, it moved five times downwards and four times upwards. As the ascending and descending lines do not coincide, the petiole manifestly circumnutates; the great evening fall and nocturnal rise being an exaggeration of one of the circumnutations. It should, however, be observed that the petiole fell much lower down in the evenings than could be seen on the vertical glass or is represented in the diagram. After 7 P.M. on the 3rd (when the last dot in Fig. 157 was made) the pot was carried into a bed-room, and the petiole was found at 12.50 A.M. (i.e. after midnight) standing almost upright, and much more highly inclined than it was at 10.40 P.M. When observed again at 4 A.M. it had begun to fall, and continued falling till 6.15 A.M., after which hour it zigzagged and again circumnutated. Similar observations were made on another petiole, with nearly the same result.

Fig. 157 Mimosa pudica: circumnutation and nyctitropic movement of main petiole, traced during 34 h. 30 m.

On two other occasions the movement of the main petiole was observed every two or three minutes, the plants being kept at a rather high temperature, viz., on the first occasion at 77°–81° F., and the filament then described 2½ ellipses in 69 m. On the second occasion, when the temperature was 81°–86° F., it made rather more than 3 ellipses in 67 m. therefore, Fig. 157, though now sufficiently complex, would have been incomparably more so, if dots had been made on the glass every 2 or 3 minutes, instead of every hour or half-hour. Although the main petiole is continually and rapidly describing small ellipses during the day, yet after the great nocturnal rising movement has commenced, if dots are made every 2 or 3 minutes, as was done for an hour between 9.30 and 10.30 P.M. (temp. 84° F.), and the dots are then joined, an almost absolutely straight line is the result.

To show that the movement of the petiole is in all probability due to the varying turgescence of the pulvinus, and not to growth (in accordance with the conclusions of Pfeffer), a very old leaf, with some of its leaflets yellowish and hardly at all sensitive, was selected for observation, and the plant was kept at the highly favourable temp. of 80° F. The petiole fell from 8 A.M. till 10.15 A.M., it then rose a little in a somewhat zigzag line, often remaining stationary, till 5 P.M., when the great evening fall commenced, which was continued till at least 10 P.M. By 7 A.M. on the following morning it had risen to the same level as on the previous morning, and then descended in a zigzag line. But from 10.30 A.M. till 4.15 P.M. it remained almost motionless, all power of movement being now lost. The petiole, therefore, of this very old leaf, which must have long ceased growing, moved periodically; but instead of circumnutating several times during the day, it moved only twice down and twice up in the course of 24 h., with the ascending and descending lines not coincident.

It has already been stated that the pinnae move independently of the main petiole. The petiole of a leaf was fixed to a cork support, close to the point whence the four pinnae diverge, with a short fine filament cemented longitudinally to one of the two terminal pinnae, and a graduated semicircle was placed close beneath it. By looking vertically down, its angular or lateral movements could be measured with accuracy. Between noon and 4.15 P.M. the pinna changed its position to one side by only 7°; but not continuously in the same direction, as it moved four times to one side, and three times to the opposite side, in one instance to the extent of 16°. This pinna, therefore circumnutated. Later in the evening the four pinnae approach each other, and the one which was observed moved inwards 59° between noon and 6.45 P.M. Ten observations were made in the course of 2 h. 20 m. (at average intervals of 14 m.), between 4.25 and 6.45 P.M.; and there was now, when the leaf was going to sleep, no swaying from side to side, but a steady inward movement. Here therefore there is in the evening the same conversion of a circumnutating into a steady movement in one direction, as in the case of the main petiole.

It has also been stated that each separate leaflet circumnutates. A pinna was cemented with shellac on the summit of a little stick driven firmly into the ground, immediately beneath a pair of leaflets, to the midribs of both of which excessively fine glass filaments were attached. This treatment did not injure the leaflets, for they went to sleep in the usual manner, and long retained their sensitiveness. the movements of one of them were traced during 49 h., as shown in Fig. 158. On the first day the leaflet sank down till 11.30 A.M., and then rose till late in the evening in a zigzag line, indicating circumnutation. On the second day, when more accustomed to its new state, it oscillated twice up and twice down during the 24 h. This plant was subjected to a rather low temperature, viz., 62°–64° F.; had it been kept warmer, no doubt the movements of the leaflet would have been much more rapid and complicated. It may be seen in the diagram that the ascending and descending lines do not coincide; but the large amount of lateral movement in the evening is the result of the leaflets bending towards the apex of the leaf when going to sleep. Another leaflet was casually observed, and found to be continually circumnutating during the same length of time.

The circumnutation of the leaves is not destroyed by their being subjected to moderately long continued darkness; but the proper periodicity of their movements is lost. Some very young seedlings were kept during two days in the dark (temp. 57°–59° F.) except when the circumnutation of their stems was occasionally observed; and on the evening of the second day the leaflets did not fully and properly go to sleep. The pot was then placed for three days in a dark cupboard, under nearly the same temperature, and at the close of this period the leaflets showed no signs of sleeping, and were only slightly sensitive to a touch. On the following day the stem was cemented to a stick, and the movements of two leaves were traced on a vertical glass during 72 h. The plants were still kept in the dark, excepting that at each observation, which lasted 3 or 4 minutes, they were illuminated by two candles. On the third day the leaflets still exhibited a vestige of sensitiveness when forcibly pressed, but in the evening they showed no signs of sleep. Nevertheless, their petioles continued to circumnutate distinctly, although the proper order of their movements in relation to the day and night was wholly lost. Thus, one leaf descended during the first two nights (i.e. between 10 P.M. and 7 A.M. next morning) instead of ascending, and on the third night it moved chiefly in a lateral direction. The second leaf behaved in an equally abnormal manner, moving laterally during the first night, descending greatly during the second, and ascending to an unusual height during the third night.

Fig 158. Mimosa pudica: circumnutation and nyctitropic movement of a leaflet (with pinna secured), traced on a vertical glass, from 8 A.M. Sept. 14th to 9 A.M. 16th.

With plants kept at a high temperature and exposed to the light, the most rapid circumnutating movement of the apex of a leaf which was observed, amounted to 1/500 of an inch in one second; and this would have equalled 1/8 of an inch in a minute, had not the leaf occasionally stood still. The actual distance travelled by the apex (as ascertained by a measure placed close to the leaf) was on one occasion nearly 3/4 of an inch in a vertical direction in 15 m.; and on another occasion 5/8 of an inch in 60 m.; but there was also some lateral movement.

Mimosa albida.[19]—The leaves of this plant, one of which is here figured (Fig. 159) reduced to 2/3 of the natural size, present some interesting peculiarities. It consists of a long petiole bearing only two pinnae (here represented as rather more divergent than is usual), each with two pairs of leaflets. But the inner basal leaflets are greatly reduced in size, owing probably to the want of space for their full development, so that they may be considered as almost rudimentary. They vary somewhat in size, and both occasionally disappear, or only one. Nevertheless, they are not in the least rudimentary in function, for they are sensitive, extremely heliotropic, circumnutate at nearly the same rate as the fully developed leaflets, and assume when asleep exactly the same position. With M. pudica the inner leaflets at the base and between the pinnae are likewise much shortened and obliquely truncated; this fact was well seen in some seedlings of M. pudica, in which the third leaf above the cotyledons bore only two pinnae, each with only 3 or 4 pairs of leaflets, of which the inner basal one was less than half as long as its fellow; so that the whole leaf resembled pretty closely that of M. albida. In this latter species the main petiole terminates in a little point, and on each side of this there is a pair of minute, flattened, lancet-shaped projections, hairy on their margins, which drop off and disappear soon after the leaf is fully developed. There can hardly be a doubt that these little projections are the last and fugacious representatives of an additional pair of leaflets to each pinna; for the outer one is twice as broad as the inner one, and a little longer, viz. 7/100 of an inch, whilst the inner one is only 5/100–6/100 long. Now if the basal pair of leaflets of the existing leaves were to become rudimentary, we should expect that the rudiments would still exhibit some trace of their present great inequality of size. The conclusion that the pinnae of the parent-form of M. albida possessed at least three pairs of leaflets, instead of, as at present, only two, is supported by the structure of the first true leaf; for this consists of a simple petiole, often bearing three pairs of leaflets. This latter fact, as well as the presence of the rudiments, both lead to the conclusion that M. albida is descended from a form the leaves of which bore more than two pairs of leaflets. The second leaf above the cotyledons resembles in all respects the leaves on fully developed plants.

[19] Mr. Thiselton Dyer informs us that this Peruvian plant (which was sent to us from Kew) is considered by Mr. Bentham ('Trans. Linn. Soc.,' vol. xxx. p. 390) to be "the species or variety which most commonly represents the M. sensitiva of our gardens."

Fig. 159. Mimosa albida: leaf seen from vertically above.

When the leaves go to sleep, each leaflet twists half round, so as to present its edge to the zenith, and comes into close contact with its fellow. The pinnae also approach each other closely, so that the four terminal leaflets come together. The large basal leaflets (with the little rudimentary ones in contact with them) move inwards and forwards, so as to embrace the outside of the united terminal leaflets, and thus all eight leaflets (the rudimentary ones included) form together a single vertical packet. The two pinnae at the same time that they approach each other sink downwards, and thus instead of extending horizontally in the same line with the main petiole, as during the day, they depend at night at about 45°, or even at a greater angle, beneath the horizon. The movement of the main petiole seems to be variable; we have seen it in the evening 27° lower than during the day; but sometimes in nearly the same position. Nevertheless, a sinking movement in the evening and a rising one during the night is probably the normal course, for this was well-marked in the petiole of the first-formed true leaf.

The circumnutation of the main petiole of a young leaf was traced during 2 3/4 days, and was considerable in extent, but less complex than that of M. pudica. The movement was much more lateral than is usual with circumnutating leaves, and this was the sole peculiarity which it presented. The apex of one of the terminal leaflets was seen under the microscope to travel 1/50 of an inch in 3 minutes.

Mimosa marginata.—The opposite leaflets rise up and approach each other at night, but do not come into close contact, except in the case of very young leaflets on vigorous shoots. Full-grown leaflets circumnutate during the day slowly and on a small scale.

Schrankia uncinata (Tribe 20).—A leaf consists of two or three pairs of pinnae, each bearing many small leaflets. These, when the plant is asleep, are directed forwards and become imbricated. The angle between the two terminal pinnae was diminished at night, in one case by 15°; and they sank almost vertically downwards. The hinder pairs of pinnae likewise sink downwards, but do not converge, that is, move towards the apex of the leaf. The main petiole does not become depressed, at least during the evening. In this latter respect, as well as in the sinking of the pinnae, there is a marked difference between the nyctitropic movements of the present plant and of Mimosa pudica. It should, however, be added that our specimen was not in a very vigorous condition. The pinnae of Schrankia aculeata also sink at night.

Acacia Farnesiana (Tribe 22).—The different appearance presented by a bush of this plant when asleep and awake is wonderful. The same leaf in the two states is shown in the following figure (Fig. 160). The leaflets move towards the apex of the pinna and become imbricated, and the pinnae then look like bits of dangling string. The following remarks and measurements do not fully apply to the small leaf here figured. The pinnae move forwards and at the same time sink downwards, whilst the main petiole rises considerably. With respect to the degree of movement: the two terminal pinnae of one specimen formed together an angle of 100° during the day, and at night of only 38°, so each had moved 31° forwards. The penultimate pinnae during the day formed together an angle of 180°, that is, they stood in a straight line opposite one another, and at night each had moved 65° forwards. The basal pair of pinnae were directed during the day, each about 21° backwards, and at night 38° forwards, so each had moved 59° forwards. But the pinnae at the same time sink greatly, and sometimes hang almost perpendicularly downwards. The main petiole, on the other hand, rises much: by 8.30 P.M. one stood 34° higher than at noon, and by 6.40 A.M. on the following morning it was still higher by 10°; shortly after this hour the diurnal sinking movement commenced. The course of a nearly full-grown leaf was traced during 14 h.; it was strongly zigzag, and apparently represented five ellipses, with their longer axes differently directed.

Fig. 160. Acacia Farnesiana: A, leaf during the day; B, the same leaf at night.

Albizzia lophantha (Tribe 23).—The leaflets at night come into contact with one another, and are directed towards the apex of the pinna. The pinnae approach one another, but remain in the same plane as during the day; and in this respect they differ much from those of the above Schrankia and Acacia. The main petiole rises but little. The first-formed leaf above the cotyledons bore 11 leaflets on each side, and these slept like those on the subsequently formed leaves; but the petiole of this first leaf was curved downwards during the day and at night straightened itself, so that the chord of its arc then stood 16° higher than in the day-time.

Melaleuca ericaefolia (Myrtaceae).—According to Bouché ('Bot. Zeit.,' 1874, p. 359) the leaves sleep at night, in nearly the same manner as those of certain species of Pimelia.

Œnothera mollissima (Onagrarieae).—According to Linnæus ('Somnus Plantarum'), the leaves rise up vertically at night.

Passiflora gracilis (Passifloracae).—The young leaves sleep by their blades hanging vertically downwards, and the whole length of the petiole then becomes somewhat curved downwards. Externally no trace of a pulvinus can be seen. The petiole of the uppermost leaf on a young shoot stood at 10.45 A.M. at 33° above the horizon; and at 10.30 P.M., when the blade was vertically dependent, at only 15°, so the petiole had fallen 18°. That of the next older leaf fell only 7°. From some unknown cause the leaves do not always sleep properly. The stem of a plant, which had stood for some time before a north-east window, was secured to a stick at the base of a young leaf, the blade of which was inclined at 40° below the horizon. From its position the leaf had to be viewed obliquely, consequently the vertically ascending and descending movements appeared when traced oblique. On the first day (Oct. 12th) the leaf descended in a zigzag line until late in the evening; and by 8.15 A.M. on the 13th had risen to nearly the same level as on the previous morning. A new tracing was now begun (Fig. 161). The leaf continued to rise until 8.50 A.M., then moved a little to the right, and afterwards descended. Between 11 A.M. and 5 P.M. it circumnutated, and after the latter hour the great nocturnal fall commenced. At 7.15 P.M. it depended vertically. The dotted line ought to have been prolonged much lower down in the figure. By 6.50 A.M. on the following morning (14th) the leaf had risen greatly, and continued to rise till 7.50 A.M., after which hour it redescended. It should be observed that the lines traced on this second morning would have coincided with and confused those previously traced, had not the pot been slided a very little to the left. In the evening (14th) a mark was placed behind the filament attached to the apex of the leaf, and its movement was carefully traced from 5 P.M. to 10.15 P.M. Between 5 and 7.15 P.M. the leaf descended in a straight line, and at the latter hour it appeared vertically dependent. But between 7.15 and 10.15 P.M. the line consisted of a succession of steps, the cause of which we could not understand; it was, however, manifest that the movement was no longer a simple descending one.

Fig. 161. Passiflora gracilis: circumnutation and nyctitropic movement of leaf, traced on vertical glass, from 8.20 A.M. Oct. 13th to 10 A.M. 14th. Figure reduced to two-thirds of original scale.

Siegesbeckia orientalis (Compositæ).—Some seedlings were raised in the middle of winter and kept in the hot-house; they flowered, but did not grow well, and their leaves never showed any signs of sleep. The leaves on other seedlings raised in May were horizontal at noon (June 22nd), and depended at a considerable angle beneath the horizon at 10 P.M. In the case of four youngish leaves which were from 2 to 2½ inches in length, these angles were found to be 50°, 56°, 60°, and 65°. At the end of August when the plants had grown to a height of 10 to 11 inches, the younger leaves were so much curved downwards at night that they might truly be said to be asleep. This is one of the species which must be well illuminated during the day in order to sleep, for on two occasions when plants were kept all day in a room with north-east windows, the leaves did not sleep at night. The same cause probably accounts for the leaves on our seedlings raised in the dead of the winter not sleeping. Professor Pfeffer informs us that the leaves of another species (S. Jorullensis ?) hang vertically down at night.

Fig. 162. Nicotiana glauca: shoots with leaves expanded during the day, and asleep at night. Figures copied from photographs, and reduced.

Ipomœa caerulea and purpurea (Convolvulaceae).—The leaves on very young plants, a foot or two in height, are depressed at night to between 68° and 80° beneath the horizon; and some hang quite vertically downwards. On the following morning they again rise into a horizontal position. The petioles become at night downwardly curved, either through their entire length or in the upper part alone; and this apparently causes the depression of the blade. It seems necessary that the leaves should be well illuminated during the day in order to sleep, for those which stood on the back of a plant before a north-east window did not sleep.

Nicotiana tabacum (var. Virginian) and glauca (Solaneae).—The young leaves of both these species sleep by bending vertically upwards. Figures of two shoots of N. glauca, awake and asleep (Fig. 162), are given on p. 385: one of the shoots, from which the photographs were taken, was accidentally bent to one side.

Fig. 163. Nicotiana tabacum: circumnutation and nyctitropic movement of a leaf (5 inches in length), traced on a vertical glass, from 3 P.M. July 10th to 8.10 A.M. 13th. Apex of leaf 4 inches from glass. Temp. 17½°–18½° C. Figure reduced to one-half original scale.

At the base of the petiole of N. tabacum, on the outside, there is a mass of cells, which are rather smaller than elsewhere, and have their longer axes differently directed from the cells of the parenchyma, and may therefore be considered as forming a sort of pulvinus. A young plant of N. tabacum was selected, and the circumnutation of the fifth leaf above the cotyledons was observed during three days. On the first morning (July 10th) the leaf fell from 9 to 10 A.M., which is its normal course, but rose during the remainder of the day; and this no doubt was due to its being illuminated exclusively from above; for properly the evening rise does not commence until 3 or 4 P.M. In the figure as given on p. 386 (Fig. 163) the first dot was made at 3 P.M.; and the tracing was continued for the following 65 h. When the leaf pointed to the dot next above that marked 3 P.M. it stood horizontally. The tracing is remarkable only from its simplicity and the straightness of the lines. The leaf each day described a single great ellipse; for it should be observed that the ascending and descending lines do not coincide. On the evening of the 11th the leaf did not descend quite so low as usual, and it now zigzagged a little. The diurnal sinking movement had already commenced each morning by 7 A.M. The broken lines at the top of the figure, representing the nocturnal vertical position of the leaf, ought to be prolonged much higher up.

Mirabilis longiflora and jalapa (Nyctagineae).—The first pair of leaves above the cotyledons, produced by seedlings of both these species, were considerably divergent during the day, and at night stood up vertically in close contact with one another. The two upper leaves on an older seedling were almost horizontal by day, and at night stood up vertically, but were not in close contact, owing to the resistance offered by the central bud.

Polygonum aviculare (Polygoneae).—Professor Batalin informs us that the young leaves rise up vertically at night. This is likewise the case, according to Linnæus, with several species of Amaranthus (Amaranthaceae); and we observed a sleep movement of this kind in one member of the genus. Again, with Chenopodium album (Chenopodieae), the upper young leaves of some seedlings, about 4 inches in height, were horizontal or sub-horizontal during the day, and at 10 P.M. on March 7th were quite, or almost quite, vertical. Other seedlings raised in the greenhouse during the winter (Jan. 28th) were observed day and night, and no difference could be perceived in the position of their leaves. According to Bouché ('Bot. Zeitung,' 1874, p. 359) the leaves of Pimelia linoides and spectabilis (Thymeleae) sleep at night.

Euphorbia jacquiniaeflora (Euphorbiaceae).—Mr. Lynch called our attention to the fact that the young leaves of this plant sleep by depending vertically. The third leaf from the summit (March 11th) was inclined during the day 30° beneath the horizon, and at night hung vertically down, as did some of the still younger leaves. It rose up to its former level on the following morning. The fourth and fifth leaves from the summit stood horizontally during the day, and sank down at night only 38°. The sixth leaf did not sensibly alter its position. The sinking movement is due to the downward curvature of the petiole, no part of which exhibits any structure like that of a pulvinus. Early on the morning of June 7th a filament was fixed longitudinally to a young leaf (the third from the summit, and 2 5/8 inches in length), and its movements were traced on a vertical glass during 72 h., the plant being illuminated from above through a skylight. Each day the leaf fell in a nearly straight line from 7 A.M. to 5 P.M., after which hour it was so much inclined downwards that the movement could no longer be traced; and during the latter part of each night, or early in the morning, the leaf rose. It therefore circumnutated in a very simple manner, making a single large ellipse every 24 h., for the ascending and descending lines did not coincide. On each successive morning it stood at a less height than on the previous one, and this was probably due partly to the increasing age of the leaf, and partly to the illumination being insufficient; for although the leaves are very slightly heliotropic, yet, according to Mr. Lynch's and our own observations, their inclination during the day is determined by the intensity of the light. On the third day, by which time the extent of the descending movement had much decreased, the line traced was plainly much more zigzag than on any previous day, and it appeared as if some of its powers of movement were thus expended. At 10 P.M. on June 7th, when the leaf depended vertically, its movements were observed by a mark being placed behind it, and the end of the attached filament was seen to oscillate slowly and slightly from side to side, as well as upwards and downwards.

Phyllanthus Niruri (Euphorbiaceae).—The leaflets of this plant sleep, as described by Pfeffer,[20] in a remarkable manner, apparently like those of Cassia, for they sink downwards at night and twist round, so that their lower surfaces are turned outwards. They are furnished as might have been expected from this complex kind of movement, with a pulvinus.

[20] 'Die Period. Beweg.,' p. 159.

GYMNOSPERMS.

Pinus Nordmanniana (Coniferæ).—M. Chatin states[21] that the leaves, which are horizontal during the day, rise up at night, so as to assume a position almost perpendicular to the branch from which they arise; we presume that he here refers to a horizontal branch. He adds: "En même temps, ce mouvement d'érection est accompangé d'un mouvement de torsion imprimé à la partie basilaire de la feuille, et pouvant souvent parcourir un arc de 90 degrés." As the lower surfaces of the leaves are white, whilst the upper are dark green, the tree presents a widely different appearance by day and night. The leaves on a small tree in a pot did not exhibit with us any nyctitropic movements. We have seen in a former chapter that the leaves of Pinus pinaster and Austriaca are continually circumnutating.

[21] 'Comptes Rendus,' Jan. 1876, p. 171.

MONOCOTYLEDONS.

Thalia dealbata (Cannaceae).—the leaves of this plant sleep by turning vertically upwards; they are furnished with a well-developed pulvinus. It is the only instance known to us of a very large leaf sleeping. The blade of a young leaf, which was as yet only 13 1/4 inches in length and 6½ in breadth, formed at noon an angle with its tall petiole of 121°, and at night stood vertically in a line with it, and so had risen 59°. The actual distance travelled by the apex (as measured by an orthogonic tracing) of another large leaf, between 7.30 A.M. and 10 P.M., was 10½ inches. The circumnutation of two young and dwarfed leaves, arising amongst the taller leaves at the base of the plant, was traced on a vertical glass during two days. On the first day the apex of one, and on the second day the apex of the other leaf, described between 6.40 A.M. and 4 P.M. two ellipses, the longer axes of which were extended in very different directions from the lines representing the great diurnal sinking and nocturnal rising movement.

Maranta arundinacea (Cannaceae).—The blades of the leaves, which are furnished with a pulvinus, stand horizontally during the day or between 10° and 20° above the horizon, and at night vertically upwards. They therefore rise between 70° and 90° at night. The plant was placed at noon in the dark in the hot-house, and on the following day the movements of the leaves were traced. Between 8.40 and 10.30 A.M. they rose, and then fell greatly till 1.37 P.M. But by 3 P.M. they had again risen a little, and continued to rise during the rest of the afternoon and night; on the following morning they stood at the same level as on the previous day. Darkness, therefore, during a day and a half does not interfere with the periodicity of their movements. On a warm but stormy evening, the plant whilst being brought into the house, had its leaves violently shaken, and at night not one went to sleep. On the next morning the plant was taken back to the hot-house, and again at night the leaves did not sleep; but on the ensuing night they rose in the usual manner between 70° and 80°. This fact is analogous with what we have observed with climbing plants, namely, that much agitation checks for a time their power of circumnutation; but the effect in this instance was much more strongly marked and prolonged.

Colocasia antiquorum (Caladium esculentum, Hort.) (Aroideae).—The leaves of this plant sleep by their blades sinking in the evening, so as to stand highly inclined, or even quite vertically with their tips pointing to the ground. They are not provided with a pulvinus. The blade of one stood at noon 1 degree beneath the horizon; at 4.20 P.M., 20°; at 6 P.M. 43°; at 7.20 P.M., 69°; and at 8.30 P.M., 68°; so it had now begun to rise; at 10.15 P.M. it stood at 65°, and on the following early morning at 11° beneath the horizon. The circumnutation of another young leaf (with its petiole only 3 1/4 inches, and the blade 4 inches in length), was traced on a vertical glass during 48 h.; it was dimly illuminated through a skylight, and this seemed to disturb the proper periodicity of its movements. Nevertheless, the leaf fell greatly during both afternoons, till either 7.10 P.M. or 9 P.M., when it rose a little and moved laterally. By an early hour on both mornings, it had assumed its diurnal position. The well-marked lateral movement for a short time in the early part of the night, was the only interesting fact which it presented, as this caused the ascending and descending lines not to coincide, in accordance with the general rule with circumnutating organs. The movements of the leaves of this plant are thus of the most simple kind; and the tracing is not worth giving. We have seen that in another genus of the Aroideae, namely, Pistia, the leaves rise so much at night that they may almost be said to sleep.

Strephium floribundum[22] (Gramineæ).—The oval leaves are provided with a pulvinus, and are extended horizontally or declined a little beneath the horizon during the day. Those on the upright culms simply rise up vertically at night, so that their tips are directed towards the zenith. (Fig. 164.) Horizontally extended leaves arising from much inclined or almost horizontal culms, move at night so that their tips point towards the apex of the culm, with one lateral margin directed towards the zenith; and in order to assume this position the leaves have to twist on their own axes through an angle of nearly 90°. Thus the surface of the blade always stands vertically, whatever may be the position of the midrib or of the leaf as a whole.

[22] A. Brongniart first observed that the leaves of this plant and of Marsilea sleep: see 'Bull. de la Soc. Bot. de France,' tom. vii. 1860, p. 470.

Fig. 164. Strephium floribundum: culms with leaves during the day, and when asleep at night. Figures reduced.

The circumnutation of a young leaf (2.3 inches in length) was traced during 48 h. (Fig. 165). The movement was remarkably simple; the leaf descended from before 6.40 A.M. until 2 or 2.50 P.M., and then rose so as to stand vertically at about 6 P.M., descending again late in the night or in the very early morning. On the second day the descending line zigzagged slightly. As usual, the ascending and descending lines did not coincide. On another occasion, when the temperature was a little higher, viz., 24°–26½° C., a leaf was observed 17 times between 8.50 A.M. and 12.16 P.M.; it changed its course by as much as a rectangle six times in this interval of 3 h. 26 m., and described two irregular triangles and a half. The leaf, therefore, on this occasion circumnutated rapidly and in a complex manner.

Fig. 165. Strephium floribundum: circumnutation and nyctitropic movement of a leaf, traced from 9 A.M. June 26th to 8.45 A.M. 27th;

filament fixed along the midrib. Apex of leaf 8 1/4 inches from the vertical glass; plant illuminated from above. Temp. 23½°–24½° C.

ACOTYLEDONS.

Marsilea quadrifoliata (Marsileaceae).—The shape of a leaf, expanded horizontally during the day, is shown at A (Fig. 166). Each leaflet is provided with a well-developed pulvinus. When the leaves sleep, the two terminal leaflets rise up, twist half round and come into contact with one another (B), and are afterwards embraced by the two lower leaflets (C); so that the four leaflets with their lower surfaces turned outwards form a vertical packet. The curvature of the summit of the petiole of the leaf figured asleep, is merely accidental. The plant was brought into a room, where the temperature was only a little above 60° F., and the movement of one of the leaflets (the petiole having been secured) was traced during 24 h. (Fig. 167). The leaf fell from the early morning till 1.50 P.M., and then rose till 6 P.M., when it was asleep. A vertically dependent glass filament was now fixed to one of the terminal and inner leaflets; and part of the tracing in Fig. 167, after 6 P.M., shows that it continued to sink, making one zigzag, until 10.40 P.M. At 6.45 A.M. on the following morning, the leaf was awaking, and the filament pointed above the vertical glass, but by 8.25 A.M. it occupied the position shown in the figure. The diagram differs greatly in appearance from most of those previously given; and this is due to the leaflet twisting and moving laterally as it approaches and comes into contact with its fellow. The movement of another leaflet, when asleep, was traced between 6 P.M. and 10.35 P.M., and it clearly circumnutated, for it continued for two hours to sink, then rose, and then sank still lower than it was at 6 P.M. It may be seen in the preceding figure (167) that the leaflet, when the plant was subjected to a rather low temperature in the house, descended and ascended during the middle of the day in a somewhat zigzag line; but when kept in the hot-house from 9 A.M. to 3 P.M. at a high but varying temperature (viz., between 72° and 83° F.) a leaflet (with the petiole secured) circumnutated rapidly, for it made three large vertical ellipses in the course of the six hours. According to Brongniart, Marsilea pubescens sleeps like the present species. These plants are the sole cryptogamic ones known to sleep.

Fig. 166. Marsilea quadrifoliata: A, leaf during the day, seen from vertically above; B, leaf beginning to go to sleep, seen laterally; C, the same asleep. Figures reduced to one-half of natural scale.

Fig. 167. Marsilea quadrifoliata: circumnutation and nyctitropic movement of leaflet traced on vertical glass, during nearly 24 h. Figure reduced to two-thirds of original scale. Plant kept at rather too low a temperature.

Summary and Concluding Remarks on the Nyctitropic or Sleep-movements of Leaves.—That these movements are in some manner of high importance to the plants which exhibit them, few will dispute who have observed how complex they sometimes are. Thus with Cassia, the leaflets which are horizontal during the day not only bend at night vertically downwards with the terminal pair directed considerably backwards, but they also rotate on their own axes, so that their lower surfaces are turned outwards. The terminal leaflet of Melilotus likewise rotates, by which movement one of its lateral edges is directed upwards, and at the same time it moves either to the left or to the right, until its upper surface comes into contact with that of the lateral leaflet on the same side, which has likewise rotated on its own axis. With Arachis, all four leaflets form together during the night a single vertical packet; and to the effect this the two anterior leaflets have to move upwards and the two posterior ones forwards, besides all twisting on their own axes. In the genus Sida the leaves of some species move at night through an angle of 90° upwards, and of others through the same angle downwards. We have seen a similar difference in the nyctitropic movements of the cotyledons in the genus Oxalis. In Lupinus, again, the leaflets move either upwards or downwards; and in some species, for instance L. luteus, those on one side of the star-shaped leaf move up, and those on the opposite side move down; the intermediate ones rotating on their axes; and by these varied movements, the whole leaf forms at night a vertical star instead of a horizontal one, as during the day. Some leaves and leaflets, besides moving either upwards or downwards, become more or less folded at night, as in Bauhinia and in some species of Oxalis. The positions, indeed, which leaves occupy when asleep are almost infinitely diversified; they may point either vertically upwards or downwards, or, in the case of leaflets, towards the apex or towards the base of the leaf, or in any intermediate position. They often rotate at least as much as 90° on their own axes. The leaves which arise from upright and from horizontal or much inclined branches on the same plant, move in some few cases in a different manner, as with Porlieria and Strephium. The whole appearance of many plants is wonderfully changed at night, as may be seen with Oxalis, and still more plainly with Mimosa. A bush of Acacia Farnesiana appears at night as if covered with little dangling bits of string instead of leaves. Excluding a few genera not seen by ourselves, about which we are in doubt, and excluding a few others the leaflets of which rotate at night, and do not rise or sink much, there are 37 genera in which the leaves or leaflets rise, often moving at the same time towards the apex or towards the base of the leaf, and 32 genera in which they sink at night.

The nyctitropic movements of leaves, leaflets, and petioles are effected in two different ways; firstly, by alternately increased growth on their opposite sides, preceded by increased turgescence of the cells; and secondly by means of a pulvinus or aggregate of small cells, generally destitute of chlorophyll, which become alternately more turgescent on nearly opposite sides; and this turgescence is not followed by growth except during the early age of the plant. A pulvinus seems to be formed (as formerly shown) by a group of cells ceasing to grow at a very early age, and therefore does not differ essentially from the surrounding tissues. The cotyledons of some species of Trifolium are provided with a pulvinus, and others are destitute of one, and so it is with the leaves in the genus Sida. We see also in this same genus gradations in the state of the development of the pulvinus; and in Nicotiana we have what may probably be considered as the commencing development of one. The nature of the movement is closely similar, whether a pulvinus is absent or present, as is evident from many of the diagrams given in this chapter. It deserves notice that when a pulvinus is present, the ascending and descending lines hardly ever coincide, so that ellipses are habitually described by the leaves thus provided, whether they are young or so old as to have quite ceased growing. This fact of ellipses being described, shows that the alternately increased turgescence of the cells does not occur on exactly opposite sides of the pulvinus, any more than the increased growth which causes the movements of leaves not furnished with pulvini. When a pulvinus is present, the nyctitropic movements are continued for a very much longer period than when such do not exist. This has been amply proved in the case of cotyledons, and Pfeffer has given observations to the same effect with respect to leaves. We have seen that a leaf of Mimosa pudica continued to move in the ordinary manner, though somewhat more simply, until it withered and died. It may be added that some leaflets of Trifolium pratense were pinned open during 10 days, and on the first evening after being released they rose up and slept in the usual manner. Besides the long continuance of the movements when effected by the aid of a pulvinus (and this appears to be the final cause of its development), a twisting movement at night, as Pfeffer has remarked, is almost confined to leaves thus provided.

It is a very general rule that the first true leaf, though it may differ somewhat in shape from the leaves on the mature plant, yet sleeps like them; and this occurs quite independently of the fact whether or not the cotyledons themselves sleep, or whether they sleep in the same manner. But with Phaseolus Roxburghii the first unifoliate leaves rise at night almost sufficiently to be said to sleep, whilst the leaflets of the secondary trifoliate leaves sink vertically at night. On young plants of Sida rhombaefolia, only a few inches in height, the leaves did not sleep, though on rather older plants they rose up vertically at night. On the other hand, the leaves on very young plants of Cytisus fragrans slept in a conspicuous manner, whilst on old and vigorous bushes kept in the greenhouse, the leaves did not exhibit any plain nyctitropic movement. In the genus Lotus the basal stipule-like leaflets rise up vertically at night, and are provided with pulvini.

As already remarked, when leaves or leaflets change their position greatly at night and by complicated movements, it can hardly be doubted that these must be in some manner beneficial to the plant. If so, we must extend the same conclusion to a large number of sleeping plants; for the most complicated and the simplest nyctitropic movements are connected together by the finest gradations. But owing to the causes specified in the beginning of this chapter, it is impossible in some few cases to determine whether or not certain movements should be called nyctitropic. Generally, the position which the leaves occupy at night indicates with sufficient clearness, that the benefit thus derived, is the protection of their upper surfaces from radiation into the open sky, and in many cases the mutual protection of all the parts from cold by their being brought into close approximation. It should be remembered that it was proved in the last chapter, that leaves compelled to remain extended horizontally at night, suffered much more from radiation than those which were allowed to assume their normal vertical position.

The fact of the leaves of several plants not sleeping unless they have been well illuminated during the day, made us for a time doubt whether the protection of their upper surfaces from radiation was in all cases the final cause of their well-pronounced nyctitropic movements. But we have no reason to suppose that the illumination from the open sky, during even the most clouded day, is insufficient for this purpose; and we should bear in mind that leaves which are shaded from being seated low down on the plant, and which sometimes do not sleep, are likewise protected at night from full radiation. Nevertheless, we do not wish to deny that there may exist cases in which leaves change their position considerably at night, without their deriving any benefit from such movements.

Although with sleeping plants the blades almost always assume at night a vertical, or nearly vertical position, it is a point of complete indifference whether the apex, or the base, or one of the lateral edges, is directed to the zenith. It is a rule of wide generality, that whenever there is any difference in the degree of exposure to radiation between the upper and the lower surfaces of leaves and leaflets, it is the upper which is the least exposed, as may be seen in Lotus, Cytisus, Trifolium, and other genera. In several species of Lupinus the leaflets do not, and apparently from their structure cannot, place themselves vertically at night, and consequently their upper surfaces, though highly inclined, are more exposed than the lower; and here we have an exception to our rule. But in other species of this genus the leaflets succeed in placing themselves vertically; this, however, is effected by a very unusual movement, namely, by the leaflets on the opposite sides of the same leaf moving in opposite directions.

It is again a very common rule that when leaflets come into close contact with one another, they do so by their upper surfaces, which are thus best protected. In some cases this may be the direct result of their rising vertically; but it is obviously for the protection of the upper surfaces that the leaflets of Cassia rotate in so wonderful a manner whilst sinking downwards; and that the terminal leaflet of Melilotus rotates and moves to one side until it meets the lateral leaflet on the same side. When opposite leaves or leaflets sink vertically down without any twisting, their lower surfaces approach each other and sometimes come into contact; but this is the direct and inevitable result of their position. With many species of Oxalis the lower surfaces of the adjoining leaflets are pressed together, and are thus better protected than the upper surfaces; but this depends merely on each leaflet becoming folded at night so as to be able to sink vertically downwards. The torsion or rotation of leaves and leaflets, which occurs in so many cases, apparently always serves to bring their upper surfaces into close approximation with one another, or with other parts of the plant, for their mutual protection. We see this best in such cases as those of Arachis, Mimosa albida, and Marsilea, in which all the leaflets form together at night a single vertical packet. If with Mimosa pudica the opposite leaflets had merely moved upwards, their upper surfaces would have come into contact and been well protected; but as it is, they all successively move towards the apex of the leaf; and thus not only their upper surfaces are protected, but the successive pairs become imbricated and mutually protect one another as well as the petioles. This imbrication of the leaflets of sleeping plants is a common phenomenon.

The nyctitropic movement of the blade is generally effected by the curvature of the uppermost part of the petiole, which has often been modified into a pulvinus; or the whole petiole, when short, may be thus modified. But the blade itself sometimes curves or moves, of which fact Bauhinia offers a striking instance, as the two halves rise up and come into close contact at night. Or the blade and the upper part of the petiole may both move. Moreover, the petiole as a whole commonly either rises or sinks at night. This movement is sometimes large: thus the petioles of Cassia pubescens stand only a little above the horizon during the day, and at night rise up almost, or quite, perpendicularly. The petioles of the younger leaves of Desmodium gyrans also rise up vertically at night. On the other hand, with Amphicarpæa, the petioles of some leaves sank down as much as 57° at night; with Arachis they sank 39°, and then stood at right angles to the stem. Generally, when the rising or sinking of several petioles on the same plant was measured, the amount differed greatly. This is largely determined by the age of the leaf: for instance, the petiole of a moderately old leaf of Desmodium gyrans rose only 46°, whilst the young ones rose up vertically; that of a young leaf of Cassia floribunda rose 41°, whilst that of an older leaf rose only 12°. It is a more singular fact that the age of the plant sometimes influences greatly the amount of movement; thus with some young seedlings of a Bauhinia the petioles rose at night 30° and 34°, whereas those on these same plants, when grown to a height of 2 or 3 feet, hardly moved at all. The position of the leaves on the plant as determined by the light, seems also to influence the amount of movement of the petiole; for no other cause was apparent why the petioles of some leaves of Melilotus officinalis rose as much as 59°, and others only 7° and 9° at night.

In the case of many plants, the petioles move at night in one direction and the leaflets in a directly opposite one. Thus, in three genera of Phaseoleae the leaflets moved vertically downwards at night, and the petioles rose in two of them, whilst in the third they sank. Species in the same genus often differ widely in the movements of their petioles. Even on the same plant of Lupinus pubescens some of the petioles rose 30°, others only 6°, and others sank 4° at night. The leaflets of Cassia Barclayana moved so little at night that they could not be said to sleep, yet the petioles of some young leaves rose as much as 34°. These several facts apparently indicate that the movements of the petioles are not performed for any special purpose; though a conclusion of this kind is generally rash. When the leaflets sink vertically down at night and the petioles rise, as often occurs, it is certain that the upward movement of the latter does not aid the leaflets in placing themselves in their proper position at night, for they have to move through a greater angular space than would otherwise have been necessary.

Notwithstanding what has just been said, it may be strongly suspected that in some cases the rising of the petioles, when considerable, does beneficially serve the plant by greatly reducing the surface exposed to radiation at night. If the reader will compare the two drawings (Fig. 155, p. 371) of Cassia pubescens, copied from photographs, he will see that the diameter of the plant at night is about one-third of what it is by day, and therefore the surface exposed to radiation is nearly nine times less. A similar conclusion may be deduced from the drawings (Fig. 149, p. 358) of a branch awake and asleep of Desmodium gyrans. So it was in a very striking manner with young plants of Bauhinia, and with Oxalis Ortegesii.

We are led to an analogous conclusion with respect to the movements of the secondary petioles of certain pinnate leaves. The pinnae of Mimosa pudica converge at night; and thus the imbricated and closed leaflets on each separate pinna are all brought close together into a single bundle, and mutually protect one another, with a somewhat smaller surface exposed to radiation. With Albizzia lophantha the pinnae close together in the same manner. Although the pinnae of Acacia Farnesiana do not converge much, they sink downwards. Those of Neptunia oleracea likewise move downwards, as well as backwards, towards the base of the leaf, whilst the main petiole rises. With Schrankia, again, the pinnae are depressed at night. Now in these three latter cases, though the pinnae do not mutually protect one another at night, yet after having sunk down they expose, as does a dependent sleeping leaf, much less surface to the zenith and to radiation than if they had remained horizontal.

Any one who had never observed continuously a sleeping plant, would naturally suppose that the leaves moved only in the evening when going to sleep, and in the morning when awaking; but he would be quite mistaken, for we have found no exception to the rule that leaves which sleep continue to move during the whole twenty-four hours; they move, however, more quickly when going to sleep and when awaking than at other times. That they are not stationary during the day is shown by all the diagrams given, and by the many more which were traced. It is troublesome to observe the movements of leaves in the middle of the night, but this was done in a few cases; and tracings were made during the early part of the night of the movements in the case of Oxalis, Amphicarpæa, two species of Erythrina, a Cassia, Passiflora, Euphorbia and Marsilea; and the leaves after they had gone to sleep, were found to be in constant movement. When, however, opposite leaflets come into close contact with one another or with the stem at night, they are, as we believe, mechanically prevented from moving, but this point was not sufficiently investigated.

When the movements of sleeping leaves are traced during twenty-four hours, the ascending and descending lines do not coincide, except occasionally and by accident for a short space; so that with many plants a single large ellipse is described during each twenty-four hours. Such ellipses are generally narrow and vertically directed, for the amount of lateral movement is small. That there is some lateral movement is shown by the ascending and descending lines not coinciding, and occasionally, as with Desmodium gyrans and Thalia dealbata, it was strongly marked. In the case of Melilotus the ellipses described by the terminal leaflet during the day are laterally extended, instead of vertically, as is usual; and this fact evidently stands in relation with the terminal leaflet moving laterally when it goes to sleep. With the majority of sleeping plants the leaves oscillate more than once up and down in the twenty-four hours; so that frequently two ellipses, one of moderate size, and one of very large size which includes the nocturnal movement, are described within the twenty-four hours. For instance, a leaf which stands vertically up during the night will sink in the morning, then rise considerably, again sink in the afternoon, and in the evening reascend and assume its vertical nocturnal position. It will thus describe, in the course of the twenty-four hours, two ellipses of unequal sizes. Other plants describe within the same time, three, four, or five ellipses. Occasionally the longer axes of the several ellipses extend in different directions, of which Acacia Farnesiana offered a good instance. The following cases will give an idea of the rate of movement: Oxalis acetosella completed two ellipses at the rate of 1 h. 25 m. for each; Marsilea quadrifoliata, at the rate of 2 h.; Trifolium subterraneum, one in 3 h. 30 m.; and Arachis hypogaea, in 4 h. 50 m. But the number of ellipses described within a given time depends largely on the state of the plant and on the conditions to which it is exposed. It often happens that a single ellipse may be described during one day, and two on the next. Erythrina corallodendron made four ellipses on the first day of observation and only a single one on the third, apparently owing to having been kept not sufficiently illuminated and perhaps not warm enough. But there seems likewise to be an innate tendency in different species of the same genus to make a different number of ellipses in the twenty-four hours: the leaflets of Trifolium repens made only one; those of T. resupinatum two, and those of T. subterraneum three in this time. Again, the leaflets of Oxalis Plumierii made a single ellipse; those of O. bupleurifolia, two; those of O. Valdiviana, two or three; and those of O. acetosella, at least five in the twenty-four hours.

The line followed by the apex of a leaf or leaflet, whilst describing one or more ellipses during the day, is often zigzag, either throughout its whole course or only during the morning or evening: Robinia offered an instance of zigzagging confined to the morning, and a similar movement in the evening is shown in the diagram (Fig. 126) given under Sida. The amount of the zigzag movement depends largely on the plant being placed under highly favourable conditions. But even under such favourable conditions, if the dots which mark the position of the apex are made at considerable intervals of time, and the dots are then joined, the course pursued will still appear comparatively simple, although the number of the ellipses will be increased; but if dots are made every two or three minutes and these are joined, the result often is that all the lines are strongly zigzag, many small loops, triangles, and other figures being also formed. This fact is shown in two parts of the diagram (Fig. 150) of the movements of Desmodium gyrans. Strephium floribundum, observed under a high temperature, made several little triangles at the rate of 43 m. for each. Mimosa pudica, similarly observed, described three little ellipses in 67 m.; and the apex of a leaflet crossed 1/500 of an inch in a second, or 0.12 inch in a minute. The leaflets of Averrhoa made a countless number of little oscillations when the temperature was high and the sun shining. The zigzag movement may in all cases be considered as an attempt to form small loops, which are drawn out by a prevailing movement in some one direction. The rapid gyrations of the little lateral leaflets of Desmodium belong to the same class of movements, somewhat exaggerated in rapidity and amplitude. The jerking movements, with a small advance and still smaller retreat, apparently not exactly in the same line, of the hypocotyl of the cabbage and of the leaves of Dionaea, as seen under the microscope, all probably come under this same head. We may suspect that we here see the energy which is freed during the incessant chemical changes in progress in the tissues, converted into motion. Finally, it should be noted that leaflets and probably some leaves, whilst describing their ellipses, often rotate slightly on their axes; so that the plane of the leaf is directed first to one and then to another side. This was plainly seen to be the case with the large terminal leaflets of Desmodium, Erythrina and Amphicarpæa, and is probably common to all leaflets provided with a pulvinus.

With respect to the periodicity of the movements of sleeping leaves, Pfeffer[23] has so clearly shown that this depends on the daily alternations of light and darkness, that nothing farther need be said on this head. But we may recall the behaviour of Mimosa in the North, where the sun does not set, and the complete inversion of the daily movements by artificial light and darkness. It has also been shown by us, that although leaves subjected to darkness for a moderately long time continue to circumnutate, yet the periodicity of their movements is soon greatly disturbed, or quite annulled. The presence of light or its absence cannot be supposed to be the direct cause of the movements, for these are wonderfully diversified even with the leaflets of the same leaf, although all have of course been similarly exposed. The movements depend on innate causes, and are of an adaptive nature. The alternations of light and darkness merely give notice to the leaves that the period has arrived for them to move in a certain manner. We may infer from the fact of several plants (Tropaeolum, Lupinus, etc.) not sleeping unless they have been well illuminated during the day, that it is not the actual decrease of light in the evening, but the contrast between the amount at this hour and during the early part of the day, which excites the leaves to modify their ordinary mode of circumnutation.

[23] 'Die Periodischen Bewegungen der Blattorgane,' 1875, p. 30, et passim.

As the leaves of most plants assume their proper diurnal position in the morning, although light be excluded, and as the leaves of some plants continue to move in the normal manner in darkness during at least a whole day, we may conclude that the periodicity of their movements is to a certain extent inherited.[24] The strength of such inheritance differs much in different species, and seems never to be rigid; for plants have been introduced from all parts of the world into our gardens and greenhouses;

and if their movements had been at all strictly fixed in relation to the alternations of day and night, they would have slept in this country at very different hours, which is not the case. Moreover, it has been observed that sleeping plants in their native homes change their times of sleep with the changing seasons.[25]

[24] Pfeffer denies such inheritance; he attributes ('Die Period. Bewegungen,' pp. 30–56) the periodicity when prolonged for a day or two in darkness, to "Nachwirkung," or the after-effects of light and darkness. But we are unable to follow his train of reasoning. There does not seem to be any more reason for attributing such movements to this cause than, for instance, the inherited habit of winter and summer wheat to grow best at different seasons; for this habit is lost after a few years, like the movements of leaves in darkness after a few days. No doubt some effect must be produced on the seeds by the long-continued cultivation of the parent-plants under different climates, but no one probably would call this the "Nachwirkung" of the climates.

[25] Pfeffer, ibid., p. 46.

We may now turn to the systematic list. This contains the names of all the sleeping plants known to us, though the list undoubtedly is very imperfect. It may be premised that, as a general rule, all the species in the same genus sleep in nearly the same manner. But there are some exceptions; in several large genera including many sleeping species (for instance, Oxalis), some do not sleep. One species of Melilotus sleeps like a Trifolium, and therefore very differently from its congeners; so does one species of Cassia. In the genus Sida, the leaves either rise or fall at night; and with Lupinus they sleep in three different methods. Returning to the list, the first point which strikes us, is that there are many more genera amongst the Leguminosae (and in almost every one of the Leguminous tribes) than in all the other families put together; and we are tempted to connect this fact with the great mobility of the stems and leaves in this family, as shown by the large number of climbing species which it contains. Next to the Leguminosae come the Malvaceae, together with some closely allied families. But by far the most important point in the list, is that we meet with sleeping plants in 28 families, in all the great divisions of the Phanerogamic series, and in one Cryptogam. Now, although it is probable that with the Leguminosae the tendency to sleep may have been inherited from one or a few progenitors, and possibly so in the cohorts of the Malvales and Chenopodiales, yet it is manifest that the tendency must have been acquired by the several genera in the other families, quite independently of one another. Hence the question naturally arises, how has this been possible? and the answer, we cannot doubt is that leaves owe their nyctitropic movements to their habit of circumnutating,—a habit common to all plants, and everywhere ready for any beneficial development or modification.

It has been shown in the previous chapters that the leaves and cotyledons of all plants are continually moving up and down, generally to a slight but sometimes to a considerable extent, and that they describe either one or several ellipses in the course of twenty-four hours; they are also so far affected by the alternations of day and night that they generally, or at least often, move periodically to a small extent; and here we have a basis for the development of the greater nyctitropic movements. That the movements of leaves and cotyledons which do not sleep come within the class of circumnutating movements cannot be doubted, for they are closely similar to those of hypocotyls, epicotyls, the stems of mature plants, and of various other organs. Now, if we take the simplest case of a sleeping leaf, we see that it makes a single ellipse in the twenty-four hours, which resembles one described by a non-sleeping leaf in every respect, except that it is much larger. In both cases the course pursued is often zigzag. As all non-sleeping leaves are incessantly circumnutating, we must conclude that a part at least of the upward and downward movement of one that sleeps, is due to ordinary circumnutation; and it seems altogether gratuitous to rank the remainder of the movement under a wholly different head. With a multitude of climbing plants the ellipses which they describe have been greatly increased for another purpose, namely, catching hold of a support. With these climbing plants, the various circumnutating organs have been so far modified in relation to light that, differently from all ordinary plants, they do not bend towards it. with sleeping plants the rate and amplitude of the movements of the leaves have been so far modified in relation to light, that they move in a certain direction with the waning light of the evening and with the increasing light of the morning more rapidly, and to a greater extent, than at other hours.

But the leaves and cotyledons of many non-sleeping plants move in a much more complex manner than in the cases just alluded to, for they describe two, three, or more ellipses in the course of a day. Now, if a plant of this kind were converted into one that slept, one side of one of the several ellipses which each leaf daily describes, would have to be greatly increased in length in the evening, until the leaf stood vertically, when it would go on circumnutating about the same spot. On the following morning, the side of another ellipse would have to be similarly increased in length so as to bring the leaf back again into its diurnal position, when it would again circumnutate until the evening. If the reader will look, for instance, at the diagram (Fig. 142, p. 351), representing the nyctitropic movements of the terminal leaflet of Trifolium subterraneum, remembering that the curved broken lines at the top ought to be prolonged much higher up, he will see that the great rise in the evening and the great fall in the morning together form a large ellipse like one of those described during the daytime, differing only in size. Or, he may look at the diagram (Fig. 103, p. 236) of the 3½ ellipses described in the course of 6 h. 35 m. by a leaf of Lupinus speciosus, which is one of the species in this genus that does not sleep; and he will see that by merely prolonging upwards the line which was already rising late in the evening, and bringing it down again next morning, the diagram would represent the movements of a sleeping plant.

With those sleeping plants which describe several ellipses in the daytime, and which travel in a strongly zigzag line, often making in their course minute loops, triangles, etc., if as soon as one of the ellipses begins in the evening to be greatly increased in size, dots are made every 2 or 3 minutes and these are joined, the line then described is almost strictly rectilinear, in strong contrast with the lines made during the daytime. This was observed with Desmodium gyrans and Mimosa pudica. With this latter plant, moreover, the pinnae converge in the evening by a steady movement, whereas during the day they are continually converging and diverging to a slight extent. In all such cases it was scarcely possible to observe the difference in the movement during the day and evening, without being convinced that in the evening the plant saves the expenditure of force by not moving laterally, and that its whole energy is now expended in gaining quickly its proper nocturnal position by a direct course. In several other cases, for instance, when a leaf after describing during the day one or more fairly regular ellipses, zigzags much in the evening, it appears as if energy was being expended, so that the great evening rise or fall might coincide with the period of the day proper for this movement.

The most complex of all the movements performed by sleeping plants, is that when leaves or leaflets, after describing in the daytime several vertically directed ellipses, rotate greatly on their axes in the evening, by which twisting movement they occupy a wholly different position at night to what they do during the day. For instance, the terminal leaflets of Cassia not only move vertically downwards in the evening, but twist round, so that their lower surfaces face outwards. Such movements are wholly, or almost wholly, confined to leaflets provided with a pulvinus. But this torsion is not a new kind of movement introduced solely for the purpose of sleep; for it has been shown that some leaflets whilst describing their ordinary ellipses during the daytime rotate slightly, causing their blades to face first to one side and then to another. Although we can see how the slight periodical movements of leaves in a vertical plane could be easily converted into the greater yet simple nyctitropic movements, we do not at present know by what graduated steps the more complex movements, effected by the torsion of the pulvini, have been acquired. A probable explanation could be given in each case only after a close investigation of the movements in all the allied forms.

From the facts and considerations now advanced we may conclude that nyctitropism, or the sleep of leaves and cotyledons, is merely a modification of their ordinary circumnutating movement, regulated in its period and amplitude by the alternations of light and darkness. The object gained is the protection of the upper surfaces of the leaves from radiation at night, often combined with the mutual protection of the several parts by their close approximation. In such cases as those of the leaflets of Cassia—of the terminal leaflets of Melilotus—of all the leaflets of Arachis, Marsilea, etc.—we have ordinary circumnutation modified to the extreme extent known to us in any of the several great classes of modified circumnutation. On this view of the origin of nyctitropism we can understand how it is that a few plants, widely distributed throughout the Vascular series, have been able to acquire the habit of placing the blades of their leaves vertically at night, that is, of sleeping,—a fact otherwise inexplicable.

The leaves of some plants move during the day in a manner, which has improperly been called diurnal sleep; for when the sun shines brightly on them, they direct their edges towards it. To such cases we shall recur in the following chapter on Heliotropism. It has been shown that the leaflets of one form of Porlieria hygrometrica keep closed during the day, as long as the plant is scantily supplied with water, in the same manner as when asleep; and this apparently serves to check evaporation. There is only one other analogous case known to us, namely, that of certain Gramineæ, which fold inwards the sides of their narrow leaves, when these are exposed to the sun and to a dry atmosphere, as described by Duval-Jouve.[26] We have also observed the same phenomenon in Elymus arenareus.

[26] 'Annal. des Sc. Nat. (Bot.),' 1875, tom. i. pp. 326–329.

There is another movement, which since the time of Linnæus has generally been called sleep, namely, that of the petals of the many flowers which close at night. These movements have been ably investigated by Pfeffer, who has shown (as was first observed by Hofmeister) that they are caused or regulated more by temperature than by the alternations of light and darkness. Although they cannot fail to protect the organs of reproduction from radiation at night, this does not seem to be their chief function, but rather the protection of the organs from cold winds, and especially from rain, during the day. the latter seems probable, as Kerner[27] has shown that a widely different kind of movement, namely, the bending down of the upper part of the peduncle, serves in many cases the same end. The closure of the flowers will also exclude nocturnal insects which may be ill-adapted for their fertilisation, and the well-adapted kinds at periods when the temperature is not favourable for fertilisation. Whether these movements of the petals consist, as is probable, of modified circumnutation we do not know.

[27] 'Die Schutzmittel des Pollens,' 1873, pp. 30–39.

Embryology of Leaves.—A few facts have been incidentally given in this chapter on what may be called the embryology of leaves. With most plants the first leaf which is developed after the cotyledons, resembles closely the leaves produced by the mature plant, but this is not always the case. the first leaves produced by some species of Drosera, for instance by D. Capensis, differ widely in shape from those borne by the mature plant, and resemble closely the leaves of D. rotundifolia, as was shown to us by Prof. Williamson of Manchester. The first true leaf of the gorse, or Ulex, is not narrow and spinose like the older leaves. On the other hand, with many Leguminous plants, for instance, Cassia, Acacia lophantha, etc., the first leaf has essentially the same character as the older leaves, excepting that it bears fewer leaflets. In Trifolium the first leaf generally bears only a single leaflet instead of three, and this differs somewhat in shape from the corresponding leaflet on the older leaves. Now, with Trifolium Pannonicum the first true leaf on some seedlings was unifoliate, and on others completely trifoliate; and between these two extreme states there were all sorts of gradations, some seedlings bearing a single leaflet more or less deeply notched on one or both sides, and some bearing a single additional and perfect lateral leaflet. Here, then, we have the rare opportunity of seeing a structure proper to a more advanced age, in the act of gradually encroaching on and replacing an earlier or embryological condition.

The genus Melilotus is closely allied to Trifolium, and the first leaf bears only a single leaflet, which at night rotates on its axis so as to present one lateral edge to the zenith. Hence it sleeps like the terminal leaflet of a mature plant, as was observed in 15 species, and wholly unlike the corresponding leaflet of Trifolium, which simply bends upwards. It is therefore a curious fact that in one of these 15 species, viz., M. Taurica (and in a lesser degree in two others), leaves arising from young shoots, produced on plants which had been cut down and kept in pots during the winter in the green-house, slept like the leaves of a Trifolium, whilst the leaves on the fully-grown branches on these same plants afterwards slept normally like those of a Melilotus. If young shoots rising from the ground may be considered as new individuals, partaking to a certain extent of the nature of seedlings, then the peculiar manner in which their leaves slept may be considered as an embryological habit, probably the result of Melilotus being descended from some form which slept like a Trifolium.

This view is partially supported by the leaves on old and young branches of another species, M. Messanensis (not included in the above 15 species), always sleeping like those of a Trifolium.

The first true leaf of Mimosa albida consists of a simple petiole, often bearing three pairs of leaflets, all of which are of nearly equal size and of the same shape: the second leaf differs widely from the first, and resembles that on a mature plant (see Fig. 159, p. 379), for it consists of two pinnae, each of which bears two pairs of leaflets, of which the inner basal one is very small. But at the base of each pinna there is a pair of minute points, evidently rudiments of leaflets, for they are of unequal sizes, like the two succeeding leaflets. These rudiments are in one sense embryological, for they exist only during the youth of the leaf, falling off and disappearing as soon as it is fully grown.

With Desmodium gyrans the two lateral leaflets are very much smaller than the corresponding leaflets in most of the species in this large genus; they vary also in position and size; one or both are sometimes absent; and they do not sleep like the fully-developed leaflets. They may therefore be considered as almost rudimentary; and in accordance with the general principles of embryology, they ought to be more constantly and fully developed on very young than on old plants. But this is not the case, for they were quite absent on some young seedlings, and did not appear until from 10 to 20 leaves had been formed. This fact leads to the suspicion that D. gyrans is descended through a unifoliate form (of which some exist) from a trifoliate species; and that the little lateral leaflets reappear through reversion. However this may be, the interesting fact of the pulvini or organs of movement of these little leaflets, not having been reduced nearly so much as their blades—taking the large terminal leaflet as the standard of comparison—gives us probably the proximate cause of their extraordinary power of gyration.

CHAPTER VIII.
MODIFIED CIRCUMNUTATION: MOVEMENTS EXCITED BY LIGHT.

Distinction between heliotropism and the effects of light on the periodicity of the movements of leaves—Heliotropic movements of Beta, Solanum, Zea, and Avena—Heliotropic movements towards an obscure light in Apios, Brassica, Phalaris, Tropaeolum, and Cassia—Apheliotropic movements of tendrils of Bignonia—Of flower-peduncles of Cyclamen—Burying of the pods—Heliotropism and apheliotropism modified forms of circumnutation—Steps by which one movement is converted into the other—Transversal-heliotropismus or diaheliotropism influenced by epinasty, the weight of the part and apogeotropism—Apogeotropism overcome during the middle of the day by diaheliotropism—Effects of the weight of the blades of cotyledons—So called diurnal sleep—Chlorophyll injured by intense light—Movements to avoid intense light

Sachs first clearly pointed out the important difference between the action of light in modifying the periodic movements of leaves, and in causing them to bend towards its source.[1] The latter, or heliotropic movements are determined by the direction of the light, whilst periodic movements are affected by changes in its intensity and not by its direction. The periodicity of the circumnutating movement often continues for some time in darkness, as we have seen in the last chapter; whilst heliotropic bending ceases very quickly when the light fails. Nevertheless, plants which have ceased through long-continued darkness to move periodically, if re-exposed to the light are still, according to Sachs, heliotropic.

[1] 'Physiologie Veg.' (French Translation), 1868, pp. 42, 517, etc.

Apheliotropism, or, as usually designated, negative heliotropism, implies that a plant, when unequally illuminated on the two sides, bends from the light, instead of, as in the last sub-class of cases, towards it; but apheliotropism is comparatively rare, at least in a well-marked degree. There is a third and large sub-class of cases, namely, those of "transversal-Heliotropismus" of Frank, which we will here call diaheliotropism. Parts of plants, under this influence, place themselves more or less transversely to the direction whence the light proceeds, and are thus fully illuminated. There is a fourth sub-class, as far as the final cause of the movement is concerned; for the leaves of some plants when exposed to an intense and injurious amount of light direct themselves, by rising or sinking or twisting, so as to be less intensely illuminated. Such movements have sometimes been called diurnal sleep. If thought advisable, they might be called paraheliotropic, and this term would correspond with our other terms.

It will be shown in the present chapter that all the movements included in these four sub-classes, consist of modified circumnutation. We do not pretend to say that if a part of a plant, whilst still growing, did not circumnutate—though such a supposition is most improbable—it could not bend towards the light; but, as a matter of fact, heliotropism seems always to consist of modified circumnutation. Any kind of movement in relation to light will obviously be much facilitated by each part circumnutating or bending successively in all directions, so that an already existing movement has only to be increased in some one direction, and to be lessened or stopped in the other directions, in order that it should become heliotropic, apheliotropic, etc., as the case may be. In the next chapter some observations on the sensitiveness of plants to light, their rate of bending towards it, and the accuracy with which they point towards its source, etc., will be given. Afterwards it will be shown—and this seems to us a point of much interest—that sensitiveness to light is sometimes confined to a small part of the plant; and that this part when stimulated by light, transmits an influence to distant parts, exciting them to bend.

Heliotropism.—When a plant which is strongly heliotropic (and species differ much in this respect) is exposed to a bright lateral light, it bends quickly towards it, and the course pursued by the stem is quite or nearly straight. But if the light is much dimmed, or occasionally interrupted, or admitted in only a slightly oblique direction, the course pursued is more or less zigzag; and as we have seen and shall again see, such zigzag movement results from the elongation or drawing out of the ellipses, loops, etc., which the plant would have described, if it had been illuminated from above. On several occasions we were much struck with this fact, whilst observing the circumnutation of highly sensitive seedlings, which were unintentionally illuminated rather obliquely, or only at successive intervals of time.

Fig. 168. Beta vulgaris: circumnutation of hypocotyl, deflected by the light being slightly lateral, traced on a horizontal glass from 8.30 A.M. to 5.30 P.M. Direction of the lighted taper by which it was illuminated shown by a line joining the first and penultimate dots. Figure reduced to one-third of the original scale.

For instance two young seedlings of Beta vulgaris were placed in the middle of a room with north-east windows, and were kept covered up, except during each observation which lasted for only a minute or two; but the result was that their hypocotyls bowed themselves to the side, whence some light occasionally entered, in lines which were only slightly zigzag. Although not a single ellipse was even approximately formed, we inferred from the zigzag lines—and, as it proved, correctly—that their hypocotyls were circumnutating, for on the following day these same seedlings were placed in a completely darkened room, and were observed each time by the aid of a small wax taper held almost directly above them, and their movements were traced on a horizontal glass above; and now their hypocotyls clearly circumnutated (Fig. 168, and Fig. 39, formerly given, p. 52); yet they moved a short distance towards the side where the taper was held up. If we look at these diagrams, and suppose that the taper had been held more on one side, and that the hypocotyls, still circumnutating, had bent themselves within the same time much more towards the light, long zigzag lines would obviously have been the result.

Fig. 169. Avena sativa: heliotropic movement and circumnutation of sheath-like cotyledon (1½ inch in height) traced on horizontal glass from 8 A.M. to 10.25 P.M. Oct. 16th.

Again, two seedlings of Solanum lycopersicum were illuminated from above, but accidentally a little more light entered on one than on any other side, and their hypocotyls became slightly bowed towards the brighter side; they moved in a zigzag line and described in their course two little triangles, as seen in Fig. 37 (p. 50), and in another tracing not given. The sheath-like cotyledons of Zea mays behaved, under nearly similar circumstances, in a nearly similar manner as described in our first chapter (p. 64), for they bowed themselves during the whole day towards one side, making, however, in their course some conspicuous flexures. Before we knew how greatly ordinary circumnutation was modified by a lateral light, some seedling oats, with rather old and therefore not highly sensitive cotyledons, were placed in front of a north-east window, towards which they bent all day in a strongly zigzag course. On the following day they continued to bend in the same direction (Fig. 169), but zigzagged much less. The sky, however, became between 12.40 and 2.35 P.M. overcast with extraordinarily dark thunder-clouds, and it was interesting to note how plainly the cotyledons circumnutated during this interval.

The foregoing observations are of some value, from having been made when we were not attending to heliotropism; and they led us to experiment on several kinds of seedlings, by exposing them to a dim lateral light, so as to observe the gradations between ordinary circumnutation and heliotropism. Seedlings in pots were placed in front of, and about a yard from, a north-east window; on each side and over the pots black boards were placed; in the rear the pots were open to the diffused light of the room, which had a second north-east and a north-west window. By hanging up one or more blinds before the window where the seedlings stood, it was easy to dim the light, so that very little more entered on this side than on the opposite one, which received the diffused light of the room. Late in the evening the blinds were successively removed, and as the plants had been subjected during the day to a very obscure light, they continued to bend towards the window later in the evening than would otherwise have occurred. Most of the seedlings were selected because they were known to be highly sensitive to light, and some because they were but little sensitive, or had become so from having grown old. The movements were traced in the usual manner on a horizontal glass cover; a fine glass filament with little triangles of paper having been cemented in an upright position to the hypocotyls. Whenever the stem or hypocotyl became much bowed towards the light, the latter part of its course had to be traced on a vertical glass, parallel to the window, and at right angles to the horizontal glass cover.

Fig. 170. Apios graveolens: heliotropic movement of hypocotyl (.45 of inch in height) towards a moderately bright lateral light, traced on a horizontal glass from 8.30 A.M. to 11.30 A.M. Sept. 18th. Figure reduced to one-third of original scale.

Apios graveolens.—The hypocotyl bends in a few hours rectangularly towards a bright lateral light. In order to ascertain how straight a course it would pursue when fairly well illuminated on one side, seedlings were first placed before a south-west window on a cloudy and rainy morning; and the movement of two hypocotyls were traced for 3 h., during which time they became greatly bowed towards the light. One of these tracings is given on p. 422 (Fig. 170), and the course may be seen to be almost straight. But the amount of light on this occasion was superfluous, for two seedlings were placed before a north-east window, protected by an ordinary linen and two muslin blinds, yet their hypocotyls moved towards this rather dim light in only slightly zigzag lines; but after 4 P.M., as the light waned, the lines became distinctly zigzag. One of these seedlings, moreover, described in the afternoon an ellipse of considerable size, with its longer axis directed towards the window.

We now determined that the light should be made dim enough, so we began by exposing several seedlings before a north-east window, protected by one linen blind, three muslin blinds, and a towel. But so little light entered that a pencil cast no perceptible shadow on a white card, and the hypocotyls did not bend at all towards the window. During this time, from 8.15 to 10.50 A.M., the hypocotyls zigzagged or circumnutated near the same spot, as may be seen at A, in Fig. 171. The towel, therefore, was removed at 10.50 A.M., and replaced by two muslin blinds, and now the light passed through one ordinary linen and four muslin blinds. When a pencil was held upright on a card close to the seedlings, it cast a shadow (pointing from the window) which could only just be distinguished. Yet this very slight excess of light on one side sufficed to cause the hypocotyls of all the seedlings immediately to begin bending in zigzag lines towards the window. The course of one is shown at A (Fig. 171): after moving towards the window from 10.50 A.M. to 12.48 P.M. it bent from the window, and then returned in a nearly parallel line; that is, it almost completed between 12.48 and 2 P.M. a narrow ellipse. Late in the evening, as the light waned, the hypocotyl ceased to bend towards the window, and circumnutated on a small scale round the same spot; during the night it moved considerably backwards, that is, became more upright, through the action of apogeotropism. At B, we have a tracing of the movements of another seedling from the hour (10.50 A.M.) when the towel was removed; and it is in all essential respects similar to the previous one. In these two cases there could be no doubt that the ordinary circumnutating movement of the hypocotyl was modified and rendered heliotropic.

Fig. 171. Apios graveolens: heliotropic movement and circumnutation of the hypocotyls of two seedlings towards a dim lateral light, traced on a horizontal glass during the day. The broken lines show their return nocturnal courses. Height of hypocotyl of A .5, and of B .55 inch. Figure reduced to one-half of original scale.

Brassica oleracea.—The hypocotyl of the cabbage, when not disturbed by a lateral light, circumnutates in a complicated manner over nearly the same space, and a figure formerly given is here reproduced (Fig. 172). If the hypocotyl is exposed to a moderately strong lateral light it moves quickly towards this side, travelling in a straight, or nearly straight, line. But when the lateral light is very dim its course is extremely tortuous, and evidently consists of modified circumnutation. Seedlings were placed before a north-east window, protected by a linen and muslin blind and by a towel. The sky was cloudy, and whenever the clouds grew a little lighter an additional muslin blind was temporarily suspended. The light from the window was thus so much obscured that, judging by the unassisted eye, the seedlings appeared to receive more light from the interior of the room than from the window; but this was not really the case, as was shown by a very faint shadow cast by a pencil on a card. Nevertheless, this extremely small excess of light on one side caused the hypocotyls, which in the morning had stood upright, to bend at right angles towards the window, so that in the evening (after 4.23 P.M.) their course had to be traced on a vertical glass parallel to the window. It should be stated that at 3.30 P.M., by which time the sky had become darker, the towel was removed and replaced by an additional muslin blind, which itself was removed at 4 P.M., the other two blinds being left suspended. In Fig. 173 the course pursued, between 8.9 A.M. and 7.10 P.M., by one of the hypocotyls thus exposed is shown. It may be observed that during the first 16 m. the hypocotyl moved obliquely from the light, and this, no doubt, was due to its then circumnutating in this direction. Similar cases were repeatedly observed, and a dim light rarely or never produced any effect until from a quarter to three-quarters of an hour had elapsed. After 5.15 P.M., by which time the light had become obscure, the hypocotyl began to circumnutate about the same spot. The contrast between the two figures (172 and 173) would have been more striking, if they had been originally drawn on the same scale, and had been equally reduced. But the movements shown in Fig. 172 were at first more magnified, and have been reduced to only one-half of the original scale; whereas those in Fig. 173 were at first less magnified, and have been reduced to a one-third scale. A tracing made at the same time with the last of the movements of a second hypocotyl, presented a closely analogous appearance; but it did not bend quite so much towards the light, and it circumnutated rather more plainly.

Fig. 172. Brassica oleracea: ordinary circumnutating movement of the hypocotyl of a seedling plant.

Fig. 173. Brassica oleracea: heliotropic movement and circumnutation of a hypocotyl towards a very dim lateral light, traced during 11 hours, on a horizontal glass in the morning, and on a vertical glass in the evening. Figure reduced to one-third of the original scale.

Phalaris Canariensis.—The sheath-like cotyledons of this monocotyledonous plant were selected for trial, because they are very sensitive to light and circumnutate well, as formerly shown (see Fig. 49, p. 63). Although we felt no doubt about the result, some seedlings were first placed before a south-west window on a moderately bright morning, and the movements of one were traced. As is so common, it moved for the first 45 m. in a zigzag line; it then felt the full influence of the light, and travelled towards it for the next 2 h. 30 m. in an almost straight line. The tracing has not been given, as it was almost identical with that of Apios under similar circumstances (Fig. 170). By noon it had bowed itself to its full extent; it then circumnutated about the same spot and described two ellipses; by 5 P.M. it had retreated considerably from the light, through the action of apogeotropism. After some preliminary trials for ascertaining the right degree of obscurity, some seedlings were placed (Sept. 16th) before a north-east window, and light was admitted through an ordinary linen and three muslin blinds. A pencil held close by the pot now cast a very faint shadow on a white card, pointing from the window. In the evening, at 4.30 and again at 6 P.M., some of the blinds were removed. In Fig. 174 we see the course pursued under these circumstances by a rather old and not very sensitive cotyledon, 1.9 inch in height, which became much bowed, but was never rectangularly bent towards the light. From 11 A.M., when the sky became rather duller, until 6.30 P.M., the zigzagging was conspicuous, and evidently consisted of drawn-out ellipses. After 6.30 P.M. and during the night, it retreated in a crooked line from the window. Another and younger seedling moved during the same time much more quickly and to a much greater distance, in an only slightly zigzag line towards the light; by 11 A.M. it was bent almost rectangularly in this direction, and now circumnutated about the same place.

Fig. 174. Phalaris Canariensis: heliotropic movement and circumnutation of a rather old cotyledon, towards a dull lateral light, traced on a horizontal glass from 8.15 A.M. Sept. 16th to 7.45 A.M. 17th. Figure reduced to one-third of original scale.

Tropaeolum majus.—Some very young seedlings, bearing only two leaves, and therefore not as yet arrived at the climbing stage of growth, were first tried before a north-east window without any blind. The epicotyls bowed themselves towards the light so rapidly that in little more than 3 h. their tips pointed rectangularly towards it. The lines traced were either nearly straight or slightly zigzag; and in this latter case we see that a trace of circumnutation was retained even under the influence of a moderately bright light. Twice whilst these epicotyls were bending towards the window, dots were made every 5 or 6 minutes, in order to detect any trace of lateral movement, but there was hardly any; and the lines formed by their junction were nearly straight, or only very slightly zigzag, as in the other parts of the figures. After the epicotyls had bowed themselves to the full extent towards the light, ellipses of considerable size were described in the usual manner.

After having seen how the epicotyls moved towards a moderately bright light, seedlings were placed at 7.48 A.M. (Sept. 7th) before a north-east window, covered by a towel, and shortly afterwards by an ordinary linen blind, but the epicotyls still moved towards the window. At 9.13 A.M. two additional muslin blinds were suspended, so that the seedlings received very little more light from the window than from the interior of the room. The sky varied in brightness, and the seedlings occasionally received for a short time less light from the window than from the opposite side (as ascertained by the shadow cast), and then one of the blinds was temporarily removed. In the evening the blinds were taken away, one by one. the course pursued by an epicotyl under these circumstances is shown in Fig. 175. During the whole day, until 6.45 P.M., it plainly bowed itself towards the light; and the tip moved over a considerable space. After 6.45 P.M. it moved backwards, or from the window, till 10.40 P.M., when the last dot was made. Here, then, we have a distinct heliotropic movement, effected by means of six elongated figures (which if dots had been made every few minutes would have been more or less elliptic) directed towards the light, with the apex of each successive ellipse nearer to the window than the previous one. Now, if the light had been only a little brighter, the epicotyl would have bowed itself more to the light, as we may safely conclude from the previous trials; there would also have been less lateral movement, and the ellipses or other figures would have been drawn out into a strongly marked zigzag line, with probably one or two small loops still formed. If the light had been much brighter, we should have had a slightly zigzag line, or one quite straight, for there would have been more movement in the direction of the light, and much less from side to side.

Fig. 175. Tropaeolum majus: heliotropic movement and circumnutation of the epicotyl of a young seedling towards a dull lateral light, traced on a horizontal glass from 7.48 A.M. to 10.40 P.M. Figure reduced to one-half of the original scale.

Fig. 176. Tropaeolum majus: heliotropic movement and circumnutation of an old internode towards a lateral light, traced on a horizontal glass from 8 A.M. Nov. 2nd to 10.20 A.M. Nov. 4th. Broken lines show the nocturnal course.

Sachs states that the older internodes of this Tropaeolum are apheliotropic; we therefore placed a plant, 11 3/4 inches high, in a box, blackened within, but open on one side in front of a north-east window without any blind. A filament was fixed to the third internode from the summit on one plant, and to the fourth internode of another. These internodes were either not old enough, or the light was not sufficiently bright, to induce apheliotropism, for both plants bent slowly towards, instead of from the window during four days. The course, during two days of the first-mentioned internode, is given in Fig. 176; and we see that it either circumnutated on a small scale, or travelled in a zigzag line towards the light. We have thought this case of feeble heliotropism in one of the older internodes of a plant, which, whilst young, is so extremely sensitive to light, worth giving.

Fig. 177. Cassia tora: heliotropic movement and circumnutation of a hypocotyl (1½ inch in height) traced on a horizontal glass from 8 A.M. to 10.10 P.M. Oct. 7th. Also its circumnutation in darkness from 7 A.M. Oct. 8th to 7.45 A.M. Oct. 9th.

Cassia tora.—The cotyledons of this plant are extremely sensitive to light, whilst the hypocotyls are much less sensitive than those of most other seedlings, as we had often observed with surprise. It seemed therefore worth while to trace their movements. They were exposed to a lateral light before a north-east window, which was at first covered merely by a muslin blind, but as the sky grew brighter about 11 A.M., an additional linen blind was suspended. After 4 P.M. one blind and then the other was removed.

The seedlings were protected on each side and above, but were open to the diffused light of the room in the rear. Upright filaments were fixed to the hypocotyls of two seedlings, which stood vertically in the morning. The accompanying figure (Fig. 177) shows the course pursued by one of them during two days; but it should be particularly noticed that during the second day the seedlings were kept in darkness, and they then circumnutated round nearly the same small space. On the first day (Oct. 7th) the hypocotyl moved from 8 A.M. to 12.23 P.M., toward the light in a zigzag line, then turned abruptly to the left and afterwards described a small ellipse. Another irregular ellipse was completed between 3 P.M. and about 5.30 P.M., the hypocotyl still bending towards the light. The hypocotyl was straight and upright in the morning, but by 6 P.M. its upper half was bowed towards the light, so that the chord of the arc thus formed stood at an angle of 20° with the perpendicular. After 6 P.M. its course was reversed through the action of apogeotropism, and it continued to bend from the window during the night, as shown by the broken line. On the next day it was kept in the dark (excepting when each observation was made by the aid of a taper), and the course followed from 7 A.M. on the 8th to 7.45 A.M. on the 9th is here likewise shown. The difference between the two parts of the figure (177), namely that described during the daytime on the 7th, when exposed to a rather dim lateral light, and that on the 8th in darkness, is striking. The difference consists in the lines during the first day having been drawn out in the direction of the light. The movements of the other seedling, traced under the same circumstances, were closely similar.

Apheliotropism.—We succeeded in observing only two cases of apheliotropism, for these are somewhat rare; and the movements are generally so slow that they would have been very troublesome to trace.

Fig. 178. Bignonia capreolata: apheliotropic movement of a tendril, traced on a horizontal glass from 6.45 A.M. July 19th to 10 A.M. 20th. Movements as originally traced, little magnified, here reduced to two-thirds of the original scale.

Bignonia capreolata.—No organ of any plant, as far as we have seen, bends away so quickly from the light as do the tendrils of this Bignonia. They are also remarkable from circumnutating much less regularly than most other tendrils, often remaining stationary; they depend on apheliotropism for coming into contact with the trunks of trees.[2] The stem of a young plant was tied to a stick at the base of a pair of fine tendrils, which projected almost vertically upwards; and it was placed in front of a north-east window, being protected on all other sides from the light. The first dot was made at 6.45 A.M., and by 7.35 A.M. both tendrils felt the full influence of the light, for they moved straight away from it until 9.20 A.M., when they circumnutated for a time, still moving, but only a little, from the light (see Fig. 178 of the left-hand tendril). After 3 P.M. they again moved rapidly away from the light in zigzag lines. By a late hour in the evening both had moved so far, that they pointed in a direct line from the light. During the night they returned a little in a nearly opposite direction. On the following morning they again moved from the light and converged, so that by the evening they had become interlocked, still pointing from the light. The right-hand tendril, whilst converging, zigzagged much more than the one figured. Both tracings showed that the apheliotropic movement was a modified form of circumnutation.

[2] 'The Movements and Habits of Climbing Plants,' 1875, p. 97.

Cyclamen Persicum.—Whilst this plant is in flower the peduncles stand upright, but their uppermost part is hooked so that the flower itself hangs downwards. As soon as the pods begin to swell, the peduncles increase much in length and slowly curve downwards, but the short, upper, hooked part straightens itself. Ultimately the pods reach the ground, and if this is covered with moss or dead leaves, they bury themselves. We have often seen saucer-like depressions formed by the pods in damp sand or sawdust; and one pod (.3 of inch in diameter) buried itself in sawdust for three-quarters of its length.[3] We shall have occasion hereafter to consider the object gained by this burying process. The peduncles can change the direction of their curvature, for if a pot, with plants having their peduncles already bowed downwards, be placed horizontally, they slowly bend at right angles to their former direction towards the centre of the earth. We therefore at first attributed the movement to geotropism; but a pot which had lain horizontally with the pods all pointing to the ground, was reversed, being still kept horizontal, so that the pods now pointed directly upwards; it was then placed in a dark cupboard, but the pods still pointed upwards after four days and nights. The pot, in the same position, was next brought back into the light, and after two days there was some bending downwards of the peduncles, and on the fourth day two of them pointed to the centre of the earth, as did the others after an additional day or two. Another plant, in a pot which had always stood upright, was left in the dark cupboard for six days; it bore 3 peduncles, and only one became within this time at all bowed downwards, and that doubtfully. The weight, therefore, of the pods is not the cause of the bending down. This pot was then brought back into the light, and after three days the peduncles were considerably bowed downwards. We are thus led to infer that the downward curvature is due to apheliotropism; though more trials ought to have been made.

[3] The peduncles of several other species of Cyclamen twist themselves into a spire, and according to Erasmus Darwin ('Botanic Garden,' Canto.,

iii. p. 126), the pods forcibly penetrate the earth. See also Grenier and Godron, 'Flore de France,' tom. ii. p. 459.

Fig. 179. Cyclamen Persicum: downward apheliotropic movement of a flower-peduncle, greatly magnified (about 47 times?), traced on a horizontal glass from 1 P.M. Feb. 18th to 8 A.M. 21st.

In order to observe the nature of this movement, a peduncle bearing a large pod which had reached and rested on the ground, was lifted a little up and secured to a stick. A filament was fixed across the pod with a mark beneath, and its movement, greatly magnified, was traced on a horizontal glass during 67 h. The plant was illuminated during the day from above. A copy of the tracing is given on p. 434 (Fig. 179); and there can be no doubt that the descending movement is one of modified circumnutation, but on an extremely small scale. The observation was repeated on another pod, which had partially buried itself in sawdust, and which was lifted up a quarter of an inch above the surface; it described three very small circles in 24 h. Considering the great length and thinness of the peduncles and the lightness of the pods, we may conclude that they would not be able to excavate saucer-like depressions in sand or sawdust, or bury themselves in moss, etc., unless they were aided by their continued rocking or circumnutating movement.

Relation between Circumnutation and Heliotropism.—Any one who will look at the foregoing diagrams, showing the movements of the stems of various plants towards a lateral and more or less dimmed light, will be forced to admit that ordinary circumnutation and heliotropism graduate into one another. When a plant is exposed to a dim lateral light and continues during the whole day bending towards it, receding late in the evening, the movement unquestionably is one of heliotropism. Now, in the case of Tropaeolum (Fig. 175) the stem or epicotyl obviously circumnutated during the whole day, and yet it continued at the same time to move heliotropically; this latter movement being effected by the apex of each successive elongated figure or ellipse standing nearer to the light than the previous one. In the case of Cassia (Fig. 177) the comparison of the movement of the hypocotyl, when exposed to a dim lateral light and to darkness, is very instructive; as is that between the ordinary circumnutating movement of a seedling Brassica (Figs. 172, 173), or that of Phalaris (Figs. 49, 174), and their heliotropic movement towards a window protected by blinds. In both these cases, and in many others, it was interesting to notice how gradually the stems began to circumnutate as the light waned in the evening. We have therefore many kinds of gradations from a movement towards the light, which must be considered as one of circumnutation very slightly modified and still consisting of ellipses or circles,—though a movement more or less strongly zigzag, with loops or ellipses occasionally formed,—to a nearly straight, or even quite straight, heliotropic course.

A plant, when exposed to a lateral light, though this may be bright, commonly moves at first in a zigzag line, or even directly from the light; and this no doubt is due to its circumnutating at the time in a direction either opposite to the source of the light, or more or less transversely to it. As soon, however, as the direction of the circumnutating movement nearly coincides with that of the entering light, the plant bends in a straight course towards the light, if this is bright. The course appears to be rendered more and more rapid and rectilinear, in accordance with the degree of brightness of the light—firstly, by the longer axes of the elliptical figures, which the plant continues to describe as long as the light remains very dim, being directed more or less accurately towards its source, and by each successive ellipse being described nearer to the light. Secondly, if the light is only somewhat dimmed, by the acceleration and increase of the movement towards it, and by the retardation or arrestment of that from the light, some lateral movement being still retained, for the light will interfere less with a movement at right angles to its direction, than with one in its own direction.[4] The result is that the course is rendered more or less zigzag and unequal in rate. Lastly, when the light is very bright all lateral movement is lost; and the whole energy of the plant is expended in rendering the circumnutating movement rectilinear and rapid in one direction alone, namely, towards the light.

[4] In his paper, 'Ueber orthotrope und plagiotrope Pflanzentheile' ('Arbeiten des Bot. Inst. in Würzburg,' Band ii. Heft ii. 1879), Sachs has discussed the manner in which geotropism and heliotropism are affected by differences in the angles at which the organs of plants stand with respect to the direction of the incident force.

The common view seems to be that heliotropism is a quite distinct kind of movement from circumnutation; and it may be urged that in the foregoing diagrams we see heliotropism merely combined with, or superimposed on, circumnutation. But if so, it must be assumed that a bright lateral light completely stops circumnutation, for a plant thus exposed moves in a straight line towards it, without describing any ellipses or circles. If the light be somewhat obscured, though amply sufficient to cause the plant to bend towards it, we have more or less plain evidence of still-continued circumnutation. It must further be assumed that it is only a lateral light which has this extraordinary power of stopping circumnutation, for we know that the several plants above experimented on, and all the others which were observed by us whilst growing, continue to circumnutate, however bright the light may be, if it comes from above. Nor should it be forgotten that in the life of each plant, circumnutation precedes heliotropism, for hypocotyls, epicotyls, and petioles circumnutate before they have broken through the ground and have ever felt the influence of light.

We are therefore fully justified, as it seems to us, in believing that whenever light enters laterally, it is the movement of circumnutation which gives rise to, or is converted into, heliotropism and apheliotropism. On this view we need not assume against all analogy that a lateral light entirely stops circumnutation; it merely excites the plant to modify its movement for a time in a beneficial manner. The existence of every possible gradation, between a straight course towards a lateral light and a course consisting of a series of loops or ellipses, becomes perfectly intelligible. Finally, the conversion of circumnutation into heliotropism or apheliotropism, is closely analogous to what takes place with sleeping plants, which during the daytime describe one or more ellipses, often moving in zigzag lines and making little loops; for when they begin in the evening to go to sleep, they likewise expend all their energy in rendering their course rectilinear and rapid. In the case of sleep-movements, the exciting or regulating cause is a difference in the intensity of the light, coming from above, at different periods of the twenty-four hours; whilst with heliotropic and apheliotropic movements, it is a difference in the intensity of the light on the two sides of the plant.

Transversal-heliotropismus (of Frank[5]*) or Diaheliotropism.*—The cause of leaves placing themselves more or less transversely to the light, with their upper surfaces directed towards it, has been of late the subject of much controversy. We do not here refer to the object of the movement, which no doubt is that their upper surfaces may be fully illuminated, but the means by which this position is gained. Hardly a better or more simple instance can be given of diaheliotropism than that offered by many seedlings, the cotyledons of which are extended horizontally. When they first burst from their seed-coats they are in contact and stand in various positions, often vertically upwards; they soon diverge, and this is effected by epinasty, which, as we have seen, is a modified form of circumnutation. After they have diverged to their full extent, they retain nearly the same position, though brightly illuminated all day long from above, with their lower surfaces close to the ground and thus much shaded. There is therefore a great contrast in the degree of illumination of their upper and lower surfaces, and if they were heliotropic they would bend quickly upwards. It must not, however, be supposed that such cotyledons are immovably fixed in a horizontal position. When seedlings are exposed before a window, their hypocotyls, which are highly heliotropic, bend quickly towards it, and the upper surfaces of their cotyledons still remain exposed at right angles to the light; but if the hypocotyl is secured so that it cannot bend, the cotyledons themselves change their position. If the two are placed in the line of the entering light, the one furthest from it rises up and that nearest to it often sinks down; if placed transversely to the light, they twist a little laterally; so that in every case they endeavour to place their upper surfaces at right angles to the light. So it notoriously is with the leaves on plants nailed against a wall, or grown in front of a window. A moderate amount of light suffices to induce such movements; all that is necessary is that the light should steadily strike the plants in an oblique direction. With respect to the above twisting movement of cotyledons, Frank has given many and much more striking instances in the case of the leaves on branches which had been fastened in various positions or turned upside down.

[5] 'Die natürliche Wagerechte Richtung von Pflanzentheilen,' 1870. See also some interesting articles by the same author, "Zur Frage über Transversal-Geo-und Heliotropismus," 'Bot. Zeitung,' 1873, p. 17 et seq.

In our observations on the cotyledons of seedling plants, we often felt surprise at their persistent horizontal position during the day, and were convinced before we had read Frank's essay, that some special explanation was necessary. De Vries has shown[6] that the more or less horizontal position of leaves is in most cases influenced by epinasty, by their own weight, and by apogeotropism. A young cotyledon or leaf after bursting free is brought down into its proper position, as already remarked, by epinasty, which, according to De Vries, long continues to act on the midribs and petioles. Weight can hardly be influential in the case of cotyledons, except in a few cases presently to be mentioned, but must be so with large and thick leaves. With respect to apogeotropism, De Vries maintains that it generally comes into play, and of this fact we shall presently advance some indirect evidence. But over these and other constant forces we believe that there is in many cases, but we do not say in all, a preponderant tendency in leaves and cotyledons to place themselves more or less transversely with respect to the light.

[6] 'Arbeiten des Bot. Instituts in Würzburg,' Heft. ii. 1872, pp. 223–277.

In the cases above alluded to of seedlings exposed to a lateral light with their hypocotyls secured, it is impossible that epinasty, weight and apogeotropism, either in opposition or combined, can be the cause of the rising of one cotyledon, and of the sinking of the other, since the forces in question act equally on both; and since epinasty, weight and apogeotropism all act in a vertical plane, they cannot cause the twisting of the petioles, which occurs in seedlings under the above conditions of illumination. All these movements evidently depend in some manner on the obliquity of the light, but cannot be called heliotropic, as this implies bending towards the light; whereas the cotyledon nearest to the light bends in an opposed direction or downwards, and both place themselves as nearly as possible at right angles to the light. The movement, therefore, deserves a distinct name. As cotyledons and leaves are continually oscillating up and down, and yet retain all day long their proper position with their upper surfaces directed transversely to the light, and if displaced reassume this position, diaheliotropism must be considered as a modified form of circumnutation. This was often evident when the movements of cotyledons standing in front of a window were traced. We see something analogous in the case of sleeping leaves or cotyledons, which after oscillating up and down during the whole day, rise into a vertical position late in the evening, and on the following morning sink down again into their horizontal or diaheliotropic position, in direct opposition to heliotropism. This return into their diurnal position, which often requires an angular movement of 90°, is analogous to the movement of leaves on displaced branches, which recover their former positions. It deserves notice that any force such as apogeotropism, will act with different degrees of power[7] in the different positions of those leaves or cotyledons which oscillate largely up and down during the day; and yet they recover their horizontal or diaheliotropic position.

[7] See former note, in reference to Sachs' remarks on this subject.

We may therefore conclude that diaheliotropic movements cannot be fully explained by the direct action of light, gravitation, weight, etc., any more than can the nyctitropic movements of cotyledons and leaves. In the latter case they place themselves so that their upper surfaces may radiate at night as little as possible into open space, with the upper surfaces of the opposite leaflets often in contact. These movements, which are sometimes extremely complex, are regulated, though not directly caused, by the alternations of light and darkness. In the case of diaheliotropism, cotyledons and leaves place themselves so that their upper surfaces may be exposed to the light, and this movement is regulated, though not directly caused, by the direction whence the light proceeds. In both cases the movement consists of circumnutation modified by innate or constitutional causes, in the same manner as with climbing plants, the circumnutation of which is increased in amplitude and rendered more circular, or again with very young cotyledons and leaves which are thus brought down into a horizontal position by epinasty.

We have hitherto referred only to those leaves and cotyledons which occupy a permanently horizontal position; but many stand more or less obliquely, and some few upright. the cause of these differences of position is not known; but in accordance with Wiesner's views, hereafter to be given, it is probable that some leaves and cotyledons would suffer, if they were fully illuminated by standing at right angles to the light.

We have seen in the second and fourth chapters that those cotyledons and leaves which do not alter their positions at night sufficiently to be said to sleep, commonly rise a little in the evening and fall again on the next morning, so that they stand during the night at a rather higher inclination than during the middle of the day. It is incredible that a rising movement of 2° or 3°, or even of 10° or 20°, can be of any service to the plant, so as to have been specially acquired. It must be the result of some periodical change in the conditions to which they are subjected, and there can hardly be a doubt that this is the daily alternations of light and darkness. De Vries states in the paper before referred to, that most petioles and midribs are apogeotropic;[8] and apogeotropism would account for the above rising movement, which is common to so many widely distinct species, if we suppose it to be conquered by diaheliotropism during the middle of the day, as long as it is of importance to the plant that its cotyledons and leaves should be fully exposed to the light. The exact hour in the afternoon at which they begin to bend slightly upwards, and the extent of the movement, will depend on their degree of sensitiveness to gravitation and on their power of resisting its action during the middle of the day, as well as on the amplitude of their ordinary circumnutating movements; and as these qualities differ much in different species, we might expect that the hour in the afternoon at which they begin to rise would differ much in different species, as is the case. Some other agency, however, besides apogeotropism, must come into play, either directly or indirectly, in this upward movement. Thus a young bean (Vicia faba), growing in a small pot, was placed in front of a window in a klinostat; and at night the leaves rose a little, although the action of apogeotropism was quite eliminated. Nevertheless, they did not rise nearly so much at night, as when subjected to apogeotropism. Is it not possible, or even probable, that leaves and cotyledons, which have moved upwards in the evening through the action of apogeotropism during countless generations, may inherit a tendency to this movement? We have seen that the hypocotyls of several Leguminous plants have from a remote period inherited a tendency to arch themselves; and we know that the sleep-movements of leaves are to a certain extent inherited, independently of the alternations of light and darkness.

[8] According to Frank ('Die nat. Wagerechte Richtung von Pflanzentheilen,' 1870, p. 46) the root-leaves of many plants, kept in darkness, rise up and even become vertical; and so it is in some cases with shoots. (See Rauwenhoff, 'Archives Néerlandaises,' tom. xii. p. 32.) These movements indicate apogeotropism; but when organs have been long kept in the dark, the amount of water and of mineral matter which they contain is so much altered, and their regular growth is so much disturbed, that it is perhaps rash to infer from their movements what would occur under normal conditions. (See Godlewski, 'Bot. Zeitung,' Feb. 14th, 1879.)

In our observations on the circumnutation of those cotyledons and leaves which do not sleep at night, we met with hardly any distinct cases of their sinking a little in the evening, and rising again in the morning,—that is, of movements the reverse of those just discussed. We have no doubt that such cases occur, inasmuch as the leaves of many plants sleep by sinking vertically downwards. How to account for the few cases which were observed must be left doubtful. The young leaves of Cannabis sativa sink at night between 30° and 40° beneath the horizon; and Kraus attributes this to epinasty in conjunction with the absorption of water. Whenever epinastic growth is vigorous, it might conquer diaheliotropism in the evening, at which time it would be of no importance to the plant to keep its leaves horizontal. The cotyledons of Anoda Wrightii, of one variety of Gossypium, and of several species of Ipomœa, remain horizontal in the evening whilst they are very young; as they grow a little older they curve a little downwards, and when large and heavy sink so much that they come under our definition of sleep. In the case of the Anoda and of some species of Ipomœa, it was proved that the downward movement did not depend on the weight of the cotyledons; but from the fact of the movement being so much more strongly pronounced after the cotyledons have grown large and heavy, we may suspect that their weight aboriginally played some part in determining that the modification of the circumnutating movement should be in a downward direction.

The so-called Diurnal Sleep of Leaves, Or Paraheliotropism.—This is another class of movements, dependent on the action of light, which supports to some extent the belief that the movements above described are only indirectly due to its action. We refer to the movements of leaves and cotyledons which when moderately illuminated are diaheliotropic; but which change their positions and present their edges to the light, when the sun shines brightly on them. These movements have sometimes been called diurnal sleep, but they differ wholly with respect to the object gained from those properly called nyctitropic; and in some cases the position occupied during the day is the reverse of that during the night.

It has long been known[9] that when the sun shines brightly on the leaflets of Robinia, they rise up and present their edges to the light; whilst their position at night is vertically downwards. We have observed the same movement, when the sun shone brightly on the leaflets of an Australian Acacia. Those of Amphicarpæa monoica turned their edges to the sun; and an analogous movement of the little almost rudimentary basal leaflets of Mimosa albida was on one occasion so rapid that it could be distinctly seen through a lens. the elongated, unifoliate, first leaves of Phaseolus Roxburghii stood at 7 A.M. at 20° above the horizon, and no doubt they afterwards sank a little lower. At noon, after having been exposed for about

2 h. to a bright sun, they stood at 56° above the horizon; they were then protected from the rays of the sun, but were left well illuminated from above, and after 30 m. they had fallen 40°, for they now stood at only 16° above the horizon. Some young plants of Phaseolus Hernandesii had been exposed to the same bright sunlight, and their broad, unifoliate, first leaves now stood up almost or quite vertically, as did many of the leaflets on the trifoliate secondary leaves; but some of the leaflets had twisted round on their own axes by as much as 90° without rising, so as to present their edges to the sun. The leaflets on the same leaf sometimes behaved in these two different manners, but always with the result of being less intensely illuminated. These plants were then protected from the sun, and were looked at after 1½ h.; and now all the leaves and leaflets had reassumed their ordinary sub-horizontal positions. The copper-coloured cotyledons of some seedlings of Cassia mimosoides were horizontal in the morning, but after the sun had shone on them, each had risen 45½° above the horizon. the movement in these several cases must not be confounded with the sudden closing of the leaflets of Mimosa pudica, which may sometimes be noticed when a plant which has been kept in an obscure place is suddenly exposed to the sun; for in this case the light seems to act, as if it were a touch.

[9] Pfeffer gives the names and dates of several ancient writers in his 'Die Periodischen Bewegungen,' 1875, p. 62.

From Prof. Wiesner's interesting observations, it is probable that the above movements have been acquired for a special purpose. the chlorophyll in leaves is often injured by too intense a light, and Prof. Wiesner[10] believes that it is protected by the most diversified means, such as the presence of hairs, colouring matter, etc., and amongst other means by the leaves presenting their edges to the sun, so that the blades then receive much less light. He experimented on the young leaflets of Robinia, by fixing them in such a position that they could not escape being intensely illuminated, whilst others were allowed to place themselves obliquely; and the former began to suffer from the light in the course of two days.

[10] 'Die Näturlichen Einrichtungen zum Schutze des Chlorophylls,' etc., 1876. Pringsheim has recently observed under the microscope the destruction of chlorophyll in a few minutes by the action of concentrated light from the sun, in the presence of oxygen. See, also, Stahl on the protection of chlorophyll from intense light, in 'Bot. Zeitung,' 1880.

In the cases above given, the leaflets move either upwards or twist laterally, so as to place their edges in the direction of the sun's light; but Cohn long ago observed that the leaflets of Oxalis bend downwards when fully exposed to the sun. We witnessed a striking instance of this movement in the very large leaflets of O. Ortegesii. A similar movement may frequently be observed with the leaflets of Averrhoa bilimbi (a member of the Oxalidæ); and a leaf is here represented (Fig. 180) on which the sun had shone. A diagram (Fig. 134) was given in the last chapter, representing the oscillations by which a leaflet rapidly descended under these circumstances; and the movement may be seen closely to resemble that (Fig. 133) by which it assumed its nocturnal position. It is an interesting fact in relation to our present subject that, as Prof. Batalin informs us in a letter, dated February, 1879, the leaflets of Oxalis acetosella may be daily exposed to the sun during many weeks, and they do not suffer if they are allowed to depress themselves; but if this be prevented, they lose their colour and wither in two or three days. Yet the duration of a leaf is about two months, when subjected only to diffused light; and in this case the leaflets never sink downwards during the day.

Fig. 180. Averrhoa bilimbi: leaf with leaflets depressed after exposure to sunshine; but the leaflets are sometimes more depressed than is here shown. Figure much reduced.

As the upward movements of the leaflets of Robinia, and the downward movements of those of Oxalis, have been proved to be highly beneficial to these plants when subjected to bright sunshine, it seems probable that they have been acquired for the special purpose of avoiding too intense an illumination. As it would have been very troublesome in all the above cases to have watched for a fitting opportunity and to have traced the movement of the leaves whilst they were fully exposed to the sunshine, we did not ascertain whether paraheliotropism always consisted of modified circumnutation; but this certainly was the case with the Averrhoa, and probably with the other species, as their leaves were continually circumnutating.

CHAPTER IX.
SENSITIVENESS OF PLANTS TO LIGHT: ITS TRANSMITTED EFFECTS.

Uses of heliotropism—Insectivorous and climbing plants not heliotropic—Same organ heliotropic at one age and not at another—Extraordinary sensitiveness of some plants to light—The effects of light do not correspond with its intensity—Effects of previous illumination—Time required for the action of light—After-effects of light—Apogeotropism acts as soon as light fails—Accuracy with which plants bend to the light—This dependent on the illumination of one whole side of the part—Localised sensitiveness to light and its transmitted effects—Cotyledons of Phalaris, manner of bending—Results of the exclusion of light from their tips—Effects transmitted beneath the surface of the ground—Lateral illumination of the tip determines the direction of the curvature of the base—Cotyledons of Avena, curvature of basal part due to the illumination of upper part—Similar results with the hypocotyls of Brassica and Beta—Radicles of Sinapis apheliotropic, due to the sensitiveness of their tips—Concluding remarks and summary of chapter—Means by which circumnutation has been converted into heliotropism or apheliotropism.

No one can look at the plants growing on a bank or on the borders of a thick wood, and doubt that the young stems and leaves place themselves so that the leaves may be well illuminated. They are thus enabled to decompose carbonic acid. But the sheath-like cotyledons of some Gramineæ, for instance, those of Phalaris, are not green and contain very little starch; from which fact we may infer that they decompose little or no carbonic acid. Nevertheless, they are extremely heliotropic; and this probably serves them in another way, namely, as a guide from the buried seeds through fissures in the ground or through overlying masses of vegetation, into the light and air. This view is strengthened by the fact that with Phalaris and Avena the first true leaf, which is bright green and no doubt decomposes carbonic acid, exhibits hardly a trace of heliotropism. The heliotropic movements of many other seedlings probably aid them in like manner in emerging from the ground; for apogeotropism by itself would blindly guide them upwards, against any overlying obstacle.

Heliotropism prevails so extensively among the higher plants, that there are extremely few, of which some part, either the stem, flower-peduncle, petiole, or leaf, does not bend towards a lateral light. Drosera rotundifolia is one of the few plants the leaves of which exhibit no trace of heliotropism. Nor could we see any in Dionaea, though the plants were not so carefully observed. Sir J. Hooker exposed the pitchers of Sarracenia for some time to a lateral light, but they did not bend towards it.[1] We can understand the reason why these insectivorous plants should not be heliotropic, as they do not live chiefly by decomposing carbonic acid; and it is much more important to them that their leaves should occupy the best position for capturing insects, than that they should be fully exposed to the light.

[1] According to F. Kurtz ('Verhandl. des Bot. Vereins der Provinz Brandenburg,' Bd. xx. 1878) the leaves or pitchers of Darlingtonia Californica are strongly apheliotropic. We failed to detect this movement in a plant which we possessed for a short time.

Tendrils, which consist of leaves or of other organs modified, and the stems of twining plants, are, as Mohl long ago remarked, rarely heliotropic; and here again we can see the reason why, for if they had moved towards a lateral light they would have been drawn away from their supports. But some tendrils are apheliotropic, for instance those of Bignonia capreolata and of Smilax aspera; and the stems of some plants which climb by rootlets, as those of the Ivy and Tecoma radicans, are likewise apheliotropic, and they thus find a support. The leaves, on the other hand, of most climbing plants are heliotropic; but we could detect no signs of any such movement in those of Mutisia clematis.

As heliotropism is so widely prevalent, and as twining plants are distributed throughout the whole vascular series, the apparent absence of any tendency in their stems to bend towards the light, seemed to us so remarkable a fact as to deserve further investigation, for it implies that heliotropism can be readily eliminated. When twining plants are exposed to a lateral light, their stems go on revolving or circumnutating about the same spot, without any evident deflection towards the light; but we thought that we might detect some trace of heliotropism by comparing the average rate at which the stems moved to and from the light during their successive revolutions.[2] Three young plants (about a foot in height) of Ipomœa caerulea and four of I. purpurea, growing in separate pots, were placed on a bright day before a north-east window in a room otherwise darkened, with the tips of their revolving stems fronting the window. When the tip of each plant pointed directly from the window, and when again towards it, the times were recorded. This was continued from 6.45 A.M. till a little after 2 P.M. on June 17th. After a few observations we concluded that we could safely estimate the time taken by each semicircle, within a limit of error of at most 5 minutes. Although the rate of movement in different parts of the same revolution varied greatly, yet 22 semicircles to the light were completed, each on an average in 73.95 minutes; and 22 semicircles from the light each in 73.5 minutes. It may, therefore, be said that they travelled to and from the light at exactly the same average rate; though probably the accuracy of the result was in part accidental. In the evening the stems were not in the least deflected towards the window. Nevertheless, there appears to exist a vestige of heliotropism, for with 6 out of the 7 plants, the first semicircle from the light, described in the early morning after they had been subjected to darkness during the night and thus probably rendered more sensitive, required rather more time, and the first semicircle to the light considerably less time, than the average. Thus with all 7 plants, taken together, the mean time of the first semicircle in the morning from the light, was 76.8 minutes, instead of 73.5 minutes, which is the mean of all the semicircles during the day from the light; and the mean of the first semicircle to the light was only 63.1, instead of 73.95 minutes, which was the mean of all the semicircles during the day to the light.

[2] Some erroneous statements are unfortunately given on this subject, in 'The Movements and Habits of Climbing Plants,' 1875, pp. 28, 32, 40, and 53. Conclusions were drawn from an insufficient number of observations, for we did not then know at how unequal a rate the stems and tendrils of climbing plants sometimes travel in different parts of the same revolution.

Similar observations were made on Wistaria Sinensis, and the mean of 9 semicircles from the light was 117 minutes, and of 7 semicircles to the light 122 minutes, and this difference does not exceed the probable limit of error. During the three days of exposure, the shoot did not become at all bent towards the window before which it stood. In this case the first semicircle from the light in the early morning of each day, required rather less time for its performance than did the first semicircle to the light; and this result, if not accidental, appears to indicate that the shoots retain a trace of an original apheliotropic tendency. With Lonicera brachypoda the semicircles from and to the light differed considerably in time; for 5 semicircles from the light required on a mean 202.4 minutes, and 4 to the light, 229.5 minutes; but the shoot moved very irregularly, and under these circumstances the observations were much too few.

It is remarkable that the same part on the same plant may be affected by light in a widely different manner at different ages, and as it appears at different seasons. The hypocotyledonous stems of Ipomœa caerulea and purpurea are extremely heliotropic, whilst the stems of older plants, only about a foot in height, are, as we have just seen, almost wholly insensible to light. Sachs states (and we have observed the same fact) that the hypocotyls of the Ivy (Hedera helix) are slightly heliotropic; whereas the stems of plants grown to a few inches in height become so strongly apheliotropic, that they bend at right angles away from the light. Nevertheless, some young plants which had behaved in this manner early in the summer again became distinctly heliotropic in the beginning of September; and the zigzag courses of their stems, as they slowly curved towards a north-east window, were traced during 10 days. The stems of very young plants of Tropaeolum majus are highly heliotropic, whilst those of older plants, according to Sachs, are slightly apheliotropic. In all these cases the heliotropism of the very young stems serves to expose the cotyledons, or when the cotyledons are hypogean the first true leaves, fully to the light; and the loss of this power by the older stems, or their becoming apheliotropic, is connected with their habit of climbing.

Most seedling plants are strongly heliotropic, and it is no doubt a great advantage to them in their struggle for life to expose their cotyledons to the light as quickly and as fully as possible, for the sake of obtaining carbon. It has been shown in the first chapter that the greater number of seedlings circumnutate largely and rapidly; and as heliotropism consists of modified circumnutation, we are tempted to look at the high development of these two powers in seedlings as intimately connected. Whether there are any plants which circumnutate slowly and to a small extent, and yet are highly heliotropic, we do not know; but there are several, and there is nothing surprising in this fact, which circumnutate largely and are not at all, or only slightly, heliotropic. Of such cases Drosera rotundifolia offers an excellent instance. The stolons of the strawberry circumnutate almost like the stems of climbing plants, and they are not at all affected by a moderate light; but when exposed late in the summer to a somewhat brighter light they were slightly heliotropic; in sunlight, according to De Vries, they are apheliotropic. Climbing plants circumnutate much more widely than any other plants, yet they are not at all heliotropic.

Although the stems of most seedling plants are strongly heliotropic, some few are but slightly heliotropic, without our being able to assign any reason. This is the case with the hypocotyl of Cassia tora, and we were struck with the same fact with some other seedlings, for instance, those of Reseda odorata. With respect to the degree of sensitiveness of the more sensitive kinds, it was shown in the last chapter that seedlings of several species, placed before a north-east window protected by several blinds, and exposed in the rear to the diffused light of the room, moved with unerring certainty towards the window, although it was impossible to judge, excepting by the shadow cast by an upright pencil on a white card, on which side most light entered, so that the excess on one side must have been extremely small.

A pot with seedlings of Phalaris Canariensis, which had been raised in darkness, was placed in a completely darkened room, at 12 feet from a very small lamp. After 3 h. the cotyledons were doubtfully curved towards the light, and after 7 h. 40 m. from the first exposure, they were all plainly, though slightly, curved towards the lamp. Now, at this distance of 12 feet, the light was so obscure that we could not see the seedlings themselves, nor read the large Roman figures on the white face of a watch, nor see a pencil line on paper, but could just distinguish a line made with Indian ink. It is a more surprising fact that no visible shadow was cast by a pencil held upright on a white card; the seedlings, therefore, were acted on by a difference in the illumination of their two sides, which the human eye could not distinguish. On another occasion even a less degree of light acted, for some cotyledons of Phalaris became slightly curved towards the same lamp at a distance of 20 feet; at this distance we could not see a circular dot 2.29 mm. (.09 inch) in diameter made with Indian ink on white paper, though we could just see a dot 3.56 mm. (.14 inch) in diameter; yet a dot of the former size appears large when seen in the light.[3]

[3] Strasburger says ('Wirkung des Lichtes auf Schwärmsporen,' 1878, p. 52), that the spores of Haematococcus moved to a light which only just sufficed to allow middle-sized type to be read.

We next tried how small a beam of light would act; as this bears on light serving as a guide to seedlings whilst they emerge through fissured or encumbered ground. A pot with seedlings of Phalaris was covered by a tin-vessel, having on one side a circular hole 1.23 mm. in diameter (i.e. a little less than the 1/20th of an inch); and the box was placed in front of a paraffin lamp and on another occasion in front of a window; and both times the seedlings were manifestly bent after a few hours towards the little hole.

A more severe trial was now made; little tubes of very thin glass, closed at their upper ends and coated with black varnish, were slipped over the cotyledons of Phalaris (which had germinated in darkness) and just fitted them. Narrow stripes of the varnish had been previously scraped off one side, through which alone light could enter; and their dimensions were afterwards measured under the microscope. As a control experiment, similar unvarnished and transparent tubes were tried, and they did not prevent the cotyledons bending towards the light. Two cotyledons were placed before a south-west window, one of which was illuminated by a stripe in the varnish, only .004 inch (0.1 mm.) in breadth and .016 inch (0.4 mm.) in length; and the other by a stripe .008 inch in breadth and .06 inch in length. The seedlings were examined after an exposure of 7 h. 40 m., and were found to be manifestly bowed towards the light. Some other cotyledons were at the same time treated similarly, excepting that the little stripes were directed not to the sky, but in such a manner that they received only the diffused light from the room; and these cotyledons did not become at all bowed. Seven other cotyledons were illuminated through narrow, but comparatively long, cleared stripes in the varnish—namely, in breadth between .01 and .026 inch, and in length between .15 and .3 inch; and these all became bowed to the side, by which light entered through the stripes, whether these were directed towards the sky or to one side of the room. That light passing through a hole only .004 inch in breadth by .016 in length, should induce curvature, seems to us a surprising fact.

Before we knew how extremely sensitive the cotyledons of Phalaris were to light, we endeavoured to trace their circumnutation in darkness by the aid of a small wax taper, held for a minute or two at each observation in nearly the same position, a little on the left side in front of the vertical glass on which the tracing was made. The seedlings were thus observed seventeen times in the course of the day, at intervals of from half to three-quarters of an hour; and late in the evening we were surprised to find that all the 29 cotyledons were greatly curved and pointed towards the vertical glass, a little to the left where the taper had been held. The tracings showed that they had travelled in zigzag lines. Thus, an exposure to a feeble light for a very short time at the above specified intervals, sufficed to induce well-marked heliotropism. An analogous case was observed with the hypocotyls of Solanum lycopersicum. We at first attributed this result to the after-effects of the light on each occasion; but since reading Wiesner's observations,[4] which will be referred to in the last chapter, we cannot doubt that an intermittent light is more efficacious than a continuous one, as plants are especially sensitive to any contrast in its amount.

[4] 'Sitz. der k. Akad. der Wissensch.' (Vienna), Jan. 1880, p. 12.

The cotyledons of Phalaris bend much more slowly towards a very obscure light than towards a bright one. Thus, in the experiments with seedlings placed in a dark room at 12 feet from a very small lamp, they were just perceptibly and doubtfully curved towards it after 3 h., and only slightly, yet certainly, after 4 h. After 8 h. 40 m. the chords of their arcs were deflected from the perpendicular by an average angle of only 16°. Had the light been bright, they would have become much more curved in between 1 and 2 h. Several trials were made with seedlings placed at various distances from a small lamp in a dark room; but we will give only one trial. Six pots were placed at distances of 2, 4, 8, 12, 16, and 20 feet from the lamp, before which they were left for 4 h. As light decreases in a geometrical ratio, the seedlings in the 2nd pot received 1/4th, those in the 3rd pot 1/16th, those in the 4th 1/36th, those in the 5th 1/64th, and those in the 6th 1/100th of the light received by the seedlings in the first or nearest pot. Therefore it might have been expected that there would have been an immense difference in the degree of their heliotropic curvature in the several pots; and there was a well-marked difference between those which stood nearest and furthest from the lamp, but the difference in each successive pair of pots was extremely small. In order to avoid prejudice, we asked three persons, who knew nothing about the experiment, to arrange the pots in order according to the degree of curvature of the cotyledons. The first person arranged them in proper order, but doubted long between the 12 feet and 16 feet pots; yet these two received light in the proportion of 36 to 64. The second person also arranged them properly, but doubted between the 8 feet and 12 feet pots, which received light in the proportion of 16 to 36. The third person arranged them in wrong order, and doubted about four of the pots. This evidence shows conclusively how little the curvature of the seedlings differed in the successive pots, in comparison with the great difference in the amount of light which they received; and it should be noted that there was no excess of superfluous light, for the cotyledons became but little and slowly curved even in the nearest pot. Close to the 6th pot, at the distance of 20 feet from the lamp, the light allowed us just to distinguish a dot 3.56 mm. (.14 inch) in diameter, made with Indian ink on white paper, but not a dot 2.29 mm. (.09 inch) in diameter.

The degree of curvature of the cotyledons of Phalaris within a given time, depends not merely on the amount of lateral light which they may then receive, but on that which they have previously received from above and on all sides. Analogous facts have been given with respect to the nyctitropic and periodic movements of plants. Of two pots containing seedlings of Phalaris which had germinated in darkness, one was still kept in the dark, and the other was exposed (Sept. 26th) to the light in a greenhouse during a cloudy day and on the following bright morning. On this morning (27th), at 10.30 A.M., both pots were placed in a box, blackened within and open in front, before a north-east window, protected by a linen and muslin blind and by a towel, so that but little light was admitted, though the sky was bright. Whenever the pots were looked at, this was done as quickly as possible, and the cotyledons were then held transversely with respect to the light, so that their curvature could not have been thus increased or diminished. After 50 m. the seedlings which had previously been kept in darkness, were perhaps, and after 70 m. were certainly, curved, though very slightly, towards the window. After 85 m. some of the seedlings, which had previously been illuminated, were perhaps a little affected, and after 100 m. some of the younger ones were certainly a little curved towards the light. At this time (i.e. after 100 m.) there was a plain difference in the curvature of the seedlings in the two pots. After 2 h. 12 m. the chords of the arcs of four of the most strongly curved seedlings in each pot were measured, and the mean angle from the perpendicular of those which had previously been kept in darkness was 19°, and of those which had previously been illuminated only 7°. Nor did this difference diminish during two additional hours. As a check, the seedlings in both pots were then placed in complete darkness for two hours, in order that apogeotropism should act on them; and those in the one pot which were little curved became in this time almost completely upright, whilst the more curved ones in the other pot still remained plainly curved.

Two days afterwards the experiment was repeated, with the sole difference that even less light was admitted through the window, as it was protected by a linen and muslin blind and by two towels; the sky, moreover, was somewhat less bright. The result was the same as before, excepting that everything occurred rather slower. The seedlings which had been previously kept in darkness were not in the least curved after 54 m., but were so after 70 m. Those which had previously been illuminated were not at all affected until 130 m. had elapsed, and then only slightly. After 145 m. some of the seedlings in this latter pot were certainly curved towards the light; and there was now a plain difference between the two pots. After 3 h. 45 m. the chords of the arcs of 3 seedlings in each pot were measured, and the mean angle from the perpendicular was 16° for those in the pot which had previously been kept in darkness, and only 5° for those which had previously been illuminated.

The curvature of the cotyledons of Phalaris towards a lateral light is therefore certainly influenced by the degree to which they have been previously illuminated. We shall presently see that the influence of light on their bending continues for a short time after the light has been extinguished. These facts, as well as that of the curvature not increasing or decreasing in nearly the same ratio with that of the amount of light which they receive, as shown in the trials with the plants before the lamp, all indicate that light acts on them as a stimulus, in somewhat the same manner as on the nervous system of animals, and not in a direct manner on the cells or cell-walls which by their contraction or expansion cause the curvature.

It has already been incidentally shown how slowly the cotyledons of Phalaris bend towards a very dim light; but when they were placed before a bright paraffin lamp their tips were all curved rectangularly towards it in 2 h. 20 m. The hypocotyls of Solanum lycopersicum had bent in the morning at right angles towards a north-east window. At 1 P.M. (Oct. 21st) the pot was turned round, so that the seedlings now pointed from the light, but by 5 P.M. they had reversed their curvature and again pointed to the light. They had thus passed through 180° in 4 h., having in the morning previously passed through about 90°. But the reversal of the first half of the curvature will have been aided by apogeotropism. Similar cases were observed with other seedlings, for instance, with those of Sinapis alba.

We attempted to ascertain in how short a time light acted on the cotyledons of Phalaris, but this was difficult on account of their rapid circumnutating movement; moreover, they differ much in sensibility, according to age; nevertheless, some of our observations are worth giving. Pots with seedlings were placed under a microscope provided with an eye-piece micrometer, of which each division equalled 1/500th of an inch (0.051 mm.); and they were at first illuminated by light from a paraffin lamp passing through a solution of bichromate of potassium, which does not induce heliotropism. Thus the direction in which the cotyledons were circumnutating could be observed independently of any action from the light; and they could be made, by turning round the pots, to circumnutate transversely to the line in which the light would strike them, as soon as the solution was removed. The fact that the direction of the circumnutating movement might change at any moment, and thus the plant might bend either towards or from the lamp independently of the action of the light, gave an element of uncertainty to the results. After the solution had been removed, five seedlings which were circumnutating transversely to the line of light, began to move towards it, in 6, 4, 7½, 6, and 9 minutes. In one of these cases, the apex of the cotyledon crossed five of the divisions of the micrometer (i.e. 1/100th of an inch, or 0.254 mm.) towards the light in 3 m. Of two seedlings which were moving directly from the light at the time when the solution was removed, one began to move towards it in 13 m., and the other in 15 m. This latter seedling was observed for more than an hour and continued to move towards the light; it crossed at one time 5 divisions of the micrometer (0.254 mm.) in 2 m. 30 s. In all these cases, the movement towards the light was extremely unequal in rate, and the cotyledons often remained almost stationary for some minutes, and two of them retrograded a little. Another seedling which was circumnutating transversely to the line of light, moved towards it in 4 m. after the solution was removed; it then remained almost stationary for 10 m.; then crossed 5 divisions of the micrometer in 6 m.; and then 8 divisions in 11m. This unequal rate of movement, interrupted by pauses, and at first with occasional retrogressions, accords well with our conclusion that heliotropism consists of modified circumnutation.

In order to observe how long the after-effects of light lasted, a pot with seedlings of Phalaris, which had germinated in darkness, was placed at 10.40 A.M. before a north-east window, being protected on all other sides from the light; and the movement of a cotyledon was traced on a horizontal glass. It circumnutated about the same space for the first 24 m., and during the next 1 h. 33 m. moved rapidly towards the light. The light was now (i.e. after 1 h. 57 m.) completely excluded, but the cotyledon continued bending in the same direction as before, certainly for more than 15 m., probably for about 27 m. The doubt arose from the necessity of not looking at the seedlings often, and thus exposing them, though momentarily, to the light. This same seedling was now kept in the dark, until 2.18 P.M., by which time it had reacquired through apogeotropism its original upright position, when it was again exposed to the light from a clouded sky. By 3 P.M. it had moved a very short distance towards the light, but during the next 45 m. travelled quickly towards it. After this exposure of 1 h. 27 m. to a rather dull sky, the light was again completely excluded, but the cotyledon continued to bend in the same direction as before for 14 m. within a very small limit of error. It was then placed in the dark, and it now moved backwards, so that after 1 h. 7 m. it stood close to where it had started from at 2.18 P.M. These observations show that the cotyledons of Phalaris, after being exposed to a lateral light, continue to bend in the same direction for between a quarter and half an hour.

In the two experiments just given, the cotyledons moved backwards or from the window shortly after being subjected to darkness; and whilst tracing the circumnutation of various kinds of seedlings exposed to a lateral light, we repeatedly observed that late in the evening, as the light waned, they moved from it. This fact is shown in some of the diagrams given in the last chapter. We wished therefore to learn whether this was wholly due to apogeotropism, or whether an organ after bending towards the light tended from any other cause to bend from it, as soon as the light failed. Accordingly, two pots of seedling Phalaris and one pot of seedling Brassica were exposed for 8 h. before a paraffin lamp, by which time the cotyledons of the former and the hypocotyls of the latter were bent rectangularly towards the light. The pots were now quickly laid horizontally, so that the upper parts of the cotyledons and of the hypocotyls of 9 seedlings projected vertically upwards, as proved by a plumb-line. In this position they could not be acted on by apogeotropism, and if they possessed any tendency to straighten themselves or to bend in opposition to their former heliotropic curvature, this would be exhibited, for it would be opposed at first very slightly by apogeotropism. They were kept in the dark for 4 h., during which time they were twice looked at; but no uniform bending in opposition to their former heliotropic curvature could be detected. We have said uniform bending, because they circumnutated in their new position, and after 2 h. were inclined in different directions (between 4° and 11°) from the perpendicular. Their directions were also changed after two additional hours, and again on the following morning. We may therefore conclude that the bending back of plants from a light, when this becomes obscure or is extinguished, is wholly due to apogeotropism.[5]

[5] It appears from a reference in Wiesner ('Die Undulirende Nutation der Internodien,' p. 7), that H. Müller of Thurgau found that a stem which is

bending heliotropically is at the same time striving, through apogeotropism, to raise itself into a vertical position.

In our various experiments we were often struck with the accuracy with which seedlings pointed to a light although of small size. To test this, many seedlings of Phalaris, which had germinated in darkness in a very narrow box several feet in length, were placed in a darkened room near to and in front of a lamp having a small cylindrical wick. The cotyledons at the two ends and in the central part of the box, would therefore have to bend in widely different directions in order to point to the light. After they had become rectangularly bent, a long white thread was stretched by two persons, close over and parallel, first to one and then to another cotyledon; and the thread was found in almost every case actually to intersect the small circular wick of the now extinguished lamp. The deviation from accuracy never exceeded, as far as we could judge, a degree or two. This extreme accuracy seems at first surprising, but is not really so, for an upright cylindrical stem, whatever its position may be with respect to the light, would have exactly half its circumference illuminated and half in shadow; and as the difference in illumination of the two sides is the exciting cause of heliotropism, a cylinder would naturally bend with much accuracy towards the light. The cotyledons, however, of Phalaris are not cylindrical, but oval in section; and the longer axis was to the shorter axis (in the one which was measured) as 100 to 70. Nevertheless, no difference could be detected in the accuracy of their bending, whether they stood with their broad or narrow sides facing the light, or in any intermediate position; and so it was with the cotyledons of Avena sativa, which are likewise oval in section. Now, a little reflection will show that in whatever position the cotyledons may stand, there will be a line of greatest illumination, exactly fronting the light, and on each side of this line an equal amount of light will be received; but if the oval stands obliquely with respect to the light, this will be diffused over a wider surface on one side of the central line than on the other. We may therefore infer that the same amount of light, whether diffused over a wider surface or concentrated on a smaller surface, produces exactly the same effect; for the cotyledons in the long narrow box stood in all sorts of positions with reference to the light, yet all pointed truly towards it.

That the bending of the cotyledons to the light depends on the illumination of one whole side or on the obscuration of the whole opposite side, and not on a narrow longitudinal zone in the line of the light being affected, was shown by the effects of painting longitudinally with Indian ink one side of five cotyledons of Phalaris. These were then placed on a table near to a south-west window, and the painted half was directed either to the right or left. The result was that instead of bending in a direct line towards the window, they were deflected from the window and towards the unpainted side, by the following angles, 35°, 83°, 31°, 43°, and 39°. It should be remarked that it was hardly possible to paint one-half accurately, or to place all the seedlings which are oval in section in quite the same position relatively to the light; and this will account for the differences in the angles. Five cotyledons of Avena were also painted in the same manner, but with greater care; and they were laterally deflected from the line of the window, towards the unpainted side, by the following angles, 44°, 44°, 55°, 51°, and 57°. This deflection of the cotyledons from the window is intelligible, for the whole unpainted side must have received some light, whereas the opposite and painted side received none; but a narrow zone on the unpainted side directly in front of the window will have received most light, and all the hinder parts (half an oval in section) less and less light in varying degrees; and we may conclude that the angle of deflection is the resultant of the action of the light over the whole of the unpainted side.

It should have been premised that painting with Indian ink does not injure plants, at least within several hours; and it could injure them only by stopping respiration. To ascertain whether injury was thus soon caused, the upper halves of 8 cotyledons of Avena were thickly coated with transparent matter,—4 with gum, and 4 with gelatine; they were placed in the morning before a window, and by the evening they were normally bowed towards the light, although the coatings now consisted of dry crusts of gum and gelatine. Moreover, if the seedlings which were painted longitudinally with Indian ink had been injured on the painted side, the opposite side would have gone on growing, and they would consequently have become bowed towards the painted side; whereas the curvature was always, as we have seen, in the opposite direction, or towards the unpainted side which was exposed to the light. We witnessed the effects of injuring longitudinally one side of the cotyledons of Avena and Phalaris; for before we knew that grease was highly injurious to them, several were painted down one side with a mixture of oil and lamp-black, and were then exposed before a window; others similarly treated were afterwards tried in darkness. These cotyledons soon became plainly bowed towards the blackened side, evidently owing to the grease on this side having checked their growth, whilst growth continued on the opposite side. But it deserves notice that the curvature differed from that caused by light, which ultimately becomes abrupt near the ground. These seedlings did not afterwards die, but were much injured and grew badly.

LOCALISED SENSITIVENESS TO LIGHT, AND ITS TRANSMITTED EFFECTS.

Phalaris Canariensis.—Whilst observing the accuracy with which the cotyledons of this plant became bent towards the light of a small lamp, we were impressed with the idea that the uppermost part determined the direction of the curvature of the lower part. When the cotyledons are exposed to a lateral light, the upper part bends first, and afterwards the bending gradually extends down to the base, and, as we shall presently see, even a little beneath the ground. This holds good with cotyledons from less than .1 inch (one was observed to act in this manner which was only .03 in height) to about .5 of an inch in height; but when they have grown to nearly an inch in height, the basal part, for a length of .15 to .2 of an inch above the ground, ceases to bend. As with young cotyledons the lower part goes on bending, after the upper part has become well arched towards a lateral light, the apex would ultimately point to the ground instead of to the light, did not the upper part reverse its curvature and straighten itself, as soon as the upper convex surface of the bowed-down portion received more light than the lower concave surface. The position ultimately assumed by young and upright cotyledons, exposed to light entering obliquely from above through a window, is shown in the accompanying figure (Fig. 181); and here it may be seen that the whole upper part has become very nearly straight. When the cotyledons were exposed before a bright lamp, standing on the same level with them, the upper part, which was at first greatly arched towards the light, became straight and strictly parallel with the surface of the soil in the pots; the basal part being now rectangularly bent. All this great amount of curvature, together with the subsequent straightening of the upper part, was often effected in a few hours.

Fig. 181. Phalaris Canariensis: cotyledons after exposure in a box open on one side in front of a south-west window during 8 h. Curvature towards the light accurately traced. The short horizontal lines show the level of the ground.

After the uppermost part has become bowed a little to the light, its overhanging weight must tend to increase the curvature of the lower part; but any such effect was shown in several ways to be quite insignificant. When little caps of tin-foil (hereafter to be described) were placed on the summits of the cotyledons, though this must have added considerably to their weight, the rate or amount of bending was not thus increased. But the best evidence was afforded by placing pots with seedlings of Phalaris before a lamp in such a position, that the cotyledons were horizontally extended and projected at right angles to the line of light. In the course of 5½ h. they were directed towards the light with their bases bent at right angles; and this abrupt curvature could not have been aided in the least by the weight of the upper part, which acted at right angles to the plane of curvature.

It will be shown that when the upper halves of the cotyledons of Phalaris and Avena were enclosed in little pipes of tin-foil or of blackened glass, in which case the upper part was mechanically prevented from bending, the lower and unenclosed part did not bend when exposed to a lateral light; and it occurred to us that this fact might be due, not to the exclusion of the light from the upper part, but to some necessity of the bending gradually travelling down the cotyledons, so that unless the upper part first became bent, the lower could not bend, however much it might be stimulated. It was necessary for our purpose to ascertain whether this notion was true, and it was proved false; for the lower halves of several cotyledons became bowed to the light, although their upper halves were enclosed in little glass tubes (not blackened), which prevented, as far as we could judge, their bending. Nevertheless, as the part within the tube might possibly bend a very little, fine rigid rods or flat splinters of thin glass were cemented with shellac to one side of the upper part of 15 cotyledons; and in six cases they were in addition tied on with threads. They were thus forced to remain quite straight. The result was that the lower halves of all became bowed to the light, but generally not in so great a degree as the corresponding part of the free seedlings in the same pots; and this may perhaps be accounted for by some slight degree of injury having been caused by a considerable surface having been smeared with shellac. It may be added, that when the cotyledons of Phalaris and Avena are acted on by apogeotropism, it is the upper part which begins first to bend; and when this part was rendered rigid in the manner just described, the upward curvature of the basal part was not thus prevented.

To test our belief that the upper part of the cotyledons of Phalaris, when exposed to a lateral light, regulates the bending of the lower part, many experiments were tried; but most of our first attempts proved useless from various causes not worth specifying. Seven cotyledons had their tips cut off for lengths varying between .1 and .16 of an inch, and these, when left exposed all day to a lateral light, remained upright. In another set of 7 cotyledons, the tips were cut off for a length of only about .05 of an inch (1.27 mm.) and these became bowed towards a lateral light, but not nearly so much as the many other seedlings in the same pots. This latter case shows that cutting off the tips does not by itself injure the plants so seriously as to prevent heliotropism; but we thought at the time, that such injury might follow when a greater length was cut off, as in the first set of experiments. Therefore, no more trials of this kind were made, which we now regret; as we afterwards found that when the tips of three cotyledons were cut off for a length of .2 inch, and of four others for lengths of .14, .12, .1, and .07 inch, and they were extended horizontally, the amputation did not interfere in the least with their bending vertically upwards, through the action of apogeotropism, like unmutilated specimens. It is therefore extremely improbable that the amputation of the tips for lengths of from .1

to .14 inch, could from the injury thus caused have prevented the lower part from bending towards the light.

We next tried the effects of covering the upper part of the cotyledons of Phalaris with little caps which were impermeable to light; the whole lower part being left fully exposed before a south-west window or a bright paraffin lamp. Some of the caps were made of extremely thin tin-foil blackened within; these had the disadvantage of occasionally, though rarely, being too heavy, especially when twice folded. The basal edges could be pressed into close contact with the cotyledons; though this again required care to prevent injuring them. Nevertheless, any injury thus caused could be detected by removing the caps, and trying whether the cotyledons were then sensitive to light. Other caps were made of tubes of the thinnest glass, which when painted black served well, with the one great disadvantage that the lower ends could not be closed. But tubes were used which fitted the cotyledons almost closely, and black paper was placed on the soil round each, to check the upward reflection of light from the soil. Such tubes were in one respect far better than caps of tin-foil, as it was possible to cover at the same time some cotyledons with transparent and others with opaque tubes; and thus our experiments could be controlled. It should be kept in mind that young cotyledons were selected for trial, and that these when not interfered with become bowed down to the ground towards the light.

We will begin with the glass-tubes. The summits of nine cotyledons, differing somewhat in height, were enclosed for rather less than half their lengths in uncoloured or transparent tubes; and these were then exposed before a south-west window on a bright day for 8 h. All of them became strongly curved towards the light, in the same degree as the many other free seedlings in the same pots; so that the glass-tubes certainly did not prevent the cotyledons from bending towards the light. Nineteen other cotyledons were, at the same time, similarly enclosed in tubes thickly painted with Indian ink. On five of them, the paint, to our surprise, contracted after exposure to the sunlight, and very narrow cracks were formed, through which a little light entered; and these five cases were rejected. Of the remaining 14 cotyledons, the lower halves of which had been fully exposed to the light for the whole time, 7 continued quite straight and upright; 1 was considerably bowed to the light, and 6 were slightly bowed, but with the exposed bases of most of them almost or quite straight. It is possible that some light may have been reflected upwards from the soil and entered the bases of these 7 tubes, as the sun shone brightly, though bits of blackened paper had been placed on the soil round them. Nevertheless, the 7 cotyledons which were slightly bowed, together with the 7 upright ones, presented a most remarkable contrast in appearance with the many other seedlings in the same pots to which nothing had been done. The blackened tubes were then removed from 10 of these seedlings, and they were now exposed before a lamp for 8 h.; 9 of them became greatly, and 1 moderately, curved towards the light, proving that the previous absence of any curvature in the basal part, or the presence of only a slight degree of curvature there, was due to the exclusion of light from the upper part.

Similar observations were made on 12 younger cotyledons with their upper halves enclosed within glass-tubes coated with black varnish, and with their lower halves fully exposed to bright sunshine. In these younger seedlings the sensitive zone seems to extend rather lower down, as was observed on some other occasions, for two became almost as much curved towards the light as the free seedlings; and the remaining ten were slightly curved, although the basal part of several of them, which normally becomes more curved than any other part, exhibited hardly a trace of curvature. These 12 seedlings taken together differed greatly in their degree of curvature from all the many other seedlings in the same pots.

Better evidence of the efficiency of the blackened tubes was incidentally afforded by some experiments hereafter to be given, in which the upper halves of 14 cotyledons were enclosed in tubes from which an extremely narrow stripe of the black varnish had been scraped off. These cleared stripes were not directed towards the window, but obliquely to one side of the room, so that only a very little light could act on the upper halves of the cotyledons. These 14 seedlings remained during eight hours of exposure before a south-west window on a hazy day quite upright; whereas all the other many free seedlings in the same pots became greatly bowed towards the light.

We will now turn to the trials with caps made of very thin tin-foil. These were placed at different times on the summits of 24 cotyledons, and they extended down for a length of between .15 and .2 of an inch. The seedlings were exposed to a lateral light for periods varying between 6 h. 30 m. and 7 h. 45 m., which sufficed to cause all the other seedlings in the same pots to become almost rectangularly bent towards the light. They varied in height from only .04 to 1.15 inch, but the greater number were about .75 inch. Of the 24 cotyledons with their summits thus protected, 3 became much bent, but not in the direction of the light, and as they did not straighten themselves through apogeotropism during the following night, either the caps were too heavy or the plants themselves were in a weak condition; and these three cases may be excluded. There are left for consideration 21 cotyledons; of these 17 remained all the time quite upright; the other 4 became slightly inclined to the light, but not in a degree comparable with that of the many free seedlings in the same pots. As the glass-tubes, when unpainted, did not prevent the cotyledons from becoming greatly bowed, it cannot be supposed that the caps of very thin tin-foil did so, except through the exclusion of the light. To prove that the plants had not been injured, the caps were removed from 6 of the upright seedlings, and these were exposed before a paraffin lamp for the same length of time as before, and they now all became greatly curved towards the light.

As caps between .15 and .2 of an inch in depth were thus proved to be highly efficient in preventing the cotyledons from bending towards the light, 8 other cotyledons were protected with caps between only .06 and .12 in depth. Of these, two remained vertical, one was considerably and five slightly curved towards the light, but far less so than the free seedlings in the same pots.

Another trial was made in a different manner, namely, by bandaging with strips of tin-foil, about .2 in breadth, the upper part, but not the actual summit, of eight moderately young seedlings a little over half an inch in height. The summits and the basal parts were thus left fully exposed to a lateral light during 8 h.; an upper intermediate zone being protected. With four of these seedlings the summits were exposed for a length of .05 inch, and in two of them this part became curved towards the light, but the whole lower part remained quite upright; whereas the entire length of the other two seedlings became slightly curved towards the light. The summits of the four other seedlings were exposed for a length of .04 inch, and of these one remained almost upright, whilst the other three became considerably curved towards the light. The many free seedlings in the same pots were all greatly curved towards the light.

From these several sets of experiments, including those with the glass-tubes, and those when the tips were cut off, we may infer that the exclusion of light from the upper part of the cotyledons of Phalaris prevents the lower part, though fully exposed to a lateral light, from becoming curved. The summit for a length of .04 or .05 of an inch, though it is itself sensitive and curves towards the light, has only a slight power of causing the lower part to bend. Nor has the exclusion of light from the summit for a length of .1 of an inch a strong influence on the curvature of the lower part. On the other hand, an exclusion for a length of between .15 and .2 of an inch, or of the whole upper half, plainly prevents the lower and fully illuminated part from becoming curved in the manner (see Fig. 181) which invariably occurs when a free cotyledon is exposed to a lateral light. With very young seedlings the sensitive zone seems to extend rather lower down relatively to their height than in older seedlings. We must therefore conclude that when seedlings are freely exposed to a lateral light some influence is transmitted from the upper to the lower part, causing the latter to bend.

This conclusion is supported by what may be seen to occur on a small scale, especially with young cotyledons, without any artificial exclusion of the light; for they bend beneath the earth where no light can enter. Seeds of Phalaris were covered with a layer one-fourth of an inch in thickness of very fine sand, consisting of extremely minute grains of silex coated with oxide of iron. A layer of this sand, moistened to the same degree as that over the seeds, was spread over a glass-plate; and when the layer was .05 of an inch in thickness (carefully measured) no light from a bright sky could be seen to pass through it, unless it was viewed through a long blackened tube, and then a trace of light could be detected, but probably much too little to affect any plant. A layer .1 of an inch in thickness was quite impermeable to light, as judged by the eye aided by the tube. It may be worth adding that the layer, when dried, remained equally impermeable to light. This sand yielded to very slight pressure whilst kept moist, and in this state did not contract or crack in the least. In a first trial, cotyledons which had grown to a moderate height were exposed for 8 h. before a paraffin lamp, and they became greatly bowed. At their bases on the shaded side opposite to the light, well-defined, crescentic, open furrows were formed, which (measured under a microscope with a micrometer) were from .02 to .03 of an inch in breadth, and these had evidently been left by the bending of the buried bases of the cotyledons towards the light. On the side of the light the cotyledons were in close contact with the sand, which was a very little heaped up. By removing with a sharp knife the sand on one side of the cotyledons in the line of the light, the bent portion and the open furrows were found to extend down to a depth of about .1 of an inch, where no light could enter. The chords of the short buried arcs formed in four cases angles of 11°, 13°, 15°, and 18°, with the perpendicular. By the following morning these short bowed portions had straightened themselves through apogeotropism.

In the next trial much younger cotyledons were similarly treated, but were exposed to a rather obscure lateral light. After some hours, a bowed cotyledon, .3 inch in height, had an open furrow on the shaded side .04 inch in breadth; another cotyledon, only .13 inch in height, had left a furrow .02 inch in breadth. But the most curious case was that of a cotyledon which had just protruded above the ground and was only .03 inch in height, and this was found to be bowed in the direction of the light to a depth of .2 of an inch beneath the surface. From what we know of the impermeability of this sand to light, the upper illuminated part in these several cases must have determined the curvature of the lower buried portions. But an apparent cause of doubt may be suggested: as the cotyledons are continually circumnutating, they tend to form a minute crack or furrow all round their bases, which would admit a little light on all sides; but this would not happen when they were illuminated laterally, for we know that they quickly bend towards a lateral light, and they then press so firmly against the sand on the illuminated side as to furrow it, and this would effectually exclude light on this side. Any light admitted on the opposite and shaded side, where an open furrow is formed, would tend to counteract the curvature towards the lamp or other source of the light. It may be added, that the use of fine moist sand, which yields easily to pressure, was indispensable in the above experiments; for seedlings raised in common soil, not kept especially damp, and exposed for 9 h. 30 m. to a strong lateral light, did not form an open furrow at their bases on the shaded side, and were not bowed beneath the surface.

Perhaps the most striking proof of the action of the upper on the lower part of the cotyledons of Phalaris, when laterally illuminated, was afforded by the blackened glass-tubes (before alluded to) with very narrow stripes of the varnish scraped off on one side, through which a little light was admitted. The breadth of these stripes or slits varied between .01 and .02 inch (.25 and .51 mm.). Cotyledons with their upper halves enclosed in such tubes were placed before a south-west window, in such a position, that the scraped stripes did not directly face the window, but obliquely to one side. The seedlings were left exposed for 8 h., before the close of which time the many free seedlings in the same pots had become greatly bowed towards the window. Under these circumstances, the whole lower halves of the cotyledons, which had their summits enclosed in the tubes, were fully exposed to the light of the sky, whilst their upper halves received exclusively or chiefly diffused light from the room, and this only through a very narrow slit on one side. Now, if the curvature of the lower part had been determined by the illumination of this part, all the cotyledons assuredly would have become curved towards the window; but this was far from being the case. Tubes of the kind just described were placed on several occasions over the upper halves of 27 cotyledons; 14 of them remained all the time quite vertical; so that sufficient diffused light did not enter through the narrow slits to produce any effect whatever; and they behaved in the same manner as if their upper halves had been enclosed in completely blackened tubes. The lower halves of the 13 other cotyledons became bowed not directly in the line of the window, but obliquely towards it; one pointed at an angle of only 18°, but the remaining 12 at angles varying between 45° and 62° from the line of the window. At the commencement of the experiment, pins had been laid on the earth in the direction towards which the slits in the varnish faced; and in this direction alone a small amount of diffused light entered. At the close of the experiment, 7 of the bowed cotyledons pointed exactly in the line of the pins, and 6 of them in a line between that of the pins and that of the window. This intermediate position is intelligible, for any light from the sky which entered obliquely through the slits would be much more efficient than the diffused light which entered directly through them. After the 8 h. exposure, the contrast in appearance between these 13 cotyledons and the many other seedlings in the same pots, which were all (excepting the above 14 vertical ones) greatly bowed in straight and parallel lines towards the window, was extremely remarkable. It is therefore certain that a little weak light striking the upper halves of the cotyledons of Phalaris, is far more potent in determining the direction of the curvature of the lower halves, than the full illumination of the latter during the whole time of exposure.

In confirmation of the above results, the effect of thickly painting with Indian ink one side of the upper part of three cotyledons of Phalaris, for a length of .2 inch from their tips, may be worth giving. These were placed so that the unpainted surface was directed not towards the window, but a little to one side; and they all became bent towards the unpainted side, and from the line of the window by angles amounting to 31°, 35°, and 83°. The curvature in this direction extended down to their bases, although the whole lower part was fully exposed to the light from the window.

Finally, although there can be no doubt that the illumination of the upper part of the cotyledons of Phalaris greatly affects the power and manner of bending of the lower part, yet some observations seemed to render it probable that the simultaneous stimulation of the lower part by light greatly favours, or is almost necessary, for its well-marked curvature; but our experiments were not conclusive, owing to the difficulty of excluding light from the lower halves without mechanically preventing their curvature.

Avena sativa.—The cotyledons of this plant become quickly bowed towards a lateral light, exactly like those of Phalaris. Experiments similar to the foregoing ones were tried, and we will give the results as briefly as possible. They are somewhat less conclusive than in the case of Phalaris, and this may possibly be accounted for by the sensitive zone varying in extension, in a species so long cultivated and variable as the common Oat. Cotyledons a little under three-quarters of an inch in height were selected for trial: six had their summits protected from light by tin-foil caps, .25 inch in depth, and two others by caps .3 inch in depth. Of these 8 cotyledons, five remained upright during 8 hours of exposure, although their lower parts were fully exposed to the light all the time; two were very slightly, and one considerably, bowed towards it. Caps only .2 or .22 inch in depth were placed over 4 other cotyledons, and now only one remained upright, one was slightly, and two considerably bowed to the light. In this and the following cases all the free seedlings in the same pots became greatly bowed to the light.

Our next trial was made with short lengths of thin and fairly transparent quills; for glass-tubes of sufficient diameter to go over the cotyledons would have been too heavy. Firstly, the summits of 13 cotyledons were enclosed in unpainted quills, and of these 11 became greatly and 2 slightly bowed to the light; so that the mere act of enclosure did not prevent the lower part from becoming bowed. Secondly, the summits of 11 cotyledons were enclosed in quills .3 inch in length, painted so as to be impermeable to light; of these, 7 did not become at all inclined towards the light, but 3 of them were slightly bent more or less transversely with respect to the line of light, and these might perhaps have been altogether excluded; one alone was slightly bowed towards the light. Painted quills, .25 inch in length, were placed over the summits of 4 other cotyledons; of these, one alone remained upright, a second was slightly bowed, and the two others as much bowed to the light as the free seedlings in the same pots. These two latter cases, considering that the caps were .25 in length, are inexplicable.

Lastly, the summits of 8 cotyledons were coated with flexible and highly transparent gold-beaters' skin, and all became as much bowed to the light as the free seedlings. The summits of 9 other cotyledons were similarly coated with gold-beaters' skin, which was then painted to a depth of between .25 and .3 inch, so as to be impermeable to light; of these 5 remained upright, and 4 were well bowed to the light, almost or quite as well as the free seedlings. These latter four cases, as well as the two in the last paragraph, offer a strong exception to the rule that the illumination of the upper part determines the curvature of the lower part. Nevertheless, 5 of these 8 cotyledons remained quite upright, although their lower halves were fully illuminated all the time; and it would almost be a prodigy to find five free seedlings standing vertically after an exposure for several hours to a lateral light.

The cotyledons of Avena, like those of Phalaris, when growing in soft, damp, fine sand, leave an open crescentric furrow on the shaded side, after bending to a lateral light; and they become bowed beneath the surface at a depth to which, as we know, light cannot penetrate. The arcs of the chords of the buried bowed portions formed in two cases angles of 20° and 21° with the perpendicular. The open furrows on the shaded side were, in four cases, .008, .016, .024, and .024 of an inch in breadth. Brassica oleracea (Common Red).—It will here be shown that the upper half of the hypocotyl of the cabbage, when illuminated by a lateral light, determines the curvature of the lower half. It is necessary to experimentise on young seedlings about half an inch or rather less in height, for when grown to an inch and upwards the basal part ceases to bend. We first tried painting the hypocotyls with Indian ink, or cutting off their summits for various lengths; but these experiments are not worth giving, though they confirm, as far as they can be trusted, the results of the following ones. These were made by folding gold-beaters' skin once round the upper halves of young hypocotyls, and painting it thickly with Indian ink or with black grease. As a control experiment, the same transparent skin, left unpainted, was folded round the upper halves of 12 hypocotyls; and these all became greatly curved to the light, excepting one, which was only moderately curved. Twenty other young hypocotyls had the skin round their upper halves painted, whilst their lower halves were left quite uncovered. These seedlings were then exposed, generally for between 7 and 8 h., in a box blackened within and open in front, either before a south-west window or a paraffin lamp. This exposure was amply sufficient, as was shown by the strongly-marked heliotropism of all the free seedlings in the same pots; nevertheless, some were left exposed to the light for a much longer time. Of the 20 hypocotyls thus treated, 14 remained quite upright, and 6 became slightly bowed to the light; but 2 of these latter cases were not really exceptions, for on removing the skin the paint was found imperfect and was penetrated by many small transparent spaces on the side which faced the light. Moreover, in two other cases the painted skin did not extend quite halfway down the hypocotyl. Although there was a wonderful contrast in the several pots between these 20 hypocotyls and the other many free seedlings, which were all greatly bowed down to their bases in the direction of the light, some being almost prostrate on the ground.

The most successful trial on any one day (included in the above results) is worth describing in detail. Six young seedlings were selected, the hypocotyls of which were nearly .45 inch, excepting one, which was .6 inch in height, measured from the bases of their petioles to the ground. Their upper halves, judged as accurately as could be done by the eye, were folded once round with gold-beaters' skin, and this was painted thickly with Indian ink. They were exposed in an otherwise darkened room before a bright paraffin lamp, which stood on a level with the two pots containing the seedlings. They were first looked at after an interval of 5 h. 10 m., and five of the protected hypocotyls were found quite erect, the sixth being very slightly inclined to the light; whereas all the many free seedlings in the same two pots were greatly bowed to the light. They were again examined after a continuous exposure to the light of 20 h. 35m.; and now the contrast between the two sets was wonderfully great; for the free seedlings had their hypocotyls extended almost horizontally in the direction of the light, and were curved down to the ground; whilst those with the upper halves protected by the painted skin, but with their lower halves fully exposed to the light, still remained quite upright, with the exception of the one which retained the same slight inclination to the light which it had before. This latter seedling was found to have been rather badly painted, for on the side facing the light the red colour of the hypocotyl could be distinguished through the paint.

We next tried nine older seedlings, the hypocotyls of which varied between 1 and 1.6 inch in height. the gold-beaters' skin round their upper parts was painted with black grease to a depth of only .3 inch, that is, from less than a third to a fourth or fifth of their total heights. They were exposed to the light for 7 h. 15 m.; and the result showed that the whole of the sensitive zone, which determines the curvature of the lower part, was not protected from the action of the light; for all 9 became curved towards it, 4 of them very slightly, 3 moderately, and 2 almost as much as the unprotected seedlings. Nevertheless, the whole 9 taken together differed plainly in their degree of curvature from the many free seedlings, and from some which were wrapped in unpainted skin, growing in the same two pots.

Seeds were covered with about a quarter of an inch of the fine sand described under Phalaris; and when the hypocotyls had grown to a height of between .4 and .55 inch, they were exposed during 9 h. before a paraffin lamp, their bases being at first closely surrounded by the damp sand. They all became bowed down to the ground, so that their upper parts lay near to and almost parallel to the surface of the soil. On the side of the light their bases were in close contact with the sand, which was here a very little heaped up; on the opposite or shaded side there were open, crescentic cracks or furrows, rather above .01 of an inch in width; but they were not so sharp and regular as those made by Phalaris and Avena, and therefore could not be so easily measured under the microscope. The hypocotyls were found, when the sand was removed on one side, to be curved to a depth beneath the surface in three cases of at least .1 inch, in a fourth case of .11, and in a fifth of .15 inch. The chords of the arcs of the short, buried, bowed portions formed angles of between 11° and 15° with the perpendicular. From what we have seen of the impermeability of this sand to light, the curvature of the hypocotyls certainly extended down to a depth where no light could enter; and the curvature must have been caused by an influence transmitted from the upper illuminated part.

The lower halves of five young hypocotyls were surrounded by unpainted gold-beaters' skin, and these, after an exposure of 8 h. before a paraffin lamp, all became as much bowed to the light as the free seedlings. The lower halves of 10 other young hypocotyls, similarly surrounded with the skin, were thickly painted with Indian ink; their upper and unprotected halves became well curved to the light, but their lower and protected halves remained vertical in all the cases excepting one, and on this the layer of paint was imperfect. This result seems to prove that the influence transmitted from the upper part is not sufficient to cause the lower part to bend, unless it be at the same time illuminated; but there remains the doubt, as in the case of Phalaris, whether the skin covered with a rather thick crust of dry Indian ink did not mechanically prevent their curvature.

Beta vulgaris.—A few analogous experiments were tried on this plant, which is not very well adapted for the purpose, as the basal part of the hypocotyl, after it has grown to above half an inch in height, does not bend much on exposure to a lateral light. Four hypocotyls were surrounded close beneath their petioles with strips of thin tin-foil, .2 inch in breadth, and they remained upright all day before a paraffin lamp; two others were surrounded with strips .15 inch in breadth, and one of these remained upright, the other becoming bowed; the bandages in two other cases were only .1 inch in breadth, and both of these hypocotyls became bowed, though one only slightly, towards the light. The free seedlings in the same pots were all fairly well curved towards the light; and during the following night became nearly upright. The pots were now turned round and placed before a window, so that the opposite sides of the seedlings were exposed to the light, towards which all the unprotected hypocotyls became bent in the course of 7 h. Seven out of the 8 seedlings with bandages of tin-foil remained upright, but one which had a bandage only .1 inch in breadth, became curved to the light. On another occasion, the upper halves of 7 hypocotyls were surrounded with painted gold-beaters' skin; of these 4 remained upright, and 3 became a little curved to the light: at the same time 4 other seedlings surrounded with unpainted skin, as well as the free ones in the same pots, all became bowed towards the lamp, before which they had been exposed during 22 hours.

Radicles of Sinapis alba.—The radicles of some plants are indifferent, as far as curvature is concerned, to the action of light; whilst others bend towards and others from it.[6] Whether these movements are of any service to the plant is very doubtful, at least in the case of subterranean roots; they probably result from the radicles being sensitive to contact, moisture, and gravitation, and as a consequence to other irritants which are never naturally encountered. The radicles of Sinapis alba, when immersed in water and exposed to a lateral light, bend from it, or are apheliotropic. They become bent for a length of about 4 mm. from their tips. To ascertain whether this movement generally occurred, 41 radicles, which had germinated in damp sawdust, were immersed in water and exposed to a lateral light; and they all, with two doubtful exceptions, became curved from the light. At the same time the tips of 54 other radicles, similarly exposed, were just touched with nitrate of silver. They were blackened for a length of from .05 to .07 mm., and probably killed; but it should be observed that this did not check materially, if at all, the growth of the upper part; for several, which were measured, increased in the course of only 8–9 h. by 5 to 7 mm. in length. Of the 54 cauterised radicles one case was doubtful, 25 curved themselves from the light in the normal manner, and 28, or more than half, were not in the least apheliotropic. There was a considerable difference, which we cannot account for, in the results of the experiments tried towards the end of April and in the middle of September. Fifteen radicles (part of the above 54) were cauterised at the former period and were exposed to sunshine, of which 12 failed to be apheliotropic, 2 were still apheliotropic, and 1 was doubtful. In September, 39 cauterised radicles were exposed to a northern light, being kept at a proper temperature; and now 23 continued to be apheliotropic in the normal manner, and only 16 failed to bend from the light. Looking at the aggregate results at both periods, there can be no doubt that the destruction of the tip for less than a millimeter in length destroyed in more than half the cases their power of moving from the light. It is probable that if the tips had been cauterised for the length of a whole millimeter, all signs of apheliotropism would have disappeared. It may be suggested that although the application of caustic does not stop growth, yet enough may be absorbed to destroy the power of movement in the upper part; but this suggestion must be rejected, for we have seen and shall again see, that cauterising one side of the tip of various kinds of radicles actually excites movement. The conclusion seems inevitable that sensitiveness to light resides in the tip of the radicle of Sinapis alba; and that the tip when thus stimulated transmits some influence to the upper part, causing it to bend. The case in this respect is parallel with that of the radicles of several plants, the tips of which are sensitive to contact and to other irritants, and, as will be shown in the eleventh chapter, to gravitation.

[6] Sachs, 'Physiologie Végétale,' 1868, p. 44.

CONCLUDING REMARKS AND SUMMARY OF CHAPTER.

We do not know whether it is a general rule with seedling plants that the illumination of the upper part determines the curvature of the lower part. But as this occurred in the four species examined by us, belonging to such distinct families as the Gramineæ, Cruciferae, and Chenopodeae, it is probably of common occurrence. It can hardly fail to be of service to seedlings, by aiding them to find the shortest path from the buried seed to the light, on nearly the same principle that the eyes of most of the lower crawling animals are seated at the anterior ends of their bodies. It is extremely doubtful whether with fully developed plants the illumination of one part ever affects the curvature of another part. The summits of 5 young plants of Asparagus officinalis (varying in height between 1.1 and 2.7 inches, and consisting of several short internodes) were covered with caps of tin-foil from 0.3 to 0.35 inch in depth; and the lower uncovered parts became as much curved towards a lateral light, as were the free seedlings in the same pots. Other seedlings of the same plant had their summits painted with Indian ink with the same negative result. Pieces of blackened paper were gummed to the edges and over the blades of some leaves on young plants of Tropaeolum majus and Ranunculus ficaria; these were then placed in a box before a window, and the petioles of the protected leaves became curved towards the light, as much as those of the unprotected leaves.

The foregoing cases with respect to seedling plants have been fully described, not only because the transmission of any effect from light is a new physiological fact, but because we think it tends to modify somewhat the current views on heliotropic movements. Until lately such movements were believed to result simply from increased growth on the shaded side. At present it is commonly admitted[7] that diminished light increases the turgescence of the cells, or the extensibility of the cell-walls, or of both together, on the shaded side, and that this is followed by increased growth. But Pfeffer has shown that a difference in the turgescence on the two sides of a pulvinus,—that is, an aggregate of small cells which have ceased to grow at an early age,—is excited by a difference in the amount of light received by the two sides; and that movement is thus caused without being followed by increased growth on the more turgescent side.[8] All observers apparently believe that light acts directly on the part which bends, but we have seen with the above described seedlings that this is not the case. Their lower halves were brightly illuminated for hours, and yet did not bend in the least towards the light, though this is the part which under ordinary circumstances bends the most. It is a still more striking fact, that the faint illumination of a narrow stripe on one side of the upper part of the cotyledons of Phalaris determined the direction of the curvature of the lower part; so that this latter part did not bend towards the bright light by which it had been fully illuminated, but obliquely towards one side where only a little light entered. These results seem to imply the presence of some matter in the upper part which is acted on by light, and which transmits its effects to the lower part. It has been shown that this transmission is independent of the bending of the upper sensitive part. We have an analogous case of transmission in Drosera, for when a gland is irritated, the basal and not the upper or intermediate part of the tentacle bends. The flexible and sensitive filament of Dionaea likewise transmits a stimulus, without itself bending; as does the stem of Mimosa.

[7] Emil Godlewski has given ('Bot. Zeitung,' 1879, Nos. 6–9) an excellent account (p. 120) of the present state of the question. See also Vines in 'Arbeiten des Bot. Inst. in Würzburg,' 1878, B. ii. pp. 114–147. Hugo de Vries has recently published a still more important article on this subject: 'Bot Zeitung,' Dec. 19th and 26th, 1879.

[8] 'Die Periodischen Bewegungen der Blattorgane,' 1875, pp. 7, 63, 123, etc. Frank has also insisted ('Die Naturliche wägerechte Richtung von Pflanzentheilen,' 1870, p. 53) on the important part which the pulvini of the leaflets of compound leaves play in placing the leaflets in a proper position with respect to the light. This holds good, especially with the

leaves of climbing plants, which are carried into all sorts of positions, ill-adapted for the action of the light.

Light exerts a powerful influence on most vegetable tissues, and there can be no doubt that it generally tends to check their growth. But when the two sides of a plant are illuminated in a slightly different degree, it does not necessarily follow that the bending towards the illuminated side is caused by changes in the tissues of the same nature as those which lead to increased growth in darkness. We know at least that a part may bend from the light, and yet its growth may not be favoured by light. This is the case with the radicles of Sinapis alba, which are plainly apheliotropic; nevertheless, they grow quicker in darkness than in light.[9] So it is with many aërial roots, according to Wiesner;[10] but there are other opposed cases. It appears, therefore, that light does not determine the growth of apheliotropic parts in any uniform manner.

[9] Francis Darwin, 'Über das Wachsthum negativ heliotropischer Wurzeln': 'Arbeiten des Bot. Inst. in Würzburg,' B. ii., Heft iii., 1880, p. 521.

[10] 'Sitzb. der k. Akad. der Wissensch' (Vienna), 1880, p. 12.

We should bear in mind that the power of bending to the light is highly beneficial to most plants. There is therefore no improbability in this power having been specially acquired. In several respects light seems to act on plants in nearly the same manner as it does on animals by means of the nervous system.[11] With seedlings the effect, as we have just seen, is transmitted from one part to another. An animal may be excited to move by a very small amount of light; and it has been shown that a difference in the illumination of the two sides of the cotyledons of Phalaris, which could not be distinguished by the human eye, sufficed to cause them to bend. It has also been shown that there is no close parallelism between the amount of light which acts on a plant and its degree of curvature; it was indeed hardly possible to perceive any difference in the curvature of some seedlings of Phalaris exposed to a light, which, though dim, was very much brighter than that to which others had been exposed. The retina, after being stimulated by a bright light, feels the effect for some time; and Phalaris continued to bend for nearly half an hour towards the side which had been illuminated. The retina cannot perceive a dim light after it has been exposed to a bright one; and plants which had been kept in the daylight during the previous day and morning, did not move so soon towards an obscure lateral light as did others which had been kept in complete darkness.

[11] Sachs has made some striking remarks to the same effect with respect to the various stimuli which excite movement in plants. See his paper

'Ueber orthotrope und plagiotrope Pflanzentheile,' 'Arb. des Bot. Inst. in Würzburg,' 1879, B. ii. p. 282.

Even if light does act in such a manner on the growing parts of plants as always to excite in them a tendency to bend towards the more illuminated side—a supposition contradicted by the foregoing experiments on seedlings and by all apheliotropic organs—yet the tendency differs greatly in different species, and is variable in degree in the individuals of the same species, as may be seen in almost any pot of seedlings of a long cultivated plant.[12] There is therefore a basis for the modification of this tendency to almost any beneficial extent. That it has been modified, we see in many cases: thus, it is of more importance for insectivorous plants to place their leaves in the best position for catching insects than to turn their leaves to the light, and they have no such power. If the stems of twining plants were to bend towards the light, they would often be drawn away from their supports; and as we have seen they do not thus bend. As the stems of most other plants are heliotropic, we may feel almost sure that twining plants, which are distributed throughout the whole vascular series, have lost a power that their non-climbing progenitors possessed. Moreover, with Ipomœa, and probably all other twiners, the stem of the young plant, before it begins to twine, is highly heliotropic, evidently in order to expose the cotyledons or the first true leaves fully to the light. With the Ivy the stems of seedlings are moderately heliotropic, whilst those of the same plants when grown a little older are apheliotropic. Some tendrils which consist of modified leaves—organs in all ordinary cases strongly diaheliotropic—have been rendered apheliotropic, and their tips crawl into any dark crevice.

[12] Strasburger has shown in his interesting work ('Wirkung des Lichtes...auf Schwärmsporen,' 1878), that the movement of the swarm-spores of various lowly organised plants to a lateral light is influenced by their stage of development, by the temperature to which they are subjected, by the degree of illumination under which they have been raised, and by other unknown causes; so that the swarm-spores of the same species may move across the field of the microscope either to or from the light. Some individuals, moreover, appear to be indifferent to the light; and those of different species behave very differently. The brighter the light, the straighter is their course. They exhibit also for a short time the after-effects of light. In all these respects they resemble the higher plants. See, also, Stahl, 'Ueber den einfluss der Lichts auf die Bewegungs-erscheinungen der Schwärmsporen' Verh. d. phys.-med. Geselsshalft in Würzburg, B. xii. 1878.

Even in the case of ordinary heliotropic movements, it is hardly credible that they result directly from the action of the light, without any special adaptation. We may illustrate what we mean by the hygroscopic movements of plants: if the tissues on one side of an organ permit of rapid evaporation, they will dry quickly and contract, causing the part to bend to this side. Now the wonderfully complex movements of the pollinia of Orchis pyramidalis, by which they clasp the proboscis of a moth and afterwards change their position for the sake of depositing the pollen-masses on the double stigma—or again the twisting movements, by which certain seeds bury themselves in the ground[13]—follow from the manner of drying of the parts in question; yet no one will suppose that these results have been gained without special adaptation. Similarly, we are led to believe in adaptation when we see the hypocotyl of a seedling, which contains chlorophyll, bending to the light; for although it thus receives less light, being now shaded by its own cotyledons, it places them—the more important organs—in the best position to be fully illuminated. The hypocotyl may therefore be said to sacrifice itself for the good of the cotyledons, or rather of the whole plant. But if it be prevented from bending, as must sometimes occur with seedlings springing up in an entangled mass of vegetation, the cotyledons themselves bend so as to face the light; the one farthest off rising up, and that nearest to the light sinking down, or both twisting laterally.[14] We may, also, suspect that the extreme sensitiveness to light of the upper part of the sheath-like cotyledons of the Gramineæ, and their power of transmitting its effects to the lower part, are specialised arrangements for finding the shortest path to the light. With plants growing on a bank, or thrown prostrate by the wind, the manner in which the leaves move, even rotating on their own axes, so that their upper surfaces may be again directed to the light, is a striking phenomenon. Such facts are rendered more striking when we remember that too intense a light injures the chlorophyll, and that the leaflets of several Leguminosae when thus exposed bend upwards and present their edges to the sun, thus escaping injury. On the other hand, the leaflets of Averrhoa and Oxalis, when similarly exposed, bend downwards.

[13] Francis Darwin, 'On the Hygroscopic Mechanism,' etc., 'Transactions Linn. Soc.,' series ii. vol. i. p. 149, 1876.

[14] Wiesner has made remarks to nearly the same effect with respect to leaves: 'Die undulirende Nutation der Internodien,' p. 6, extracted from B. lxxvii. (1878). Sitb. der k. Akad. der Wissensch. Wien.

It was shown in the last chapter that heliotropism is a modified form of circumnutation; and as every growing part of every plant circumnutates more or less, we can understand how it is that the power of bending to the light has been acquired by such a multitude of plants throughout the vegetable kingdom. The manner in which a circumnutating movement—that is, one consisting of a succession of irregular ellipses or loops—is gradually converted into a rectilinear course towards the light, has been already explained. First, we have a succession of ellipses with their longer axes directed towards the light, each of which is described nearer and nearer to its source; then the loops are drawn out into a strongly pronounced zigzag line, with here and there a small loop still formed. At the same time that the movement towards the light is increased in extent and accelerated, that in the opposite direction is lessened and retarded, and at last stopped. The zigzag movement to either side is likewise gradually lessened, so that finally the course becomes rectilinear. Thus under the stimulus of a fairly bright light there is no useless expenditure of force.

As with plants every character is more or less variable, there seems to be no great difficulty in believing that their circumnutating movements may have been increased or modified in any beneficial manner by the preservation of varying individuals. The inheritance of habitual movements is a necessary contingent for this process of selection, or the survival of the fittest; and we have seen good reason to believe that habitual movements are inherited by plants. In the case of twining species the circumnutating movements have been increased in amplitude and rendered more circular; the stimulus being here an internal or innate one. With sleeping plants the movements have been increased in amplitude and often changed in direction; and here the stimulus is the alternation of light and darkness, aided, however, by inheritance. In the case of heliotropism, the stimulus is the unequal illumination of the two sides of the plant, and this determines, as in the foregoing cases, the modification of the circumnutating movement in such a manner that the organ bends to the light. A plant which has been rendered heliotropic by the above means, might readily lose this tendency, judging from the cases already given, as soon as it became useless or injurious. A species which has ceased to be heliotropic might also be rendered apheliotropic by the preservation of the individuals which tended to circumnutate (though the cause of this and most other variations is unknown) in a direction more or less opposed to that whence the light proceeded. In like manner a plant might be rendered diaheliotropic.

CHAPTER X.
MODIFIED CIRCUMNUTATION: MOVEMENTS EXCITED BY GRAVITATION.

Means of observation—Apogeotropism—Cytisus—Verbena—Beta—Gradual conversion of the movement of circumnutation into apogeotropism in Rubus, Lilium, Phalaris, Avena, and Brassica—Apogeotropism retarded by heliotropism—Effected by the aid of joints or pulvini—Movements of flower-peduncles of Oxalis—General remarks on apogeotropism—Geotropism—Movements of radicles—Burying of seed-capsules—Use of process—Trifolium subterraneum—Arachis—Amphicarpæa—Diageotropism—Conclusion

Our object in the present chapter is to show that geotropism, apogeotropism, and diageotropism are modified forms of circumnutation. Extremely fine filaments of glass, bearing two minute triangles of paper, were fixed to the summits of young stems, frequently to the hypocotyls of seedlings, to flower-peduncles, radicles, etc., and the movements of the parts were then traced in the manner already described on vertical and horizontal glass-plates. It should be remembered that as the stems or other parts become more and more oblique with respect to the glasses, the figures traced on them necessarily become more and more magnified. The plants were protected from light, excepting whilst each observation was being made, and then the light, which was always a dim one, was allowed to enter so as to interfere as little as possible with the movement in progress; and we did not detect any evidence of such interference.

When observing the gradations between circumnutation and heliotropism, we had the great advantage of being able to lessen the light; but with geotropism analogous experiments were of course impossible. We could, however, observe the movements of stems placed at first only a little from the perpendicular, in which case geotropism did not act with nearly so much power, as when the stems were horizontal and at right angles to the force. Plants, also, were selected which were but feebly geotropic or apogeotropic, or had become so from having grown rather old. Another plan was to place the stems at first so that they pointed 30 or 40° beneath the horizon, and then apogeotropism had a great amount of work to do before the stem was rendered upright; and in this case ordinary circumnutation was often not wholly obliterated. Another plan was to observe in the evening plants which during the day had become greatly curved heliotropically; for their stems under the gradually waning light very slowly became upright through the action of apogeotropism; and in this case modified circumnutation was sometimes well displayed.

Apogeotropism.—Plants were selected for observation almost by chance, excepting that they were taken from widely different families. If the stem of a plant which is even moderately sensitive to apogeotropism be placed horizontally, the upper growing part bends quickly upwards, so as to become perpendicular; and the line traced by joining the dots successively made on a glass-plate, is generally almost straight. For instance, a young Cytisus fragrans, 12 inches in height, was placed so that the stem projected 10° beneath the horizon, and its course was traced during 72 h. At first it bent a very little downwards (Fig. 182), owing no doubt to the weight of the stem, as this occurred with most of the other plants observed, though, as they were of course circumnutating, the short downward lines were often oblique. After three-quarters of an hour the stem began to curve upwards, quickly during the first two hours, but much more slowly during the afternoon and night, and on the following day. During the second night it fell a little, and circumnutated during the following day; but it also moved a short distance to the right, which was caused by a little light having been accidentally admitted on this side. The stem was now inclined 60° above the horizon, and had therefore risen 70°. With time allowed it would probably have become upright, and no doubt would have continued circumnutating. The sole remarkable feature in the figure here given is the straightness of the course pursued. The stem, however, did not move upwards at an equable rate, and it sometimes stood almost or quite still. Such periods probably represent attempts to circumnutate in a direction opposite to apogeotropism.

Fig. 182. Cytisus fragrans: apogeotropic movement of stem from 10° beneath to 60° above horizon, traced on vertical glass, from 8.30 A.M. March 12th to 10.30 P.M. 13th. The subsequent circumnutating movement is likewise shown up to 6.45 A.M. on the 15th. Nocturnal course represented, as usual, by a broken line. Movement not greatly magnified, and tracing reduced to two-thirds of original scale.

The herbaceous stem of a Verbena melindres (?) laid horizontally, rose in 7 h. so much that it could no longer be observed on the vertical glass which stood in front of the plant. The long line which was traced was almost absolutely straight. After the 7 h. it still continued to rise, but now circumnutated slightly. On the following day it stood upright, and circumnutated regularly, as shown in Fig. 82, given in the fourth chapter. The stems of several other plants which were highly sensitive to apogeotropism rose up in almost straight lines, and then suddenly began to circumnutate. A partially etiolated and somewhat old hypocotyl of a seedling cabbage (2 3/4 inches in height) was so sensitive that when placed at an angle of only 23° from the perpendicular, it became vertical in 33 minutes. As it could not have been strongly acted upon by apogeotropism in the above slightly inclined position, we expected that it would have circumnutated, or at least have moved in a zigzag course. Accordingly, dots were made every 3 minutes; but, when these were joined, the line was nearly straight. After this hypocotyl had become upright it still moved onwards for half an hour in the same general direction, but in a zigzag manner. During the succeeding 9 h. it circumnutated regularly, and described 3 large ellipses. In this case apogeotropism, although acting at a very unfavourable angle, quite overcame the ordinary circumnutating movement.

Fig. 183. Beta vulgaris: apogeotropic movement of hypocotyl from 19° beneath horizon to a vertical position, with subsequent circumnutation, traced on a vertical and on a horizontal glass-plate, from 8.28 A.M. Sept. 28th to 8.40 A.M. 29th. Figure reduced to one-third of original scale.

The hypocotyls of Beta vulgaris are highly sensitive to apogeotropism. One was placed so as to project 19° beneath the horizon; it fell at first a very little (see Fig. 183), no doubt owing to its weight; but as it was circumnutating the line was oblique. During the next 3 h. 8 m. it rose in a nearly straight line, passing through an angle of 109°, and then (at 12.3 P.M.) stood upright. It continued for 55 m. to move in the same general direction beyond the perpendicular, but in a zigzag course. It returned also in a zigzag line, and then circumnutated regularly, describing three large ellipses during the remainder of the day. It should be observed that the ellipses in this figure are exaggerated in size, relatively to the length of the upward straight line, owing to the position of the vertical and horizontal glass-plates. Another and somewhat old hypocotyl was placed so as to stand at only 31° from the perpendicular, in which position apogeotropism acted on it with little force, and its course accordingly was slightly zigzag.

The sheath-like cotyledons of Phalaris Canariensis are extremely sensitive to apogeotropism. One was placed so as to project 40° beneath the horizon. Although it was rather old and 1.3 inch in height, it became vertical in 4 h. 30 m., having passed through an angle of 130° in a nearly straight line. It then suddenly began to circumnutate in the ordinary manner. The cotyledons of this plant, after the first leaf has begun to protrude, are but slightly apogeotropic, though they still continue to circumnutate. One at this stage of development was placed horizontally, and did not become upright even after 13 h., and its course was slightly zigzag. So, again, a rather old hypocotyl of Cassia tora (1 1/4 inch in height) required 28 h. to become upright, and its course was distinctly zigzag; whilst younger hypocotyls moved much more quickly and in a nearly straight line.

When a horizontally placed stem or other organ rises in a zigzag line, we may infer from the many cases given in our previous chapters, that we have a modified form of circumnutation; but when the course is straight, there is no evidence of circumnutation, and any one might maintain that this latter movement had been replaced by one of a wholly distinct kind. This view seems the more probable when (as sometimes occurred with the hypocotyls of Brassica and Beta, the stems of Cucurbita, and the cotyledons of Phalaris) the part in question, after bending up in a straight course, suddenly begins to circumnutate to the full extent and in the usual manner. A fairly good instance of a sudden change of this kind—that is, from a nearly straight upward movement to one of circumnutation—is shown in Fig. 183; but more striking instances were occasionally observed with Beta, Brassica, and Phalaris.

We will now describe a few cases in which it may be seen how gradually circumnutation becomes changed into apogeotropism, under circumstances to be specified in each instance.

Rubus idæus (hybrid).—A young plant, 11 inches in height, growing in a pot, was placed horizontally; and the upward movement was traced during nearly 70 h.; but the plant, though growing vigorously, was not highly sensitive to apogeotropism, or it was not capable of quick movement, for during the above time it rose only 67°. We may see in the diagram (Fig. 184) that during the first day of 12 h. it rose in a nearly straight line. When placed horizontally, it was evidently circumnutating, for it rose at first a little, notwithstanding the weight of the stem, and then sank down; so that it did not start on its permanently upward course until 1 h. 25 m. had elapsed. On the second day, by which time it had risen considerably, and when apogeotropism acted on it with somewhat less power, its course during 15½ h. was clearly zigzag, and the rate of the upward movement was not equable. During the third day, also of 15½ h., when apogeotropism acted on it with still less power, the stem plainly circumnutated, for it moved during this day 3 times up and 3 times down, 4 times to the left and 4 to the right. But the course was so complex that it could hardly be traced on the glass. We can, however, see that the successively formed irregular ellipses rose higher and higher. Apogeotropism continued to act on the fourth morning, as the stem was still rising, though it now stood only 23° from the perpendicular. In this diagram the several stages may be followed by which an almost rectilinear, upward, apogeotropic course first becomes zigzag, and then changes into a circumnutating movement, with most of the successively formed, irregular ellipses directed upwards.

Fig 184: Rubus idæus (hybrid): apogeotropic movement of stem, traced on a vertical glass during 3 days and 3 nights, from 10.40 A.M. March 18th to 8 A.M. 21st. Figure reduced to one-half of the original scale.

Lilium auratum.—A plant 23 inches in height was placed horizontally, and the upper part of the stem rose 58° in 46 h., in the manner shown in the accompanying diagram (Fig. 185). We here see that during the whole of the second day of 15½ h., the stem plainly circumnutated whilst bending upwards through apogeotropism. It had still to rise considerably, for when the last dot in the figure was made, it stood 32° from an upright position.

Fig. 185. Lilium auratum: apogeotropic movement of stem, traced on a vertical glass during 2 days and 2 nights, from 10.40 A.M. March 18th to 8 A.M. 20th. Figure reduced to one-half of the original scale.

Phalaris Canariensis.—A cotyledon of this plant (1.3 inch in height) has already been described as rising in 4 h. 30 m. from 40° beneath the horizon into a vertical position, passing through an angle of 130° in a nearly straight line, and then abruptly beginning to circumnutate. Another somewhat old cotyledon of the same height (but from which a true leaf had not yet protruded), was similarly placed at 40° beneath the horizon. For the first 4 h. it rose in a nearly straight course (Fig. 186), so that by 1.10 P.M. it was highly inclined, and now apogeotropism acted on it with much less power than before, and it began to zigzag. At 4.15 P.M. (i.e. in 7 h. from the commencement) it stood vertically, and afterwards continued to circumnutate in the usual manner about the same spot. Here then we have a graduated change from a straight upward apogeotropic course into circumnutation, instead of an abrupt change, as in the former case.

Avena sativa.—The sheath-like cotyledons, whilst young, are strongly apogeotropic; and some which were placed at 45° beneath the horizon rose 90° in 7 or 8 h. in lines almost absolutely straight. An oldish cotyledon, from which the first leaf began to protrude whilst the following observations were being made, was placed at 10° beneath the horizon, and it rose only 59° in 24h. It behaved rather differently from any other plant, observed by us, for during the first 4½ h. it rose in a line not far from straight; during the next 6½ h. it circumnutated, that is, it descended and again ascended in a strongly marked zigzag course; it then resumed its upward movement in a moderately straight line, and, with time allowed, no doubt would have become upright. In this case, after the first 4½ h., ordinary circumnutation almost completely conquered for a time apogeotropism.

Fig 186. Phalaris Canariensis: apogeotropic movement of cotyledon, traced on a vertical and horizontal glass, from 9.10 A.M. Sept. 19th to 9 A.M. 20th. Figure here reduced to one-fifth of original scale.

Brassica oleracea.—The hypocotyls of several young seedlings placed horizontally, rose up vertically in the course of 6 or 7 h. in nearly straight lines. A seedling which had grown in darkness to a height of 2 1/4 inches, and was therefore rather old and not highly sensitive, was placed so that the hypocotyl projected at between 30° and 40° beneath the horizon. The upper part alone became curved upwards, and rose during the first 3 h. 10 m. in a nearly straight line (Fig. 187); but it was not possible to trace the upward movement on the vertical glass for the first 1 h. 10 m., so that the nearly straight line in the diagram ought to have been much longer. During the next 11 h. the hypocotyl circumnutated, describing irregular figures, each of which rose a little above the one previously formed. During the night and following early morning it continued to rise in a zigzag course, so that apogeotropism was still acting. At the close of our observations, after 23 h. (represented by the highest dot in the diagram) the hypocotyl was still 32° from the perpendicular. There can be little doubt that it would ultimately have become upright by describing an additional number of irregular ellipses, one above the other.

Fig 187. Brassica oleracea: apogeotropic movement of hypocotyl, traced on vertical glass, from 9.20 A.M., Sept. 12th to 8.30 A.M. 13th. The upper part of the figure is more magnified than the lower part. If the whole course had been traced, the straight upright line would have been much longer. Figure here reduced to one-third of the original scale.

Apogeotropism retarded by Heliotropism.—When the stem of any plant bends during the day towards a lateral light, the movement is opposed by apogeotropism; but as the light gradually wanes in the evening the latter power slowly gains the upper hand, and draws the stem back into a vertical position. Here then we have a good opportunity for observing how apogeotropism acts when very nearly balanced by an opposing force. For instance, the plumule of Tropaeolum majus (see former Fig. 175) moved towards the dim evening light in a slightly zigzag line until 6.45 P.M., it then returned on its course until 10.40 P.M., during which time it zigzagged and described an ellipse of considerable size. The hypocotyl of Brassica oleracea (see former Fig. 173) moved in a straight line to the light until 5.15 P.M., and then from the light, making in its backward course a great rectangular bend, and then returned for a short distance towards the former source of the light; no observations were made after 7.10 P.M., but during the night it recovered its vertical position. A hypocotyl of Cassia tora moved in the evening in a somewhat zigzag line towards the failing light until 6 P.M., and was now bowed 20° from the perpendicular; it then returned on its course, making before 10.30 P.M. four great, nearly rectangular bends and almost completing an ellipse. Several other analogous cases were casually observed, and in all of them the apogeotropic movement could be seen to consist of modified circumnutation.

Apogeotropic Movements effected by the aid of joints or pulvini.—Movements of this kind are well known to occur in the Gramineæ, and are effected by means of the thickened bases of their sheathing leaves; the stem within being in this part thinner than elsewhere.[1] According to the analogy of all other pulvini, such joints ought to continue circumnutating for a long period, after the adjoining parts have ceased to grow. We therefore wished to ascertain whether this was the case with the Gramineæ; for if so, the upward curvature of their stems, when extended horizontally or laid prostrate, would be explained in accordance with our view—namely, that apogeotropism results from modified circumnutation. After these joints have curved upwards, they are fixed in their new position by increased growth along their lower sides.

[1] This structure has been recently described by De Vries in an interesting article, 'Ueber die Aufrichtung des gelagerten Getreides,' in 'Landwirthschaftliche Jahrbücher,' 1880, p. 473.

Lolium perenne.—A young stem, 7 inches in height, consisting of 3 internodes, with the flower-head not yet protruded, was selected for observation. A long and very thin glass filament was cemented horizontally to the stem close above the second joint, 3 inches above the ground. This joint was subsequently proved to be in an active condition, as its lower side swelled much through the action of apogeotropism (in the manner described by De Vries) after the haulm had been fastened down for 24 h. in a horizontal position. The pot was so placed that the end of the filament stood beneath the 2-inch object glass of a microscope with an eye-piece micrometer, each division of which equalled 1/500 of an inch. The end of the filament was repeatedly observed during 6 h., and was seen to be in constant movement; and it crossed 5 divisions of the micrometer (1/100 inch) in 2 h. Occasionally it moved forwards by jerks, some of which were 1/1000 inch in length, and then slowly retreated a little, afterwards again jerking forwards. These oscillations were exactly like those described under Brassica and Dionaea, but they occurred only occasionally. We may therefore conclude that this moderately old joint was continually circumnutating on a small scale.

Alopecurus pratensis.—A young plant, 11 inches in height, with the flower-head protruded, but with the florets not yet expanded, had a glass filament fixed close above the second joint, at a height of only 2 inches above the ground. The basal internode, 2 inches in length, was cemented to a stick to prevent any possibility of its circumnutating. The extremity of the filament, which projected about 50° above the horizon, was often observed during

24 h. in the same manner as in the last case. Whenever looked at, it was always in movement, and it crossed 30 divisions of the micrometer (3/50 inch) in 3½ h.; but it sometimes moved at a quicker rate, for at one time it crossed 5 divisions in 1½ h. The pot had to be moved occasionally, as the end of the filament travelled beyond the field of vision; but as far as we could judge it followed during the daytime a semicircular course; and it certainly travelled in two different directions at right angles to one another. It sometimes oscillated in the same manner as in the last species, some of the jerks forwards being as much as 1/1000 of an inch. We may therefore conclude that the joints in this and the last species of grass long continue to circumnutate; so that this movement would be ready to be converted into an apogeotropic movement, whenever the stem was placed in an inclined or horizontal position.

Movements of the Flower-peduncles of Oxalis carnosa, due to apogeotropism and other forces.—The movements of the main peduncle, and of the three or four sub-peduncles which each main peduncle of this plant bears, are extremely complex, and are determined by several distinct causes. Whilst the flowers are expanded, both kinds of peduncles circumnutate about the same spot, as we have seen (Fig. 91) in the fourth chapter. But soon after the flowers have begun to wither the sub-peduncles bend downwards, and this is due to epinasty; for on two occasions when pots were laid horizontally, the sub-peduncles assumed the same position relatively to the main peduncle, as would have been the case if they had remained upright; that is, each of them formed with it an angle of about 40°. If they had been acted on by geotropism or apheliotropism (for the plant was illuminated from above), they would have directed themselves to the centre of the earth. A main peduncle was secured to a stick in an upright position, and one of the upright sub-peduncles which had been observed circumnutating whilst the flower was expanded, continued to do so for at least 24 h. after it had withered. It then began to bend downwards, and after 36 h. pointed a little beneath the horizon. A new figure was now begun (A, Fig. 188), and the sub-peduncle was traced descending in a zigzag line from 7.20 P.M. on the 19th to 9 A.M. on the 22nd. It now pointed almost perpendicularly downwards, and the glass filament had to be removed and fastened transversely across the base of the young capsule. We expected that the sub-peduncle would have been motionless in its new position; but it continued slowly to swing, like a pendulum, from side to side, that is, in a plane at right angles to that in which it had descended. This circumnutating movement was observed from 9 A.M. on 22nd to 9 A.M. 24th, as shown at B in the diagram. We were not able to observe this particular sub-peduncle any longer; but it would certainly have gone on circumnutating until the capsule was nearly ripe (which requires only a short time), and it would then have moved upwards.

The upward movement (C, Fig. 188) is effected in part by the whole sub-peduncle rising in the same manner as it had previously descended through epinasty—namely, at the joint where united to the main peduncle. As this upward movement occurred with plants kept in the dark and in whatever position the main peduncle was fastened, it could not have been caused by heliotropism or apogeotropism, but by hyponasty. Besides this movement at the joint, there is another of a very different kind, for the sub-peduncle becomes upwardly bent in the middle part. If the sub-peduncle happens at the time to be inclined much downwards, the upward curvature is so great that the whole forms a hook. The upper end bearing the capsule, thus always places itself upright, and as this occurs in darkness, and in whatever position the main peduncle may have been secured, the upward curvature cannot be due to heliotropism or hyponasty, but to apogeotropism.

Fig. 188. Oxalis carnosa: movements of flower-peduncle, traced on a vertical glass: A, epinastic downward movement; B, circumnutation whilst depending vertically; C, subsequent upward movement, due to apogeotropism and hyponasty combined.

In order to trace this upward movement, a filament was fixed to a sub-peduncle bearing a capsule nearly ripe, which was beginning to bend upwards by the two means just described. Its course was traced (see C, Fig 188) during 53 h., by which time it had become nearly upright. The course is seen to be strongly zigzag, together with some little loops. We may therefore conclude that the movement consists of modified circumnutation.

The several species of Oxalis probably profit in the following manner by their sub-peduncles first bending downwards and then upwards. They are known to scatter their seeds by the bursting of the capsule; the walls of which are so extremely thin, like silver paper, that they would easily be permeated by rain. But as soon as the petals wither, the sepals rise up and enclose the young capsule, forming a perfect roof over it as soon as the sub-peduncle has bent itself downwards. By its subsequent upward movement, the capsule stands when ripe at a greater height above the ground by twice the length of the sub-peduncle, than it did when dependent, and is thus able to scatter its seeds to a greater distance. The sepals, which enclose the ovarium whilst it is young, present an additional adaptation by expanding widely when the seeds are ripe, so as not to interfere with their dispersal. In the case of Oxalis acetosella, the capsules are said sometimes to bury themselves under loose leaves or moss on the ground, but this cannot occur with those of O. carnosa, as the woody stem is too high.

Oxalis acetosella.—The peduncles are furnished with a joint in the middle, so that the lower part answers to the main peduncle, and the upper part to one of the sub-peduncles of O. carnosa. The upper part bends downwards, after the flower has begun to wither, and the whole peduncle then forms a hook; that this bending is due to epinasty we may infer from the case of O. carnosa. When the pod is nearly ripe, the upper part straightens itself and becomes erect; and this is due to hyponasty or apogeotropism, or both combined, and not to heliotropism, for it occurred in darkness. The short, hooked part of the peduncle of a cleistogamic flower, bearing a pod nearly ripe, was observed in the dark during three days. The apex of the pod at first pointed perpendicularly down, but in the course of three days rose 90°, so that it now projected horizontally. The course during the two latter days is shown in Fig. 189; and it may be seen how greatly the peduncle, whilst rising, circumnutated. The lines of chief movement were at right angles to the plane of the originally hooked part. The tracing was not continued any longer; but after two additional days, the peduncle with its capsule had become straight and stood upright.

Fig. 189. Oxalis acetosella: course pursued by the upper part of a peduncle, whilst rising, traced from 11 A.M. June 1st to 9 A.M. 3rd. Figure here reduced to one-half of the original scale.

Concluding Remarks on Apogeotropism.—When apogeotropism is rendered by any means feeble, it acts, as shown in the several foregoing cases, by increasing the always present circumnutating movement in a direction opposed to gravity, and by diminishing that in the direction of gravity, as well as that to either side. The upward movement thus becomes unequal in rate, and is sometimes interrupted by stationary periods. Whenever irregular ellipses or loops are still formed, their longer axes are almost always directed in the line of gravity, in an analogous manner as occurred with heliotropic movements in reference to the light. As apogeotropism acts more and more energetically, ellipses or loops cease to be formed, and the course becomes at first strongly, and then less and less zigzag, and finally rectilinear. From this gradation in the nature of the movement, and more especially from all growing parts, which alone (except when pulvini are present) are acted on by apogeotropism, continually circumnutating, we may conclude that even a rectilinear course is merely an extremely modified form of circumnutation. It is remarkable that a stem or other organ which is highly sensitive to apogeotropism, and which has bowed itself rapidly upwards in a straight line, is often carried beyond the vertical, as if by momentum. It then bends a little backwards to a point round which it finally circumnutates. Two instances of this were observed with the hypocotyls of Beta vulgaris, one of which is shown in Fig. 183, and two other instances with the hypocotyls of Brassica. This momentum-like movement probably results from the accumulated effects of apogeotropism. For the sake of observing how long such after-effects lasted, a pot with seedlings of Beta was laid on its side in the dark, and the hypocotyls in 3 h. 15 m. became highly inclined. The pot, still in the dark, was then placed upright, and the movements of the two hypocotyls were traced; one continued to bend in its former direction, now in opposition to apogeotropism, for about 37 m., perhaps for 48 m.; but after 61 m. it moved in an opposite direction. The other hypocotyl continued to move in its former course, after being placed upright, for at least 37 m.

Different species and different parts of the same species are acted on by apogeotropism in very different degrees. Young seedlings, most of which circumnutate quickly and largely, bend upwards and become vertical in much less time than do any older plants observed by us; but whether this is due to their greater sensitiveness to apogeotropism, or merely to their greater flexibility we do not know. A hypocotyl of Beta traversed an angle of 109° in 3 h. 8 m., and a cotyledon of Phalaris an angle of 130° in 4 h. 30 m. On the other hand, the stem of a herbaceous Verbena rose 90° in about 24 h.; that of Rubus 67°, in 70 h.; that of Cytisus 70°, in 72 h.; that of a young American Oak only 37°, in 72 h. The stem of a young Cyperus alternifolius rose only 11° in 96 h.; the bending being confined to near its base. Though the sheath-like cotyledons of Phalaris are so extremely sensitive to apogeotropism, the first true leaves which protrude from them exhibited only a trace of this action. Two fronds of a fern, Nephrodium molle, both of them young and one with the tip still inwardly curled, were kept in a horizontal position for 46 h., and during this time they rose so little that it was doubtful whether there was any true apogeotropic movement.

The most curious case known to us of a difference in sensitiveness to gravitation, and consequently of movement, in different parts of the same organ, is that offered by the petioles of the cotyledons of Ipomœa leptophylla. The basal part for a short length where united to the undeveloped hypocotyl and radicle is strongly geotropic, whilst the whole upper part is strongly apogeotropic. But a portion near the blades of the cotyledons is after a time acted on by epinasty and curves downwards, for the sake of emerging in the form of an arch from the ground; it subsequently straightens itself, and is then again acted on by apogeotropism.

A branch of Cucurbita ovifera, placed horizontally, moved upwards during 7 h. in a straight line, until it stood at 40° above the horizon; it then began to circumnutate, as if owing to its trailing nature it had no tendency to rise any higher. Another upright branch was secured to a stick, close to the base of a tendril, and the pot was then laid horizontally in the dark. In this position the tendril circumnutated and made several large ellipses during 14 h., as it likewise did on the following day; but during this whole time it was not in the least affected by apogeotropism. On the other hand, when branches of another Cucurbitaceous plant, Echinocytis lobata, were fixed in the dark so that the tendrils depended beneath the horizon, these began immediately to bend upwards, and whilst thus moving they ceased to circumnutate in any plain manner; but as soon as they had become horizontal they recommenced to revolve conspicuously.[2] The tendrils of Passiflora gracilis are likewise apogeotropic. Two branches were tied down so that their tendrils pointed many degrees beneath the horizon. One was observed for 8 h., during which time it rose, describing two circles, one above the other. The other tendril rose in a moderately straight line during the first 4 h., making however one small loop in its course; it then stood at about 45° above the horizon, where it circumnutated during the remaining 8 h. of observation.

[2] For details see 'The Movements and Habits of Climbing Plants,' 1875, p. 131.

A part or organ which whilst young is extremely sensitive to apogeotropism ceases to be so as it grows old; and it is remarkable, as showing the independence of this sensitiveness and of the circumnutating movement, that the latter sometimes continues for a time after all power of bending from the centre of the earth has been lost. Thus a seedling Orange bearing only 3 young leaves, with a rather stiff stem, did not curve in the least upwards during 24 h. whilst extended horizontally; yet it circumnutated all the time over a small space. The hypocotyl of a young seedling of Cassia tora, similarly placed, became vertical in 12 h.; that of an older seedling, 1 1/4 inch in height, became so in 28 h.; and that of another still older one, 1½ inch in height, remained horizontal during two days, but distinctly circumnutated during this whole time.

When the cotyledons of Phalaris or Avena are laid horizontally, the uppermost part first bends upwards, and then the lower part; consequently, after the lower part has become much curved upwards, the upper part is compelled to curve backwards in an opposite direction, in order to straighten itself and to stand vertically; and this subsequent straightening process is likewise due to apogeotropism. The upper part of 8 young cotyledons of Phalaris were made rigid by being cemented to thin glass rods, so that this part could not bend in the least; nevertheless, the basal part was not prevented from curving upward. A stem or other organ which bends upwards through apogeotropism exerts considerable force; its own weight, which has of course to be lifted, was sufficient in almost every instance to cause the part at first to bend a little downwards; but the downward course was often rendered oblique by the simultaneous circumnutating movement. The cotyledons of Avena placed horizontally, besides lifting their own weight, were able to furrow the soft sand above them, so as to leave little crescentic open spaces on the lower sides of their bases; and this is a remarkable proof of the force exerted.

As the tips of the cotyledons of Phalaris and Avena bend upwards through the action of apogeotropism before the basal part, and as these same tips when excited by a lateral light transmit some influence to the lower part, causing it to bend, we thought that the same rule might hold good with apogeotropism. Consequently, the tips of 7 cotyledons of Phalaris were cut off for a length in three cases of .2 inch and in the four other cases of .14, .12, .1, and .07 inch. But these cotyledons, after being extended horizontally, bowed themselves upwards as effectually as the unmutilated specimens in the same pots, showing that sensitiveness to gravitation is not confined to their tips.

GEOTROPISM.

This movement is directly the reverse of apogeotropism. Many organs bend downwards through epinasty or apheliotropism or from their own weight; but we have met with very few cases of a downward movement in sub-aërial organs due to geotropism. We shall however, give one good instance in the following section, in the case of Trifolium subterraneum, and probably in that of Arachis hypogaea.

On the other hand, all roots which penetrate the ground (including the modified root-like petioles of Megarrhiza and Ipomœa leptophylla) are guided in their downward course by geotropism; and so are many aërial roots, whilst others, as those of the Ivy, appear to be indifferent to its action. In our first chapter the movements of the radicles of several seedlings were described. We may there see (Fig. 1) how a radicle of the cabbage, when pointing vertically upwards so as to be very little acted on by geotropism, circumnutated; and how another (Fig. 2) which was at first placed in an inclined position bowed itself downwards in a zigzag line, sometimes remaining stationary for a time. Two other radicles of the cabbage travelled downwards in almost rectilinear courses. A radicle of the bean placed upright (Fig. 20) made a great sweep and zigzagged; but as it sank downwards and was more strongly acted on by geotropism, it moved in an almost straight course. A radicle of Cucurbita, directed upwards (Fig. 26), also zigzagged at first, and described small loops; it then moved in a straight line. Nearly the same result was observed with the radicles of Zea mays. But the best evidence of the intimate connection between circumnutation and geotropism was afforded by the radicles of Phaseolus, Vicia, and Quercus, and in a less degree by those of Zea and Æsculus (see Figs. 18, 19, 21, 41, and 52); for when these were compelled to grow and slide down highly inclined surfaces of smoked glass, they left distinctly serpentine tracks.

The Burying of Seed-capsules: Trifolium subterraneum.—The flower-heads of this plant are remarkable from producing only 3 or 4 perfect flowers, which are situated exteriorly. All the other many flowers abort, and are modified into rigid points, with a bundle of vessels running up their centres. After a time 5 long, elastic, claw-like projections, which represent the divisions of the calyx, are developed on their summits. As soon as the perfect flowers wither they bend downwards, supposing the peduncle to stand upright, and they then closely surround its upper part. This movement is due to epinasty, as is likewise the case with the flowers of T. repens. The imperfect central flowers ultimately follow, one after the other, the same course. Whilst the perfect flowers are thus bending down, the whole peduncle curves downwards and increases much in length, until the flower-head reaches the ground. Vaucher[3] says that when the plant is so placed that the heads cannot soon reach the ground, the peduncles grow to the extraordinary length of from 6 to 9 inches. In whatever position the branches may be placed, the upper part of the peduncle at first bends vertically upwards through heliotropism; but as soon as the flowers begin to wither the downward curvature of the whole peduncle commences. As this latter movement occurred in complete darkness, and with peduncles arising from upright and from dependent branches, it cannot be due to apheliotropism or to epinasty, but must be attributed to geotropism. Nineteen upright flower-heads, arising from branches in all sorts of positions, on plants growing in a warm greenhouse, were marked with thread, and after 24 h. six of them were vertically dependent; these therefore had travelled through 180° in this time. Ten were extended sub-horizontally, and these had moved through about 90°. Three very young peduncles had as yet moved only a little downwards, but after an additional 24 h. were greatly inclined.

[3] 'Hist. Phys. des Plantes d'Europe,' tom. ii. 1841, p. 106.

At the time when the flower-heads reach the ground, the younger imperfect flowers in the centre are still pressed closely together, and form a conical projection; whereas the perfect and imperfect flowers on the outside are upturned and closely surround the peduncle. They are thus adapted to offer as little resistance, as the case admits of, in penetrating the ground, though the diameter of the flower-head is still considerable. The means by which this penetration is effected will presently be described. The flower-heads are able to bury themselves in common garden mould, and easily in sand or in fine sifted cinders packed rather closely. The depth to which they penetrated, measured from the surface to the base of the head, was between 1/4 and ½ inch, but in one case rather above 0.6 inch. With a plant kept in the house, a head partly buried itself in sand in 6 h.: after 3 days only the tips of the reflexed calyces were visible, and after 6 days the whole had disappeared. But with plants growing out of doors we believe, from casual observations, that they bury themselves in a much shorter time.

After the heads have buried themselves, the central aborted flowers increase considerably in length and rigidity, and become bleached. They gradually curve, one after the other, upwards or towards the peduncle, in the same manner as did the perfect flowers at first. In thus moving, the long claws on their summits carry with them some earth. Hence a flower-head which has been buried for a sufficient time, forms a rather large ball, consisting of the aborted flowers, separated from one another by earth, and surrounding the little pods (the product of the perfect flowers) which lie close round the upper part of the peduncle. The calyces of the perfect and imperfect flowers are clothed with simple and multicellular hairs, which have the power of absorption; for when placed in a weak solution of carbonate of ammonia (2 gr. to 1 oz. of water) their protoplasmic contents immediately became aggregated and afterwards displayed the usual slow movements. This clover generally grows in dry soil, but whether the power of absorption by the hairs on the buried flower-heads is of any importance to them we do not know. Only a few of the flower-heads, which from their position are not able to reach the ground and bury themselves, yield seeds; whereas the buried ones never failed, as far as we observed, to produce as many seeds as there had been perfect flowers.

We will now consider the movements of the peduncle whilst curving down to the ground. We have seen in Chap. IV., Fig. 92, p. 225, that an upright young flower-head circumnutated conspicuously; and that this movement continued after the peduncle had begun to bend downwards. The same peduncle was observed when inclined at an angle of 19° above the horizon, and it circumnutated during two days. Another which was already curved 36° beneath the horizon, was observed from 11 A.M. July 22nd to the 27th, by which latter date it had become vertically dependent. Its course during the first 12 h. is shown in Fig. 190, and its position on the three succeeding mornings until the 25th, when it was nearly vertical. During the first day the peduncle clearly circumnutated, for it moved 4 times down and 3 times up; and on each succeeding day, as it sank downwards, the same movement continued, but was only occasionally observed and was less strongly marked. It should be stated that these peduncles were observed under a double skylight in the house, and that they generally moved downwards very much more slowly than those on plants growing out of doors or in the greenhouse.

Fig. 190. Trifolium subterraneum: downward movement of peduncle from 19° beneath the horizon to a nearly vertically dependent position, traced from 11 A.M. July 22nd to the morning of 25th. Glass filament fixed transversely across peduncle, at base of flower-head.

Fig. 191. Trifolium subterraneum: circumnutating movement of peduncle, whilst the flower-head was burying itself in sand, with the reflexed tips of the calyx still visible; traced from 8 A.M. July 26th to 9 A.M. on 27th. Glass filament fixed transversely across peduncle, near flower-head.

Fig. 192. Trifolium subterraneum: movement of same peduncle, with flower-head completely buried beneath the sand; traced from 8 A.M. to 7.15 P.M. on July 29th.

The movement of another vertically dependent peduncle with the flower-head standing half an inch above the ground, was traced, and again when it first touched the ground; in both cases irregular ellipses were described every 4 or 5 h. A peduncle on a plant which had been brought into the house, moved from an upright into a vertically dependent position in a single day; and here the course during the first 12 h. was nearly straight, but with a few well-marked zigzags which betrayed the essential nature of the movement. Lastly the circumnutation of a peduncle was traced during 51 h. whilst in the act of burying itself obliquely in a little heap of sand. After it had buried itself to such a depth that the tips of the sepals were alone visible, the above figure (Fig 191) was traced during 25 h. When the flower-head had completely disappeared beneath the sand, another tracing was made during 11 h. 45 m. (Fig. 192); and here again we see that the peduncle was circumnutating.

Any one who will observe a flower-head burying itself, will be convinced that the rocking movement, due to the continued circumnutation of the peduncle, plays an important part in the act. Considering that the flower-heads are very light, that the peduncles are long, thin, and flexible, and that they arise from flexible branches, it is incredible that an object as blunt as one of these flower-heads could penetrate the ground by means of the growing force of the peduncle, unless it were aided by the rocking movement. After a flower-head has penetrated the ground to a small depth, another and efficient agency comes into play; the central rigid aborted flowers, each terminating in five long claws, curve up towards the peduncle; and in doing so can hardly fail to drag the head down to a greater depth, aided as this action is by the circumnutating movement, which continues after the flower-head has completely buried itself. The aborted flowers thus act something like the hands of the mole, which force the earth backwards and the body forwards.

It is well known that the seed-capsules of various widely distinct plants either bury themselves in the ground, or are produced from imperfect flowers developed beneath the surface. Besides the present case, two other well-marked instances will be immediately given. It is probable that one chief good thus gained is the protection of the seeds from animals which prey on them. In the case of T. subterraneum, the seeds are not only concealed by being buried, but are likewise protected by being closely surrounded by the rigid, aborted flowers. We may the more confidently infer that protection is here aimed at, because the seeds of several species in this same genus are protected in other ways;[4] namely, by the swelling and closure of the calyx, or by the persistence and bending down of the standard-petal, etc. But the most curious instance is that of T. globosum, in which the upper flowers are sterile, as in T. subterraneum, but are here developed into large brushes of hairs which envelop and protect the seed-bearing flowers. Nevertheless, in all these cases the capsules, with their seeds, may profit, as Mr. T. Thiselton Dyer has remarked,[5] by their being kept somewhat damp; and the advantage of such dampness perhaps throws light on the presence of the absorbent hairs on the buried flower-heads of T. subterraneum. According to Mr. Bentham, as quoted by Mr. Dyer, the prostrate habit of Helianthemum prostratum "brings the capsules in contact with the surface of the ground, postpones their maturity, and so favours the seeds attaining a larger size." The capsules of Cyclamen and of Oxalis acetosella are only occasionally buried, and this only beneath dead leaves or moss. If it be an advantage to a plant that its capsules should be kept damp and cool by being laid on the ground, we have in these latter cases the first step, from which the power of penetrating the ground, with the aid of the always present movement of circumnutation, might afterwards have been gained.

[4] Vaucher, 'Hist. Phys. des Plantes d'Europe,' tom. ii. p. 110.

[5] See his interesting article in 'Nature,' April 4th, 1878, p. 446.

Arachis hypogoea.—The flowers which bury themselves, rise from stiff branches a few inches above the ground, and stand upright. After they have fallen off, the gynophore, that is the part which supports the ovarium, grows to a great length, even to 3 or 4 inches, and bends perpendicularly downwards. It resembles closely a peduncle, but has a smooth and pointed apex, which contains the ovules, and is at first not in the least enlarged. The apex after reaching the ground penetrates it, in one case observed by us to a depth of 1 inch, and in another to 0.7 inch. It there becomes developed into a large pod. Flowers which are seated too high on the plant for the gynophore to reach the ground are said[6] never to produce pods.

[6] 'Gard. Chronicle,' 1857, p. 566.

The movement of a young gynophore, rather under an inch in length and vertically dependent, was traced during 46 H. by means of a glass filament (with sights) fixed transversely a little above the apex. It plainly circumnutated (Fig. 193) whilst increasing in length and growing downwards. It was then raised up, so as to be extended almost horizontally, and the terminal part curved itself downwards, following a nearly straight course during 12 h., but with one attempt to circumnutate, as shown in Fig. 194. After 24 h. it had become nearly vertical. Whether the exciting cause of the downward movement is geotropism or apheliotropism was not ascertained; but probably it is not apheliotropism, as all the gynophores grew straight down towards the ground, whilst the light in the hot-house entered from one side as well as from above. Another and older gynophore, the apex of which had nearly reached the ground, was observed during 3 days in the same manner as the first-mentioned short one; and it was found to be always circumnutating. During the first 34 h. it described a figure which represented four ellipses. Lastly, a long gynophore, the apex of which had buried itself to the depth of about half an inch, was pulled up and extended horizontally: it quickly began to curve downwards in a zigzag line; but on the following day the terminal bleached portion was a little shrivelled. As the gynophores are rigid and arise from stiff branches, and as they terminate in sharp smooth points, it is probable that they could penetrate the ground by the mere force of growth. But this action must be aided by the circumnutating movement, for fine sand, kept moist, was pressed close round the apex of a gynophore which had reached the ground, and after a few hours it was surrounded by a narrow open crack. After three weeks this gynophore was uncovered, and the apex was found at a depth of rather above half an inch developed into a small, white, oval pod.

Fig. 193 Arachis hypogoea: circumnutation of vertically dependent young gynophore, traced on a vertical glass from 10 A.M. July 31st to 8 A.M. Aug. 2nd.

Fig. 194. Arachis hypogoea: downward movement of same young gynophore, after being extended horizontally; traced on a vertical glass from 8.30 A.M. to 8.30 P.M. Aug. 2nd.

Amphicarpoea monoica.—This plant produces long thin shoots, which twine round a support and of course circumnutate. Early in the summer shorter shoots are produced from the lower parts of the plant, which grow perpendicularly downwards and penetrate the ground. One of these, terminating in a minute bud, was observed to bury itself in sand to a depth of 0.2 inch in 24 h. It was lifted up and fixed in an inclined position about 25° beneath the horizon, being feebly illuminated from above. In this position it described two vertical ellipses in 24 h.; but on the following day, when brought into the house, it circumnutated only a very little round the same spot. Other branches were seen to penetrate the ground, and were afterwards found running like roots beneath the surface for a length of nearly two inches, and they had grown thick. One of these, after thus running, had emerged into the air. How far circumnutation aids these delicate branches in entering the ground we do not know; but the reflexed hairs with which they are clothed will assist in the work. This plant produces pods in the air, and others beneath the ground; which differ greatly in appearance. Asa Gray says[7] that it is the imperfect flowers on the creeping branches near the base of the plant which produce the subterranean pods; these flowers, therefore, must bury themselves like those of Arachis. But it may be suspected that the branches which were seen by us to penetrate the ground also produce subterranean flowers and pods.

[7] 'Manual of the Botany of the Northern United States,' 1856, p. 106.

DIAGEOTROPISM.

Besides geotropism and apogeotropism, there is, according to Frank, an allied form of movement, namely, "transverse-geotropism," or diageotropism, as we may call it for the sake of matching our other terms. Under the influence of gravitation certain parts are excited to place themselves more or less transversely to the line of its action.[8] We made no observations on this subject, and will here only remark that the position of the secondary radicles of various plants, which extend horizontally or are a little inclined downwards, would probably be considered by Frank as due to transverse-geotropism. As it has been shown in Chap. I. that the secondary radicles of Cucurbita made serpentine tracks on a smoked glass-plate, they clearly circumnutated, and there can hardly be a doubt that this holds good with other secondary radicles. It seems therefore highly probable that they place themselves in their diageotropic position by means of modified circumnutation.

[8] Elfving has lately described ('Arbeiten des Bot. Instituts in Würzburg,' B. ii. 1880, p. 489) an excellent instance of such movements in the rhizomes of certain plants.

Finally, we may conclude that the three kinds of movement which have now been described and which are excited by gravitation, consist of modified circumnutation. Different parts or organs on the same plant, and the same part in different species, are thus excited to act in a widely different manner. We can see no reason why the attraction of gravity should directly modify the state of turgescence and subsequent growth of one part on the upper side and of another part on the lower side. We are therefore led to infer that both geotropic, apogeotropic, and diageotropic movements, the purpose of which we can generally understand, have been acquired for the advantage of the plant by the modification of the ever-present movement of circumnutation. This, however, implies that gravitation produces some effect on the young tissues sufficient to serve as a guide to the plant.

CHAPTER XI.
LOCALISED SENSITIVENESS TO GRAVITATION, AND ITS TRANSMITTED EFFECTS.

General considerations—Vicia faba, effects of amputating the tips of the radicles—Regeneration of the tips—Effects of a short exposure of the tips to geotropic action and their subsequent amputation—Effects of amputating the tips obliquely—Effects of cauterising the tips—Effects of grease on the tips—Pisum sativum, tips of radicles cauterised transversely, and on their upper and lower sides—Phaseolus, cauterisation and grease on the tips—Gossypium—Cucurbita, tips cauterised transversely, and on their upper and lower sides—Zea, tips cauterised—Concluding remarks and summary of chapter—Advantages of the sensibility to geotropism being localised in the tips of the radicles.

Ciesielski states[1] that when the roots of Pisum, Lens and Vicia were extended horizontally with their tips cut off, they were not acted on by geotropism; but some days afterwards, when a new root-cap and vegetative point had been formed, they bent themselves perpendicularly downwards. He further states that if the tips are cut off, after the roots have been left extended horizontally for some little time, but before they have begun to bend downwards, they may be placed in any position, and yet will bend as if still acted on by geotropism; and this shows that some influence had been already transmitted to the bending part from the tip before it was amputated. Sachs repeated these experiments; he cut off a length of between .05 and 1 mm. (measured from the apex of the vegetative point) of the tips of the radicles of the bean (Vicia faba), and placed them horizontally or vertically in damp air, earth, and water, with the result that they became bowed in all sorts of directions.[2] He therefore disbelieved in Ciesielski's conclusions. But as we have seen with several plants that the tip of the radicle is sensitive to contact and to other irritants, and that it transmits some influence to the upper growing part causing it to bend, there seemed to us to be no a priori improbability in Ciesielski's statements. We therefore determined to repeat his experiments, and to try others on several species by different methods.

[1] 'Abwartskrümmung der Wurzel,' Inaug. Dissert. Breslau, 1871, p. 29.

[2] 'Arbeiten des Bot. Instituts in Würzburg,' Heft. iii. 1873, p. 432.

Vicia faba.—Radicles of this plant were extended horizontally either over water or with their lower surfaces just touching it. Their tips had previously been cut off, in a direction as accurately transverse as could be done, to different lengths, measured from the apex of the root-cap, and which will be specified in each case. Light was always excluded. We had previously tried hundreds of unmutilated radicles under similar circumstances, and found that every one that was healthy became plainly geotropic in under 12 h. In the case of four radicles which had their tips cut off for a length of 1.5 mm., new root caps and new vegetative points were re-formed after an interval of 3 days 20 h.; and these when placed horizontally were acted on by geotropism. On some other occasions this regeneration of the tips and reacquired sensitiveness occurred within a somewhat shorter time. Therefore, radicles having their tips amputated should be observed in from 12 to 48 h. after the operation.

Four radicles were extended horizontally with their lower surfaces touching the water, and with their tips cut off for a length of only 0.5 mm.: after 23 h. three of them were still horizontal; after 47 h. one of the three became fairly geotropic; and after 70 h. the other two showed a trace of this action. The fourth radicle was vertically geotropic after 23 h.; but by an accident the root-cap alone and not the vegetative point was found to have been amputated; so that this case formed no real exception and might have been excluded.

Five radicles were extended horizontally like the last, and had their tips cut off for a length of 1 mm.; after 22–23 h., four of them were still horizontal, and one was slightly geotropic; after 48 h. the latter had become vertical; a second was also somewhat geotropic; two remained approximately horizontal; and the last or fifth had grown in a disordered manner, for it was inclined upwards at an angle of 65° above the horizon.

Fourteen radicles were extended horizontally at a little height over the water with their tips cut off for a length of 1.5 mm.; after 12 h. all were horizontal, whilst five control or standard specimens in the same jar were all bent greatly downwards. After 24 h. several of the amputated radicles remained horizontal, but some showed a trace of geotropism, and one was plainly geotropic, for it was inclined at 40° beneath the horizon.

Seven horizontally extended radicles from which the tips had been cut off for the unusual length of 2 mm. unfortunately were not looked at until 35 h. had elapsed; three were still horizontal, but to our surprise, four were more or less plainly geotropic.

The radicles in the foregoing cases were measured before their tips were amputated, and in the course of 24 h. they had all increased greatly in length; but the measurements are not worth giving. It is of more importance that Sachs found that the rate of growth of the different parts of radicles with amputated tips was the same as with unmutilated ones. Altogether twenty-nine radicles were operated on in the manner above described, and of these only a few showed any geotropic curvature within 24 h.; whereas radicles with unmutilated tips always became, as already stated, much bent down in less than half of this time. The part of the radicle which bends most lies at the distance of from 3 to 6 mm. from the tip, and as the bending part continues to grow after the operation, there does not seem any reason why it should not have been acted on by geotropism, unless its curvature depended on some influence transmitted from the tip. And we have clear evidence of such transmission in Ciesielski's experiments, which we repeated and extended in the following manner.

Beans were embedded in friable peat with the hilum downwards, and after their radicles had grown perpendicularly down for a length of from ½ to 1 inch, sixteen were selected which were perfectly straight, and these were placed horizontally on the peat, being covered by a thin layer of it. They were thus left for an average period of 1 h. 37 m. The tips were then cut off transversely for a length of 1.5 mm., and immediately afterwards they were embedded vertically in the peat. In this position geotropism would not tend to induce any curvature, but if some influence had already been transmitted from the tip to the part which bends most, we might expect that this part would become curved in the direction in which geotropism had previously acted; for it should be noted that these radicles being now destitute of their sensitive tips, would not be prevented by geotropism from curving in any direction. The result was that of the sixteen vertically embedded radicles, four continued for several days to grow straight downwards, whilst twelve became more or less bowed laterally. In two of the twelve, a trace of curvature was perceptible in 3 h. 30 m., counting from the time when they had first been laid horizontally; and all twelve were plainly bowed in 6 h., and still more plainly in 9 h. In every one of them the curvature was directed towards the side which had been downwards whilst the radicles remained horizontal. The curvature extended for a length of from 5 to, in one instance, 8 mm., measured from the cut-off end. Of the twelve bowed radicles five became permanently bent into a right angle; the other seven were at first much less bent, and their curvature generally decreased after 24 h., but did not wholly disappear. This decrease of curvature would naturally follow, if an exposure of only 1 h. 37 m. to geotropism, served to modify the turgescence of the cells, but not their subsequent growth to the full extent. The five radicles which were rectangularly bent became fixed in this position, and they continued to grow out horizontally in the peat for a length of about 1 inch during from 4 to 6 days. By this time new tips had been formed; and it should be remarked that this regeneration occurred slower in the peat than in water, owing perhaps to the radicles being often looked at and thus disturbed. After the tips had been regenerated, geotropism was able to act on them, so that they now became bowed vertically downwards. An accurate drawing (Fig. 195) is given on the opposite page of one of these five radicles, reduced to half the natural size.

We next tried whether a shorter exposure to geotropism would suffice to produce an after-effect. Seven radicles were extended horizontally for an hour, instead of 1 h. 37 m. as in the former trial; and after their tips (1.5 mm. in length) had been amputated, they were placed vertically in damp peat. Of these, three were not in the least affected and continued for days to grow straight downwards. Four showed after 8 h. 30 m. a mere trace of curvature in the direction in which they had been acted on by geotropism; and in this respect they differed much from those which had been exposed for 1 h. 37 m., for many of the latter were plainly curved in 6 h. The curvature of one of these four radicles almost disappeared after 24 h. In the second, the curvature increased during two days and then decreased. the third radicle became permanently bent, so that its terminal part made an angle of about 45° with its original vertical direction. The fourth radicle became horizontal. These two, latter radicles continued during two more days to grow in the peat in the same directions, that is, at an angle of 45° beneath the horizon and horizontally. By the fourth morning new tips had been re-formed, and now geotropism was able to act on them again, and they became bent perpendicularly downwards, exactly as in the case of the five radicles described in the last paragraph and as is shown in (Fig. 195) here given.

Fig. 195. Vicia faba: radicle, rectangularly bent at A, after the amputation of the tip, due to the previous influence of geotropism. L, side of bean which lay on the peat, whilst geotropism acted on the radicle. A, point of chief curvature of the radicle, whilst standing vertically downwards. B, point of chief curvature after the regeneration of the tip, when geotropism again acted. C, regenerated tip.

Lastly, five other radicles were similarly treated, but were exposed to geotropism during only 45 m. After 8 h. 30 m. only one was doubtfully affected; after 24 h. two were just perceptibly curved towards the side which had been acted on by geotropism; after 48 h. the one first mentioned had a radius of curvature of 60 mm. That this curvature was due to the action of geotropism during the horizontal position of the radicle, was shown after 4 days, when a new tip had been re-formed, for it then grew perpendicularly downwards. We learn from this case that when the tips are amputated after an exposure to geotropism of only 45 m., though a slight influence is sometimes transmitted to the adjoining part of the radicle, yet this seldom suffices, and then only slowly, to induce even moderately well-pronounced curvature.

In the previously given experiments on 29 horizontally extended radicles with their tips amputated, only one grew irregularly in any marked manner, and this became bowed upwards at an angle of 65°. In Ciesielski's experiments the radicles could not have grown very irregularly, for if they had done so, he could not have spoken confidently of the obliteration of all geotropic action. It is therefore remarkable that Sachs, who experimented on many radicles with their tips amputated, found extremely disordered growth to be the usual result. As horizontally extended radicles with amputated tips are sometimes acted on slightly by geotropism within a short time, and are often acted on plainly after one or two days, we thought that this influence might possibly prevent disordered growth, though it was not able to induce immediate curvature. Therefore 13 radicles, of which 6 had their tips amputated transversely for a length of 1.5 mm., and the other 7 for a length of only 0.5 mm., were suspended vertically in damp air, in which position they would not be affected by geotropism; but they exhibited no great irregularity of growth, whilst observed during 4 to 6 days. We next thought that if care were not taken in cutting off the tips transversely, one side of the stump might be irritated more than the other, either at first or subsequently during the regeneration of the tip, and that this might cause the radicle to bend to one side. It has also been shown in Chapter III. that if a thin slice be cut off one side of the tip of the radicle, this causes the radicle to bend from the sliced side. Accordingly, 30 radicles, with tips amputated for a length of 1.5 mm., were allowed to grow perpendicularly downwards into water. Twenty of them were amputated at an angle of 20° with a line transverse to their longitudinal axes; and such stumps appeared only moderately oblique. The remaining ten radicles were amputated at an angle of about 45°. Under these circumstances no less than 19 out of the 30 became much distorted in the course of 2 or 3 days. Eleven other radicles were similarly treated, excepting that only 1 mm. (including in this and all other cases the root-cap) was amputated; and of these only one grew much, and two others slightly distorted; so that this amount of oblique amputation was not sufficient. Out of the above 30 radicles, only one or two showed in the first 24 h. any distortion, but this became plain in the 19 cases on the second day, and still more conspicuous at the close of the third day, by which time new tips had been partially or completely regenerated. When therefore a new tip is reformed on an oblique stump, it probably is developed sooner on one side than on the other: and this in some manner excites the adjoining part to bend to one side. Hence it seems probable that Sachs unintentionally amputated the radicles on which he experimented, not strictly in a transverse direction.

This explanation of the occasional irregular growth of radicles with amputated tips, is supported by the results of cauterising their tips; for often a greater length on one side than on the other was unavoidably injured or killed. It should be remarked that in the following trials the tips were first dried with blotting-paper, and then slightly rubbed with a dry stick of nitrate of silver or lunar caustic. A few touches with the caustic suffice to kill the root-cap and some of the upper layers of cells of the vegetative point. Twenty-seven radicles, some young and very short, others of moderate length, were suspended vertically over water, after being thus cauterised. Of these some entered the water immediately, and others on the second day. The same number of uncauterised radicles of the same age were observed as controls. After an interval of three or four days the contrast in appearance between the cauterised and control specimens was wonderfully great. The controls had grown straight downwards, with the exception of the normal curvature, which we have called Sachs' curvature. Of the 27 cauterised radicles, 15 had become extremely distorted; 6 of them grew upwards and formed hoops, so that their tips sometimes came into contact with the bean above; 5 grew out rectangularly to one side; only a few of the remaining 12 were quite straight, and some of these towards the close of our observations became hooked at their extreme lower ends. Radicles, extended horizontally instead of vertically, with their tips cauterised, also sometimes grew distorted, but not so commonly, as far as we could judge, as those suspended vertically; for this occurred with only 5 out of 19 radicles thus treated.

Instead of cutting off the tips, as in the first set of experiments, we next tried the effects of touching horizontally extended radicles with caustic in the manner just described. But some preliminary remarks must first be made. It may be objected that the caustic would injure the radicles and prevent them from bending; but ample evidence was given in Chapter III. that touching the tips of vertically suspended radicles with caustic on one side, does not stop their bending; on the contrary, it causes them to bend from the touched side. We also tried touching both the upper and the lower sides of the tips of some radicles of the bean, extended horizontally in damp friable earth. The tips of three were touched with caustic on their upper sides, and this would aid their geotropic bending; the tips of three were touched on their lower sides, which would tend to counteract the bending downwards; and three were left as controls. After 24 h. an independent observer was asked to pick out of the nine radicles, the two which were most and the two which were least bent; he selected as the latter, two of those which had been touched on their lower sides, and as the most bent, two of those which had been touched on the upper side. Hereafter analogous and more striking experiments with Pisum sativum and Cucurbita ovifera will be given. We may therefore safely conclude that the mere application of caustic to the tip does not prevent the radicles from bending.

In the following experiments, the tips of young horizontally extended radicles were just touched with a stick of dry caustic; and this was held transversely, so that the tip might be cauterised all round as symmetrically as possible. The radicles were then suspended in a closed vessel over water, kept rather cool, viz., 55°–59° F. This was done because we had found that the tips were more sensitive to contact under a low than under a high temperature; and we thought that the same rule might apply to geotropism. In one exceptional trial, nine radicles (which were rather too old, for they had grown to a length of from 3 to 5 cm.), were extended horizontally in damp friable earth, after their tips had been cauterised and were kept at too high a temperature, viz., of 68° F., or 20° C. The result in consequence was not so striking as in the subsequent cases for although when after 9 h. 40 m. six of them were examined, these did not exhibit any geotropic bending, yet after 24 h., when all nine were examined, only two remained horizontal, two exhibited a trace of geotropism, and five were slightly or moderately geotropic, yet not comparable in degree with the control specimens. Marks had been made on seven of these cauterised radicles at 10 mm. from the tips, which includes the whole growing portion; and after the 24 h. this part had a mean length of 37 mm., so that it had increased to more than 3½ times its original length; but it should be remembered that these beans had been exposed to a rather high temperature.

Nineteen young radicles with cauterised tips were extended at different times horizontally over water. In every trial an equal number of control specimens were observed. In the first trial, the tips of three radicles were lightly touched with the caustic for 6 or 7 seconds, which was a longer application than usual. After 23 h. 30 m. (temp. 55°–56° F.) these three radicles, A, B, C (Fig. 196), were still horizontal, whilst the three control specimens had become within 8 h. slightly geotropic, and strongly so (D, E, F) in 23 h. 30 m. A dot had been made on all six radicles at 10 mm. from their tips, when first placed horizontally. After the 23 h. 30 m. this terminal part, originally 10 mm. in length, had increased in the cauterised specimens to a mean length of 17.3 mm., and to 15.7 mm. in the control radicles, as shown in the figures by the unbroken transverse line; the dotted line being at 10 mm. from the apex. The control or uncauterised radicles, therefore, had actually grown less than the cauterised; but this no doubt was accidental, for radicles of different ages grow at different rates, and the growth of different individuals is likewise affected by unknown causes. The state of the tips of these three radicles, which had been cauterised for a rather longer time than usual, was as follows: the blackened apex, or the part which had been actually touched by the caustic, was succeeded by a yellowish zone, due probably to the absorption of some of the caustic; in A, both zones together were 1.1 mm. in length, and 1.4 mm. in diameter at the base of the yellowish zone; in B, the length of both was only 0.7 mm., and the diameter 0.7 mm.; in C, the length was 0.8 mm., and the diameter 1.2 mm.

Fig. 196. Vicia faba: state of radicles which had been extended horizontally for 23 h. 30 m.; A, B, C, tips touched with caustic; D, E, F, tips uncauterised. Lengths of radicles reduced to one-half scale, but by an accident the beans themselves not reduced in the same degree.

Three other radicles, the tips of which had been touched with caustic curing 2 or 3 seconds, remained (temp. 58°–59° F.) horizontal for 23 h.; the control radicles having, of course, become geotropic within this time. The terminal growing part, 10 mm. in length, of the cauterised radicles had increased in this interval to a mean length of 24.5 mm., and of the controls to a mean of 26 mm. A section of one of the cauterised tips showed that the blackened part was 0.5 mm. in length, of which 0.2 mm. extended into the vegetative point; and a faint discoloration could be detected even to 1.6 mm. from the apex of the root-cap.

In another lot of six radicles (temp. 55°–57° F.) the three control specimens were plainly geotropic in 8½ h.; and after 24 h. the mean length of their terminal part had increased from 10 mm. to 21 mm. When the caustic was applied to the three cauterised specimens, it was held quite motionless during 5 seconds, and the result was that the black marks were extremely minute. Therefore, caustic was again applied, after 8½ h., during which time no geotropic action had occurred. When the specimens were re-examined after an additional interval of 15½ h., one was horizontal and the other two showed, to our surprise, a trace of geotropism which in one of them soon afterwards became strongly marked; but in this latter specimen the discoloured tip was only 2/3 mm. in length. The growing part of these three radicles increased in 24 h. from 10 mm. to an average of 16.5 mm.

It would be superfluous to describe in detail the behaviour of the 10 remaining cauterised radicles. The corresponding control specimens all became geotropic in 8 h. Of the cauterised, 6 were first looked at after 8 h., and one alone showed a trace of geotropism; 4 were first looked at after 14 h., and one alone of these was slightly geotropic. After 23–24h., 5 of the 10 were still horizontal, 4 slightly, and 1 decidedly, geotropic. After 48 h. some of them became strongly geotropic. The cauterised radicles increased greatly in length, but the measurements are not worth giving.

As five of the last-mentioned cauterised radicles had become in 24 h. somewhat geotropic, these (together with three which were still horizontal) had their positions reversed, so that their tips were now a little upturned, and they were again touched with caustic. After 24 h. they showed no trace of geotropism; whereas the eight corresponding control specimens, which had likewise been reversed, in which position the tips of several pointed to the zenith, all became geotropic; some having passed in the 24 h. through an angle of 180°, others through about 135°, and others through only 90°. The eight radicles, which had been twice cauterised, were observed for an additional day (i.e. for 48 h. after being reversed), and they still showed no signs of geotropism. Nevertheless, they continued to grow rapidly; four were measured 24 h. after being reversed, and they had in this time increased in length between 8 and 11 mm.; the other four were measured 48 h. after being reversed, and these had increased by 20, 18, 23, and 28 mm.

In coming to a conclusion with respect to the effects of cauterising the tips of these radicles, we should bear in mind, firstly, that horizontally extended control radicles were always acted on by geotropism, and became somewhat bowed downwards in 8 or 9 h.; secondly, that the chief seat of the curvature lies at a distance of from 3 to 6 mm. from the tip; thirdly, that the tip was discoloured by the caustic rarely for more than 1 mm. in length; fourthly, that the greater number of the cauterised radicles, although subjected to the full influence of geotropism during the whole time, remained horizontal for 24 h., and some for twice as long; and that those which did become bowed were so only in a slight degree; fifthly, that the cauterised radicles continued to grow almost, and sometimes quite, as well as the uninjured ones along the part which bends most. And lastly, that a touch on the tip with caustic, if on one side, far from preventing curvature, actually induces it. Bearing all these facts in mind, we must infer that under normal conditions the geotropic curvature of the root is due to an influence transmitted from the apex to the adjoining part where the bending takes place; and that when the tip of the root is cauterised it is unable to originate the stimulus necessary to produce geotropic curvature.

As we had observed that grease was highly injurious to some plants, we determined to try its effects on radicles. When the cotyledons of Phalaris and Avena were covered with grease along one side, the growth of this side was quite stopped or greatly checked, and as the opposite side continued to grow, the cotyledons thus treated became bowed towards the greased side. This same matter quickly killed the delicate hypocotyls and young leaves of certain plants. The grease which we employed was made by mixing lamp-black and olive oil to such a consistence that it could be laid on in a thick layer. The tips of five radicles of the bean were coated with it for a length of 3 mm., and to our surprise this part increased in length in 23 h. to 7.1 mm.; the thick layer of grease being curiously drawn out. It thus could not have checked much, if at all, the growth of the terminal part of the radicle. With respect to geotropism, the tips of seven horizontally extended radicles were coated for a length of 2 mm., and after 24 h. no clear difference could be perceived between their downward curvature and that of an equal number of control specimens. The tips of 33 other radicles were coated on different occasions for a length of 3 mm.; and they were compared with the controls after 8 h., 24 h., and 48 h. On one occasion, after 24 h., there was very little difference in curvature between the greased and control specimens; but generally the difference was unmistakable, those with greased tips being considerably less curved downwards. The whole growing part (the greased tips included) of six of these radicles was measured and was found to have increased in 23 h. from 10 mm. to a mean length of 17.7 mm.; whilst the corresponding part of the controls had increased to 20.8 mm. It appears therefore, that although the tip itself, when greased, continues to grow, yet the growth of the whole radicle is somewhat checked, and that the geotropic curvature of the upper part, which was free from grease, was in most cases considerably lessened.

Pisum sativum.—Five radicles, extended horizontally over water, had their tips lightly touched two or three times with dry caustic. These tips were measured in two cases, and found to be blackened for a length of only half a millimeter. Five other radicles were left as controls. The part which is most bowed through geotropism lies at a distance of several millimeters from the apex. After 24 h., and again after 32 h. from the commencement, four of the cauterised radicles were still horizontal, but one was plainly geotropic, being inclined at 45° beneath the horizon. The five controls were somewhat geotropic after 7 h. 20 m., and after 24 h. were all strongly geotropic; being inclined at the following angles beneath the horizon, viz., 59°, 60°, 65°, 57°, and 43°. The length of the radicles was not measured in either set, but it was manifest that the cauterised radicles had grown greatly.

The following case proves that the action of the caustic by itself does not prevent the curvature of the radicle. Ten radicles were extended horizontally on and beneath a layer of damp friable peat-earth; and before being extended their tips were touched with dry caustic on the upper side. Ten other radicles similarly placed were touched on the lower side; and this would tend to make them bend from the cauterised side; and therefore, as now placed, upwards, or in opposition to geotropism. Lastly, ten uncauterised radicles were extended horizontally as controls. After 24 h. all the latter were geotropic; and the ten with their tips cauterised on the upper side were equally geotropic; and we believe that they became curved downwards before the controls. The ten which had been cauterised on the lower side presented a widely different appearance: No. 1, however, was perpendicularly geotropic, but this was no real exception, for on examination under the microscope, there was no vestige of a coloured mark on the tip, and it was evident that by a mistake it had not been touched with the caustic. No. 2 was plainly geotropic, being inclined at about 45° beneath the horizon; No. 3 was slightly, and No. 4 only just perceptibly geotropic; Nos. 5 and 6 were strictly horizontal; and the four remaining ones were bowed upwards, in opposition to geotropism. In these four cases the radius of the upward curvatures (according to Sachs' cyclometer) was 5 mm., 10 mm., 30 mm., and 70 mm. This curvature was distinct long before the 24 h. had elapsed, namely, after 8 h. 45 m. from the time when the lower sides of the tips were touched with the caustic.

Phaseolus multiflorus.—Eight radicles, serving as controls, were extended horizontally, some in damp friable peat and some in damp air. They all became (temp 20°–21° C.) plainly geotropic in 8 h. 30 m., for they then stood at an average angle of 63° beneath the horizon. A rather greater length of the radicle is bowed downwards by geotropism than in the case of Vicia faba, that is to say, rather more than 6 mm. as measured from the apex of the root-cap. Nine other radicles were similarly extended, three in damp peat and six in damp air, and dry caustic was held transversely to their tips during 4 or 5 seconds. Three of their tips were afterwards examined: in (1) a length of 0.68 mm. was discoloured, of which the basal 0.136 mm. was yellow, the apical part being black; in (2) the discoloration was 0.65 mm. in length, of which the basal 0.04 mm. was yellow; in (3) the discoloration was 0.6 mm. in length, of which the basal 0.13 mm. was yellow. Therefore less than 1 mm. was affected by the caustic, but this sufficed almost wholly to prevent geotropic action; for after 24 h. one alone of the nine cauterised radicles became slightly geotropic, being now inclined at 10° beneath the horizon; the eight others remained horizontal, though one was curved a little laterally.

The terminal part (10 mm. in length) of the six cauterised radicles in the damp air, had more than doubled in length in the 24 h., for this part was now on an average 20.7 mm. long. The increase in length within the same time was greater in the control specimens, for the terminal part had grown on an average from 10 mm. to 26.6 mm. But as the cauterised radicles had more than doubled their length in the 24 h., it is manifest that they had not been seriously injured by the caustic. We may here add that when experimenting on the effects of touching one side of the tip with caustic, too much was applied at first, and the whole tip (but we believe not more than 1 mm. in length) of six horizontally extended radicles was killed, and these continued for two or three days to grow out horizontally.

Many trials were made, by coating the tips of horizontally extended radicles with the before described thick grease. The geotropic curvature of 12 radicles, which were thus coated for a length of 2 mm., was delayed during the first 8 or 9 h., but after 24 h. was nearly as great as that of the control specimens. The tips of nine radicles were coated for a length of 3 mm., and after 7 h. 10 m. these stood at an average angle of 30° beneath the horizon, whilst the controls stood at an average of 54°. After 24 h. the two lots differed but little in their degree of curvature. In some other trials, however, there was a fairly well-marked difference after 24 h. between those with greased tips and the controls. The terminal part of eight control specimens increased in 24 h. from 10 mm. to a mean length of 24.3 mm., whilst the mean increase of those with greased tips was 20.7 mm. The grease, therefore, slightly checked the growth of the terminal part, but this part was not much injured; for several radicles which had been greased for a length of 2 mm. continued to grow during seven days, and were then only a little shorter than the controls. The appearance presented by these radicles after the seven days was very curious, for the black grease had been drawn out into the finest longitudinal striae, with dots and reticulations, which covered their surfaces for a length of from 26 to 44 mm., or of 1 to 1.7 inch. We may therefore conclude that grease on the tips of the radicles of this Phaseolus somewhat delays and lessens the geotropic curvature of the part which ought to bend most.

Gossypium herbaceum.—The radicles of this plant bend, through the action of geotropism, for a length of about 6 mm. Five radicles, placed horizontally in damp air, had their tips touched with caustic, and the discoloration extended for a length of from 2/3 to 1 mm. They showed, after 7 h. 45 m. and again after 23 h., not a trace of geotropism; yet the terminal portion, 9 mm. in length, had increased on an average to 15.9 mm. Six control radicles, after 7 h. 45 m., were all plainly geotropic, two of them being vertically dependent, and after 23 h. all were vertical, or nearly so.

Cucurbita ovifera.—A large number of trials proved almost useless, from the three following causes: Firstly, the tips of radicles which have grown somewhat old are only feebly geotropic if kept in damp air; nor did we succeed well in our experiments, until the germinating seeds were placed in peat and kept at a rather high temperature. Secondly, the hypocotyls of the seeds which were pinned to the lids of the jars gradually became arched; and, as the cotyledons were fixed, the movement of the hypocotyl affected the position of the radicle, and caused confusion. Thirdly, the point of the radicle is so fine that it is difficult not to cauterise it either too much or too little. But we managed generally to overcome this latter difficulty, as the following experiments show, which are given to prove that a touch with caustic on one side of the tip does not prevent the upper part of the radicle from bending. Ten radicles were laid horizontally beneath and on damp friable peat, and their tips were touched with caustic on the upper side. After 8 h. all were plainly geotropic, three of them rectangularly; after 19 h. all were strongly geotropic, most of them pointing perpendicularly downwards. Ten other radicles, similarly placed, had their tips touched with caustic on the lower side; after 8 h. three were slightly geotropic, but not nearly so much so as the least geotropic of the foregoing specimens; four remained horizontal; and three were curved upwards in opposition to geotropism. After 19 h. the three which were slightly geotropic had become strongly so. Of the four horizontal radicles, one alone showed a trace of geotropism; of the three up-curved radicles, one retained this curvature, and the other two had become horizontal.

The radicles of this plant, as already remarked, do not succeed well in damp air, but the result of one trial may be briefly given. Nine young radicles between .3 and .5 inch in length, with their tips cauterised and blackened for a length never exceeding ½ mm., together with eight control specimens, were extended horizontally in damp air. After an interval of only 4 h. 10 m. all the controls were slightly geotropic, whilst not one of the cauterised specimens exhibited a trace of this action. After 8 h. 35 m., there was the same difference between the two sets, but rather more strongly marked. By this time both sets had increased greatly in length. The controls, however, never became much more curved downwards; and after 24 h. there was no great difference between the two sets in their degree of curvature.

Eight young radicles of nearly equal length (average .36 inch) were placed beneath and on peat-earth, and were exposed to a temp. of 75°–76° F. Their tips had been touched transversely with caustic, and five of them were blackened for a length of about 0.5 mm., whilst the other three were only just visibly discoloured. In the same box there were 15 control radicles, mostly about .36 inch in length, but some rather longer and older, and therefore less sensitive. After 5 h., the 15 control radicles were all more or less geotropic: after 9 h., eight of them were bent down beneath the horizon at various angles between 45° and 90°, the remaining seven being only slightly geotropic: after 25 h. all were rectangularly geotropic. The state of the eight cauterised radicles after the same intervals of time was as follows: after 5 h. one alone was slightly geotropic, and this was one with the tip only a very little discoloured: after 9 h. the one just mentioned was rectangularly geotropic, and two others were slightly so, and these were the three which had been scarcely affected by the caustic; the other five were still strictly horizontal. After 24 h. 40 m. the three with only slightly discoloured tips were bent down rectangularly; the other five were not in the least affected, but several of them had grown rather tortuously, though still in a horizontal plane. The eight cauterised radicles which had at first a mean length of .36 inch, after 9 h. had increased to a mean length of .79 inch; and after 24 h. 40 m. to the extraordinary mean length of 2 inches. There was no plain difference in length between the five well cauterised radicles which remained horizontal, and the three with slightly cauterised tips which had become abruptly bent down. A few of the control radicles were measured after 25 h., and they were on an average only a little longer than the cauterised, viz., 2.19 inches. We thus see that killing the extreme tip of the radicle of this plant for a length of about 0.5 mm., though it stops the geotropic bending of the upper part, hardly interferes with the growth of the whole radicle.

In the same box with the 15 control specimens, the rapid geotropic bending and growth of which have just been described, there were six radicles, about .6 inch in length, extended horizontally, from which the tips had been cut off in a transverse direction for a length of barely 1 mm. These radicles were examined after 9 h. and again after 24 h. 40 m., and they all remained horizontal. They had not become nearly so tortuous as those above described which had been cauterised. The radicles with their tips cut off had grown in the 24 h. 40 m. as much, judging by the eye, as the cauterised specimens.

Zea mays.—The tips of several radicles, extended horizontally in damp air, were dried with blotting-paper and then touched in the first trial during 2 or 3 seconds with dry caustic; but this was too long a contact, for the tips were blackened for a length of rather above 1 mm. They showed no signs of geotropism after an interval of 9 h., and were then thrown away. In a second trial the tips of three radicles were touched for a shorter time, and were blackened for a length of from 0.5 to 0.75 mm.: they all remained horizontal for 4 h., but after 8 h. 30 m. one of them, in which the blackened tip was only 0.5 mm. in length, was inclined at 21° beneath the horizon. Six control radicles all became slightly geotropic in 4 h., and strongly so after 8 h. 30 m., with the chief seat of curvature generally between 6 or 7 mm. from the apex. In the cauterised specimens, the terminal growing part, 10 mm. in length, increased during the 8 h. 30 m. to a mean length of 13 mm.; and in the controls to 14.3 mm.

In a third trial the tips of five radicles (exposed to a temp. of 70°–71°) were touched with the caustic only once and very slightly; they were afterwards examined under the microscope, and the part which was in any way discoloured was on an average .76 mm. in length. After 4 h. 10 m. none were bent; after 5 h. 45 m., and again after 23 h. 30 m., they still remained horizontal, excepting one which was now inclined 20° beneath the horizon.

The terminal part, 10 mm. in length, had increased greatly in length during the 23 h. 30 m., viz., to an average of 26 mm. Four control radicles became slightly geotropic after the 4 h. 10 m., and plainly so after the 5 h. 45 m. Their mean length after the 23 h. 30 m. had increased from 10 mm. to 31 mm. Therefore a slight cauterisation of the tip checks slightly the growth of the whole radicle, and manifestly stops the bending of that part which ought to bend most under the influence of geotropism, and which still continues to increase greatly in length.]

Concluding Remarks.—Abundant evidence has now been given, showing that with various plants the tip of the radicle is alone sensitive to geotropism; and that when thus excited, it causes the adjoining parts to bend. The exact length of the sensitive part seems to be somewhat variable, depending in part on the age of the radicle; but the destruction of a length of from less than 1 to 1.5 mm. (about 1/20th of an inch), in the several species observed, generally sufficed to prevent any part of the radicle from bending within 24 h., or even for a longer period. The fact of the tip alone being sensitive is so remarkable a fact, that we will here give a brief summary of the foregoing experiments. The tips were cut off 29 horizontally extended radicles of Vicia faba, and with a few exceptions they did not become geotropic in 22 or 23 h., whilst unmutilated radicles were always bowed downwards in 8 or 9 h. It should be borne in mind that the mere act of cutting off the tip of a horizontally extended radicle does not prevent the adjoining parts from bending, if the tip has been previously exposed for an hour or two to the influence of geotropism. The tip after amputation is sometimes completely regenerated in three days; and it is possible that it may be able to transmit an impulse to the adjoining parts before its complete regeneration. The tips of six radicles of Cucurbita ovifera were amputated like those of Vicia faba; and these radicles showed no signs of geotropism in 24 h.; whereas the control specimens were slightly affected in 5 h., and strongly in 9 h.

With plants belonging to six genera, the tips of the radicles were touched transversely with dry caustic; and the injury thus caused rarely extended for a greater length than 1 mm., and sometimes to a less distance, as judged by even the faintest discoloration. We thought that this would be a better method of destroying the vegetative point than cutting it off; for we knew, from many previous experiments and from some given in the present chapter, that a touch with caustic on one side of the apex, far from preventing the adjoining part from bending, caused it to bend. In all the following cases, radicles with uncauterised tips were observed at the same time and under similar circumstances, and they became, in almost every instance, plainly bowed downwards in one-half or one-third of the time during which the cauterised specimens were observed. With Vicia faba 19 radicles were cauterised; 12 remained horizontal during 23–24 h.; 6 became slightly and 1 strongly geotropic. Eight of these radicles were afterwards reversed, and again touched with caustic, and none of them became geotropic in 24 h., whilst the reversed control specimens became strongly bowed downwards within this time. With Pisum sativum, five radicles had their tips touched with caustic, and after 32 h. four were still horizontal. The control specimens were slightly geotropic in 7 h. 20 m., and strongly so in 24 h. The tips of 9 other radicles of this plant were touched only on the lower side, and 6 of them remained horizontal for 24 h., or were upturned in opposition to geotropism; 2 were slightly, and 1 plainly geotropic. With Phaseolus multiflorus, 15 radicles were cauterised, and 8 remained horizontal for 24 h.; whereas all the controls were plainly geotropic in 8 h. 30 m. Of 5 cauterised radicles of Gossypium herbaceum, 4 remained horizontal for 23 h. and 1 became slightly geotropic; 6 control radicles were distinctly geotropic in 7 h. 45 m. Five radicles of Cucurbita ovifera remained horizontal in peat-earth during 25 h., and 9 remained so in damp air during 8½ h.; whilst the controls became slightly geotropic in 4 h. 10 m. The tips of 10 radicals of this plant were touched on their lower sides, and 6 of them remained horizontal or were upturned after 19 h., 1 being slightly and 3 strongly geotropic.

Lastly, the tips of several radicles of Vicia faba and Phaseolus multiflorus were thickly coated with grease for a length of 3 mm. This matter, which is highly injurious to most plants, did not kill or stop the growth of the tips, and only slightly lessened the rate of growth of the whole radicle; but it generally delayed a little the geotropic bending of the upper part.

The several foregoing cases would tell us nothing, if the tip itself was the part which became most bent; but we know that it is a part distant from the tip by some millimeters which grows quickest, and which, under the influence of geotropism, bends most. We have no reason to suppose that this part is injured by the death or injury of the tip; and it is certain that after the tip has been destroyed this part goes on growing at such a rate, that its length was often doubled in a day. We have also seen that the destruction of the tip does not prevent the adjoining part from bending, if this part has already received some influence from the tip. As with horizontally extended radicles, of which the tip has been cut off or destroyed, the part which ought to bend most remains motionless for many hours or days, although exposed at right angles to the full influence of geotropism, we must conclude that the tip alone is sensitive to this power, and transmits some influence or stimulus to the adjoining parts, causing them to bend. We have direct evidence of such transmission; for when a radicle was left extended horizontally for an hour or an hour and a half, by which time the supposed influence will have travelled a little distance from the tip, and the tip was then cut off, the radicle afterwards became bent, although placed perpendicularly. The terminal portions of several radicles thus treated continued for some time to grow in the direction of their newly-acquired curvature; for as they were destitute of tips, they were no longer acted on by geotropism. But after three or four days when new vegetative points were formed, the radicles were again acted on by geotropism, and now they curved themselves perpendicularly downwards. To see anything of the above kind in the animal kingdom, we should have to suppose than an animal whilst lying down determined to rise up in some particular direction; and that after its head had been cut off, an impulse continued to travel very slowly along the nerves to the proper muscles; so that after several hours the headless animal rose up in the predetermined direction.

As the tip of the radicle has been found to be the part which is sensitive to geotropism in the members of such distinct families as the Leguminosae, Malvaceae, Cucurbitaceæ and Gramineæ, we may infer that this character is common to the roots of most seedling plants. Whilst a root is penetrating the ground, the tip must travel first; and we can see the advantage of its being sensitive to geotropism, as it has to determine the course of the whole root. Whenever the tip is deflected by any subterranean obstacle, it will also be an advantage that a considerable length of the root should be able to bend, more especially as the tip itself grows slowly and bends but little, so that the proper downward course may be soon recovered. But it appears at first sight immaterial whether this were effected by the whole growing part being sensitive to geotropism, or by an influence transmitted exclusively from the tip. We should, however, remember that it is the tip which is sensitive to the contact of hard objects, causing the radicle to bend away from them, thus guiding it along the lines of least resistance in the soil. It is again the tip which is alone sensitive, at least in some cases, to moisture, causing the radicle to bend towards its source. These two kinds of sensitiveness conquer for a time the sensitiveness to geotropism, which, however, ultimately prevails. Therefore, the three kinds of sensitiveness must often come into antagonism; first one prevailing, and then another; and it would be an advantage, perhaps a necessity, for the interweighing and reconciling of these three kinds of sensitiveness, that they should be all localised in the same group of cells which have to transmit the command to the adjoining parts of the radicle, causing it to bend to or from the source of irritation.

Finally, the fact of the tip alone being sensitive to the attraction of gravity has an important bearing on the theory of geotropism. Authors seem generally to look at the bending of a radicle towards the centre of the earth, as the direct result of gravitation, which is believed to modify the growth of the upper or lower surfaces, in such a manner as to induce curvature in the proper direction. But we now know that it is the tip alone which is acted on, and that this part transmits some influence to the adjoining parts, causing them to curve downwards. Gravity does not appear to act in a more direct manner on a radicle, than it does on any lowly organised animal, which moves away when it feels some weight or pressure.

CHAPTER XII.
CONCLUDING REMARKS.

Nature of the circumnutating movement—History of a germinating seed—The radicle first protrudes and circumnutates—Its tip highly sensitive—Emergence of the hypocotyl or of the epicotyl from the ground under the form of an arch–Its circumnutation and that of the cotyledons—The seedling throws up a leaf-bearing stem—The circumnutation of all the parts or organs—Modified circumnutation—Epinasty and hyponasty—Movements of climbing plants—Nyctitropic movements—Movements excited by light and gravitation—Localised sensitiveness—Resemblance between the movements of plants and animals—The tip of the radicle acts like a brain.

It may be useful to the reader if we briefly sum up the chief conclusions, which, as far as we can judge, have been fairly well established by the observations given in this volume. All the parts or organs in every plant whilst they continue to grow, and some parts which are provided with pulvini after they have ceased to grow, are continually circumnutating. This movement commences even before the young seedling has broken through the ground. The nature of the movement and its causes, as far as ascertained, have been briefly described in the Introduction. Why every part of a plant whilst it is growing, and in some cases after growth has ceased, should have its cells rendered more turgescent and its cell-walls more extensile first on one side and then on another, thus inducing circumnutation is not known. It would appear as if the changes in the cells required periods of rest.

In some cases, as with the hypocotyls of Brassica, the leaves of Dionaea and the joints of the Gramineæ, the circumnutating movement when viewed under the microscope is seen to consist of innumerable small oscillations. The part under observation suddenly jerks forwards for a length of .002 to .001 of an inch, and then slowly retreats for a part of this distance; after a few seconds it again jerks forwards, but with many intermissions. The retreating movement apparently is due to the elasticity of the resisting tissues. How far this oscillatory movement is general we do not know, as not many circumnutating plants were observed by us under the microscope; but no such movement could be detected in the case of Drosera with a 2-inch object-glass which we used. The phenomenon is a remarkable one. The whole hypocotyl of a cabbage or the whole leaf of a

Dionaea could not jerk forwards unless a very large number of cells on one side were simultaneously affected. Are we to suppose that these cells steadily become more and more turgescent on one side, until the part suddenly yields and bends, inducing what may be called a microscopically minute earthquake in the plant; or do the cells on one side suddenly become turgescent in an intermittent manner; each forward movement thus caused being opposed by the elasticity of the tissues?

Circumnutation is of paramount importance in the life of every plant; for it is through its modification that many highly beneficial or necessary movements have been acquired. When light strikes one side of a plant, or light changes into darkness, or when gravitation acts on a displaced part, the plant is enabled in some unknown manner to increase the always varying turgescence of the cells on one side; so that the ordinary circumnutating movement is modified, and the part bends either to or from the exciting cause; or it may occupy a new position, as in the so-called sleep of leaves. The influence which modifies circumnutation may be transmitted from one part to another. Innate or constitutional changes, independently of any external agency, often modify the circumnutating movements at particular periods of the life of the plant. As circumnutation is universally present, we can understand how it is that movements of the same kind have been developed in the most distinct members of the vegetable series. But it must not be supposed that all the movements of plants arise from modified circumnutation; for, as we shall presently see, there is reason to believe that this is not the case.

Having made these few preliminary remarks, we will in imagination take a germinating seed, and consider the part which the various movements play in the life-history of the plant. The first change is the protrusion of the radicle, which begins at once to circumnutate. This movement is immediately modified by the attraction of gravity and rendered geotropic. The radicle, therefore, supposing the seed to be lying on the surface, quickly bends downwards, following a more or less spiral course, as was seen on the smoked glass-plates. Sensitiveness to gravitation resides in the tip; and it is the tip which transmits some influence to the adjoining parts, causing them to bend. As soon as the tip, protected by the root-cap, reaches the ground, it penetrates the surface, if this be soft or friable; and the act of penetration is apparently aided by the rocking or circumnutating movement of the whole end of the radicle. If the surface is compact, and cannot easily be penetrated, then the seed itself, unless it be a heavy one, is displaced or lifted up by the continued growth and elongation of the radicle. But in a state of nature seeds often get covered with earth or other matter, or fall into crevices, etc., and thus a point of resistance is afforded, and the tip can more easily penetrate the ground. But even with seeds lying loose on the surface there is another aid: a multitude of excessively fine hairs are emitted from the upper part of the radicle, and these attach themselves firmly to stones or other objects lying on the surface, and can do so even to glass; and thus the upper part is held down whilst the tip presses against and penetrates the ground. The attachment of the root-hairs is effected by the liquefaction of the outer surface of the cellulose walls, and by the subsequent setting hard of the liquefied matter. This curious process probably takes place, not for the sake of the attachment of the radicles to superficial objects, but in order that the hairs may be brought into the closest contact with the particles in the soil, by which means they can absorb the layer of water surrounding them, together with any dissolved matter.

After the tip has penetrated the ground to a little depth, the increasing thickness of the radicle, together with the root-hairs, hold it securely in its place; and now the force exerted by the longitudinal growth of the radicle drives the tip deeper into the ground. This force, combined with that due to transverse growth, gives to the radicle the power of a wedge. Even a growing root of moderate size, such as that of a seedling bean, can displace a weight of some pounds. It is not probable that the tip when buried in compact earth can actually circumnutate and thus aid its downward passage, but the circumnutating movement will facilitate the tip entering any lateral or oblique fissure in the earth, or a burrow made by an earth-worm or larva; and it is certain that roots often run down the old burrows of worms. The tip, however, in endeavouring to circumnutate, will continually press against the earth on all sides, and this can hardly fail to be of the highest importance to the plant; for we have seen that when little bits of card-like paper and of very thin paper were cemented on opposite sides of the tip, the whole growing part of the radicle was excited to bend away from the side bearing the card or more resisting substance, towards the side bearing the thin paper. We may therefore feel almost sure that when the tip encounters a stone or other obstacle in the ground, or even earth more compact on one side than the other, the root will bend away as much as it can from the obstacle or the more resisting earth, and will thus follow with unerring skill a line of least resistance.

The tip is more sensitive to prolonged contact with an object than to gravitation when this acts obliquely on the radicle, and sometimes even when it acts in the most favourable direction at right angles to the radicle. The tip was excited by an attached bead of shellac weighing less than 1/200th of a grain (0.33 mg.); it is therefore more sensitive than the most delicate tendril, namely, that of Passiflora gracilis, which was barely acted on by a bit of wire weighing 1/50th of a grain. But this degree of sensitiveness is as nothing compared with that of the glands of Drosera, for these are excited by particles weighing only 1/78740 of a grain. The sensitiveness of the tip cannot be accounted for by its being covered by a thinner layer of tissue than the other parts, for it is protected by the relatively thick root-cap. It is remarkable that although the radicle bends away, when one side of the tip is slightly touched with caustic, yet if the side be much cauterised the injury is too great, and the power of transmitting some influence to the adjoining parts causing them to bend, is lost. Other analogous cases are known to occur.

After a radicle has been deflected by some obstacle, geotropism directs the tip again to grow perpendicularly downwards; but geotropism is a feeble power, and here, as Sachs has shown, another interesting adaptive movement comes into play; for radicles at a distance of a few millimeters from the tip are sensitive to prolonged contact in such a manner that they bend towards the touching object, instead of from it as occurs when an object touches one side of the tip. Moreover, the curvature thus caused is abrupt; the pressed part alone bending. Even slight pressure suffices, such as a bit of card cemented to one side. therefore a radicle, as it passes over the edge of any obstacle in the ground, will through the action of geotropism press against it; and this pressure will cause the radicle to endeavour to bend abruptly over the edge. It will thus recover as quickly as possible its normal downward course.

Radicles are also sensitive to air which contains more moisture on one side than the other, and they bend towards its source. It is therefore probable that they are in like manner sensitive to dampness in the soil. It was ascertained in several cases that this sensitiveness resides in the tip, which transmits an influence causing the adjoining upper part to bend in opposition to geotropism towards the moist object. We may therefore infer that roots will be deflected from their downward course towards any source of moisture in the soil.

Again, most or all radicles are slightly sensitive to light, and according to Wiesner, generally bend a little from it. Whether this can be of any service to them is very doubtful, but with seeds germinating on the surface it will slightly aid geotropism in directing the radicles to the ground.[1] We ascertained in one instance that such sensitiveness resided in the tip, and caused the adjoining parts to bend from the light. The sub-aërial roots observed by Wiesner were all apheliotropic, and this, no doubt, is of use in bringing them into contact with trunks of trees or surfaces of rock, as is their habit.

[1] Dr. Karl Richter, who has especially attended to this subject ('K. Akad. der Wissenschaften in Wien,' 1879, p. 149), states that apheliotropism does not aid radicles in penetrating the ground.

We thus see that with seedling plants the tip of the radicle is endowed with diverse kinds of sensitiveness; and that the tip directs the adjoining growing parts to bend to or from the exciting cause, according to the needs of the plant. The sides of the radicle are also sensitive to contact, but in a widely different manner. Gravitation, though a less powerful cause of movement than the other above specified stimuli, is ever present; so that it ultimately prevails and determines the downward growth of the root.

The primary radicle emits secondary ones which project sub-horizontally; and these were observed in one case to circumnutate. Their tips are also sensitive to contact, and they are thus excited to bend away from any touching object; so that they resemble in these respects, as far as they were observed, the primary radicles. If displaced they resume, as Sachs has shown, their original sub-horizontal position; and this apparently is due to diageotropism. The secondary radicles emit tertiary ones, but these, in the case of the bean, are not affected by gravitation; consequently they protrude in all directions. Thus the general arrangement of the three orders of roots is excellently adapted for searching the whole soil for nutriment.

Sachs has shown that if the tip of the primary radicle is cut off (and the tip will occasionally be gnawed off with seedlings in a state of nature) one of the secondary radicles grows perpendicularly downwards, in a manner which is analogous to the upward growth of a lateral shoot after the amputation of the leading shoot. We have seen with radicles of the bean that if the primary radicle is merely compressed instead of being cut off, so that an excess of sap is directed into the secondary radicles, their natural condition is disturbed and they grow downwards. Other analogous facts have been given. As anything which disturbs the constitution is apt to lead to reversion, that is, to the resumption of a former character, it appears probable that when secondary radicles grow downwards or lateral shoots upwards, they revert to the primary manner of growth proper to radicles and shoots.

With dicotyledonous seeds, after the protrusion of the radicle, the hypocotyl breaks through the seed-coats; but if the cotyledons are hypogean, it is the epicotyl which breaks forth. These organs are at first invariably arched, with the upper part bent back parallel to the lower; and they retain this form until they have risen above the ground. In some cases, however, it is the petioles of the cotyledons or of the first true leaves which break through the seed-coats as well as the ground, before any part of the stem protrudes; and then the petioles are almost invariably arched. We have met with only one exception, and that only a partial one, namely, with the petioles of the two first leaves of Acanthus candelabrum. With Delphinium nudicaule the petioles of the two cotyledons are completely confluent, and they break through the ground as an arch; afterwards the petioles of the successively formed early leaves are arched, and they are thus enabled to break through the base of the confluent petioles of the cotyledons. In the case of Megarrhiza, it is the plumule which breaks as an arch through the tube formed by the confluence of the cotyledon-petioles. With mature plants, the flower-stems and the leaves of some few species, and the rachis of several ferns, as they emerge separately from the ground, are likewise arched.

The fact of so many different organs in plants of many kinds breaking through the ground under the form of an arch, shows that this must be in some manner highly important to them. According to Haberlandt, the tender growing apex is thus saved from abrasion, and this is probably the true explanation. But as both legs of the arch grow, their power of breaking through the ground will be much increased as long as the tip remains within the seed-coats and has a point of support. In the case of monocotyledons the plumule or cotyledon is rarely arched, as far as we have seen; but this is the case with the leaf-like cotyledon of the onion; and the crown of the arch is here strengthened by a special protuberance. In the Gramineæ the summit of the straight, sheath-like cotyledon is developed into a hard sharp crest, which evidently serves for breaking through the earth. With dicotyledons the arching of the epicotyl or hypocotyl often appears as if it merely resulted from the manner in which the parts are packed within the seed; but it is doubtful whether this is the whole of the truth in any case, and it certainly was not so in several cases, in which the arching was seen to commence after the parts had wholly escaped from the seed-coats. As the arching occurred in whatever position the seeds were placed, it is no doubt due to temporarily increased growth of the nature of epinasty or hyponasty along one side of the part.

As this habit of the hypocotyl to arch itself appears to be universal, it is probably of very ancient origin. It is therefore not surprising that it should be inherited, at least to some extent, by plants having hypogean cotyledons, in which the hypocotyl is only slightly developed and never protrudes above the ground, and in which the arching is of course now quite useless. This tendency explains, as we have seen, the curvature of the hypocotyl (and the consequent movement of the radicle) which was first observed by Sachs, and which we have often had to refer to as Sachs' curvature.

The several foregoing arched organs are continually circumnutating, or endeavouring to circumnutate, even before they break through the ground. As soon as any part of the arch protrudes from the seed-coats it is acted upon by apogeotropism, and both the legs bend upwards as quickly as the surrounding earth will permit, until the arch stands vertically. By continued growth it then forcibly breaks through the ground; but as it is continually striving to circumnutate this will aid its emergence in some slight degree, for we know that a circumnutating hypocotyl can push away damp sand on all sides. As soon as the faintest ray of light reaches a seedling, heliotropism will guide it through any crack in the soil, or through an entangled mass of overlying vegetation; for apogeotropism by itself can direct the seedling only blindly upwards. Hence probably it is that sensitiveness to light resides in the tip of the cotyledons of the Gramineæ, and in the upper part of the hypocotyls of at least some plants.

As the arch grows upwards the cotyledons are dragged out of the ground. The seed-coats are either left behind buried, or are retained for a time still enclosing the cotyledons. These are afterwards cast off merely by the swelling of the cotyledons. But with most of the Cucurbitaceæ there is a curious special contrivance for bursting the seed-coats whilst beneath the ground, namely, a peg at the base of the hypocotyl, projecting at right angles, which holds down the lower half of the seed-coats, whilst the growth of the arched part of the hypocotyl lifts up the upper half, and thus splits them in twain. A somewhat analogous structure occurs in Mimosa pudica and some other plants. Before the cotyledons are fully expanded and have diverged, the hypocotyl generally straightens itself by increased growth along the concave side, thus reversing the process which caused the arching. Ultimately not a trace of the former curvature is left, except in the case of the leaf-like cotyledons of the onion.

The cotyledons can now assume the function of leaves, and decompose carbonic acid; they also yield up to other parts of the plant the nutriment which they often contain. When they contain a large stock of nutriment they generally remain buried beneath the ground, owing to the small development of the hypocotyl; and thus they have a better chance of escaping destruction by animals. From unknown causes, nutriment is sometimes stored in the hypocotyl or in the radicle, and then one of the cotyledons or both become rudimentary, of which several instances have been given. It is probable that the extraordinary manner of germination of *Megarrhiza Californica*, *Ipomœa leptophylla* and *pandurata*, and of *Quercus virens*, is connected with the burying of the tuber-like roots, which at an early age are stocked with nutriment; for in these plants it is the petioles of the cotyledons which first protrude from the seeds, and they are then merely tipped with a minute radicle and hypocotyl. These petioles bend down geotropically like a root and penetrate the ground, so that the true root, which afterwards becomes greatly enlarged, is buried at some little depth beneath the surface. Gradations of structure are always interesting, and Asa Gray informs us that with Ipomœa Jalappa, which likewise forms huge tubers, the hypocotyl is still of considerable length, and the petioles of the cotyledons are only moderately elongated. But in addition to the advantage gained by the concealment of the nutritious matter stored within the tubers, the plumule, at least in the case of Megarrhiza, is protected from the frosts of winter by being buried.

With many dicotyledonous seedlings, as has lately been described by De Vries, the contraction of the parenchyma of the upper part of the radicle drags the hypocotyl downwards into the earth; sometimes (it is said) until even the cotyledons are buried. The hypocotyl itself of some species contracts in a like manner. It is believed that this burying process serves to protect the seedlings against the frosts of winter.

Our imaginary seedling is now mature as a seedling, for its hypocotyl is straight and its cotyledons are fully expanded. In this state the upper part of the hypocotyl and the cotyledons continue for some time to circumnutate, generally to a wide extent relatively to the size of the parts, and at a rapid rate. But seedlings profit by this power of movement only when it is modified, especially by the action of light and gravitation; for they are thus enabled to move more rapidly and to a greater extent than can most mature plants. Seedlings are subjected to a severe struggle for life, and it appears to be highly important to them that they should adapt themselves as quickly and as perfectly as possible to their conditions. Hence also it is that they are so extremely sensitive to light and gravitation. The cotyledons of some few species are sensitive to a touch; but it is probable that this is only an indirect result of the foregoing kinds of sensitiveness, for there is no reason to believe that they profit by moving when touched.

Our seedling now throws up a stem bearing leaves, and often branches, all of which whilst young are continually circumnutating. If we look, for instance, at a great acacia tree, we may feel assured that every one of the innumerable growing shoots is constantly describing small ellipses; as is each petiole, sub-petiole, and leaflet. The latter, as well as ordinary leaves, generally move up and down in nearly the same vertical plane, so that they describe very narrow ellipses. The flower-peduncles are likewise continually circumnutating. If we could look beneath the ground, and our eyes had the power of a microscope, we should see the tip of each rootlet endeavouring to sweep small ellipses or circles, as far as the pressure of the surrounding earth permitted. All this astonishing amount of movement has been going on year after year since the time when, as a seedling, the tree first emerged from the ground.

Stems are sometimes developed into long runners or stolons. These circumnutate in a conspicuous manner, and are thus aided in passing between and over surrounding obstacles. But whether the circumnutating movement has been increased for this special purpose is doubtful.

We have now to consider circumnutation in a modified form, as the source of several great classes of movement. The modification may be determined by innate causes, or by external agencies. Under the first head we see leaves which, when first unfolded, stand in a vertical position, and gradually bend downwards as they grow older. We see flower-peduncles bending down after the flower has withered, and others rising up; or again, stems with their tips at first bowed downwards, so as to be hooked, afterwards straightening themselves; and many other such cases. These changes of position, which are due to epinasty or hyponasty, occur at certain periods of the life of the plant, and are independent of any external agency. They are effected not by a continuous upward or downward movement, but by a succession of small ellipses, or by zigzag lines,—that is, by a circumnutating movement which is preponderant in some one direction.

Again, climbing plants whilst young circumnutate in the ordinary manner, but as soon as the stem has grown to a certain height, which is different for different species, it elongates rapidly, and now the amplitude of the circumnutating movement is immensely increased, evidently to favour the stem catching hold of a support. The stem also circumnutates rather more equally to all sides than in the case of non-climbing plants. This is conspicuously the case with those tendrils which consist of modified leaves, as these sweep wide circles; whilst ordinary leaves usually circumnutate nearly in the same vertical plane. Flower-peduncles when converted into tendrils have their circumnutating movement in like manner greatly increased.

We now come to our second group of circumnutating movements—those modified through external agencies. The so-called sleep or nyctitropic movements of leaves are determined by the daily alternations of light and darkness. It is not the darkness which excites them to move, but the difference in the amount of light which they receive during the day and night; for with several species, if the leaves have not been brightly illuminated during the day, they do not sleep at night. They inherit, however, some tendency to move at the proper periods, independently of any change in the amount of light. The movements are in some cases extraordinarily complex, but as a full summary has been given in the chapter devoted to this subject, we will here say but little on this head. Leaves and cotyledons assume their nocturnal position by two means, by the aid of pulvini and without such aid. In the former case the movement continues as long as the leaf or cotyledon remains in full health; whilst in the latter case it continues only whilst the part is growing. Cotyledons appear to sleep in a larger proportional number of species than do leaves. In some species, the leaves sleep and not the cotyledons; in others, the cotyledons and not the leaves; or both may sleep, and yet assume widely different positions at night.

Although the nyctitropic movements of leaves and cotyledons are wonderfully diversified, and sometimes differ much in the species of the same genus, yet the blade is always placed in such a position at night, that its upper surface is exposed as little as possible to full radiation. We cannot doubt that this is the object gained by these movements; and it has been proved that leaves exposed to a clear sky, with their blades compelled to remain horizontal, suffered much more from the cold than others which were allowed to assume their proper vertical position. Some curious facts have been given under this head, showing that horizontally extended leaves suffered more at night, when the air, which is not cooled by radiation, was prevented from freely circulating beneath their lower surfaces; and so it was, when the leaves were allowed to go to sleep on branches which had been rendered motionless. In some species the petioles rise up greatly at night, and the pinnae close together. The whole plant is thus rendered more compact, and a much smaller surface is exposed to radiation.

That the various nyctitropic movements of leaves result from modified circumnutation has, we think, been clearly shown. In the simplest cases a leaf describes a single large ellipse during the 24 h.; and the movement is so arranged that the blade stands vertically during the night, and reassumes its former position on the following morning. The course pursued differs from ordinary circumnutation only in its greater amplitude, and in its greater rapidity late in the evening and early on the following morning. Unless this movement is admitted to be one of circumnutation, such leaves do not circumnutate at all, and this would be a monstrous anomaly. In other cases, leaves and cotyledons describe several vertical ellipses during the 24 h.; and in the evening one of them is increased greatly in amplitude until the blade stands vertically either upwards or downwards. In this position it continues to circumnutate until the following morning, when it reassumes its former position. These movements, when a pulvinus is present, are often complicated by the rotation of the leaf or leaflet; and such rotation on a small scale occurs during ordinary circumnutation. The many diagrams showing the movements of sleeping and non-sleeping leaves and cotyledons should be compared, and it will be seen that they are essentially alike. Ordinary circumnutation is converted into a nyctitropic movement, firstly by an increase in its amplitude, but not to so great a degree as in the case of climbing plants, and secondly by its being rendered periodic in relation to the alternations of day and night. But there is frequently a distinct trace of periodicity in the circumnutating movements of non-sleeping leaves and cotyledons. The fact that nyctitropic movements occur in species distributed in many families throughout the whole vascular series, is intelligible, if they result from the modification of the universally present movement of circumnutation; otherwise the fact is inexplicable.

In the seventh chapter we have given the case of a Porlieria, the leaflets of which remained closed all day, as if asleep, when the plant was kept dry, apparently for the sake of checking evaporation. Something of the same kind occurs with certain Gramineæ. At the close of this same chapter, a few observations were appended on what may be called the embryology of leaves. The leaves produced by young shoots on cut-down plants of Melilotus Taurica slept like those of a Trifolium, whilst the leaves on the older branches on the same plants slept in a very different manner, proper to the genus; and from the reasons assigned we are tempted to look at this case as one of reversion to a former nyctitropic habit. So again with Desmodium gyrans, the absence of small lateral leaflets on very young plants, makes us suspect that the immediate progenitor of this species did not possess lateral leaflets, and that their appearance in an almost rudimentary condition at a somewhat more advanced age is the result of reversion to a trifoliate predecessor. However this may be, the rapid circumnutating or gyrating movements of the little lateral leaflets, seem to be due proximately to the pulvinus, or organ of movement, not having been reduced nearly so much as the blade, during the successive modifications through which the species has passed.

We now come to the highly important class of movements due to the action of a lateral light. When stems, leaves, or other organs are placed, so that one side is illuminated more brightly than the other, they bend towards the light. This heliotropic movement manifestly results from the modification of ordinary circumnutation; and every gradation between the two movements could be followed. When the light was dim, and only a very little brighter on one side than on the other, the movement consisted of a succession of ellipses, directed towards the light, each of which approached nearer to its source than the previous one. When the difference in the light on the two sides was somewhat greater, the ellipses were drawn out into a strongly-marked zigzag line, and when much greater the course became rectilinear. We have reason to believe that changes in the turgescence of the cells is the proximate cause of the movement of circumnutation; and it appears that when a plant is unequally illuminated on the two sides, the always changing turgescence is augmented along one side, and is weakened or quite arrested along the other sides. Increased turgescence is commonly followed by increased growth, so that a plant which has bent itself towards the light during the day would be fixed in this position were it not for apogeotropism acting during the night. But parts provided with pulvini bend, as Pfeffer has shown, towards the light; and here growth does not come into play any more than in the ordinary circumnutating movements of pulvini.

Heliotropism prevails widely throughout the vegetable kingdom, but whenever, from the changed habits of life of any plant, such movements become injurious or useless, the tendency is easily eliminated, as we see with climbing and insectivorous plants.

Apheliotropic movements are comparatively rare in a well-marked degree, excepting with sub-aërial roots. In the two cases investigated by us, the movement certainly consisted of modified circumnutation.

The position which leaves and cotyledons occupy during the day, namely, more or less transversely to the direction of the light, is due, according to Frank, to what we call diaheliotropism. As all leaves and cotyledons are continually circumnutating, there can hardly be a doubt that diaheliotropism results from modified circumnutation. From the fact of leaves and cotyledons frequently rising a little in the evening, it appears as if diaheliotropism had to conquer during the middle of the day a widely prevalent tendency to apogeotropism.

Lastly, the leaflets and cotyledons of some plants are known to be injured by too much light; and when the sun shines brightly on them, they move upwards or downwards, or twist laterally, so that they direct their edges towards the light, and thus they escape being injured. These paraheliotropic movements certainly consisted in one case of modified circumnutation; and so it probably is in all cases, for the leaves of all the species described circumnutate in a conspicuous manner. This movement has hitherto been observed only with leaflets provided with pulvini, in which the increased turgescence on opposite sides is not followed by growth; and we can understand why this should be so, as the movement is required only for a temporary purpose. It would manifestly be disadvantageous for the leaf to be fixed by growth in its inclined position. For it has to assume its former horizontal position, as soon as possible after the sun has ceased shining too brightly on it.

The extreme sensitiveness of certain seedlings to light, as shown in our ninth chapter, is highly remarkable. The cotyledons of Phalaris became curved towards a distant lamp, which emitted so little light, that a pencil held vertically close to the plants, did not cast any shadow which the eye could perceive on a white card. These cotyledons, therefore, were affected by a difference in the amount of light on their two sides, which the eye could not distinguish. The degree of their curvature within a given time towards a lateral light did not correspond at all strictly with the amount of light which they received; the light not being at any time in excess. They continued for nearly half an hour to bend towards a lateral light, after it had been extinguished. They bend with remarkable precision towards it, and this depends on the illumination of one whole side, or on the obscuration of the whole opposite side. The difference in the amount of light which plants at any time receive in comparison with what they have shortly before received, seems in all cases to be the chief exciting cause of those movements which are influenced by light. Thus seedlings brought out of darkness bend towards a dim lateral light, sooner than others which had previously been exposed to daylight. We have seen several analogous cases with the nyctitropic movements of leaves. A striking instance was observed in the case of the periodic movements of the cotyledons of a Cassia; in the morning a pot was placed in an obscure part of a room, and all the cotyledons rose up closed; another pot had stood in the sunlight, and the cotyledons of course remained expanded; both pots were now placed close together in the middle of the room, and the cotyledons which had been exposed to the sun, immediately began to close, while the others opened; so that the cotyledons in the two pots moved in exactly opposite directions whilst exposed to the same degree of light.

We found that if seedlings, kept in a dark place, were laterally illuminated by a small wax taper for only two or three minutes at intervals of about three-quarters of an hour, they all became bowed to the point where the taper had been held. We felt much surprised at this fact, and until we had read Wiesner's observations, we attributed it to the after-effects of the light; but he has shown that the same degree of curvature in a plant may be induced in the course of an hour by several interrupted illuminations lasting altogether for 20 m., as by a continuous illumination of 60 m. We believe that this case, as well as our own, may be explained by the excitement from light being due not so much to its actual amount, as to the difference in amount from that previously received; and in our case there were repeated alternations from complete darkness to light. In this, and in several of the above specified respects, light seems to act on the tissues of plants, almost in the same manner as it does on the nervous system of animals. There is a much more striking analogy of the same kind, in the sensitiveness to light being localised in the tips of the cotyledons of Phalaris and Avena, and in the upper part of the hypocotyls of Brassica and Beta; and in the transmission of some influence from these upper to the lower parts, causing the latter to bend towards the light. This influence is also transmitted beneath the soil to a depth where no light enters. It follows from this localisation, that the lower parts of the cotyledons of Phalaris, etc., which normally become more bent towards a lateral light than the upper parts, may be brightly illuminated during many hours, and will not bend in the least, if all light be excluded from the tip. It is an interesting experiment to place caps over the tips of the cotyledons of Phalaris, and to allow a very little light to enter through minute orifices on one side of the caps, for the lower part of the cotyledons will then bend to this side, and not to the side which has been brightly illuminated during the whole time.

In the case of the radicles of Sinapis alba, sensitiveness to light also resides in the tip, which, when laterally illuminated, causes the adjoining part of the root to bend apheliotropically.

Gravitation excites plants to bend away from the centre of the earth, or towards it, or to place themselves in a transverse position with respect to it. Although it is impossible to modify in any direct manner the attraction of gravity, yet its influence could be moderated indirectly, in the several ways described in the tenth chapter; and under such circumstances the same kind of evidence as that given in the chapter on Heliotropism, showed in the plainest manner that apogeotropic and geotropic, and probably diageotropic movements, are all modified forms of circumnutation.

Different parts of the same plant and different species are affected by gravitation in widely different degrees and manners. Some plants and organs exhibit hardly a trace of its action. Young seedlings which, as we know, circumnutate rapidly, are eminently sensitive; and we have seen the hypocotyl of Beta bending upwards through 109° in 3 h. 8 m. The after-effects of apogeotropism last for above half an hour; and horizontally-laid hypocotyls are sometimes thus carried temporarily beyond an upright position. The benefits derived from geotropism, apogeotropism, and diageotropism, are generally so manifest that they need not be specified. With the flower-peduncles of Oxalis, epinasty causes them to bend down, so that the ripening pods may be protected by the calyx from the rain. Afterwards they are carried upwards by apogeotropism in combination with hyponasty, and are thus enabled to scatter their seeds over a wider space. The capsules and flower-heads of some plants are bowed downwards through geotropism, and they then bury themselves in the earth for the protection and slow maturation of the seeds. This burying process is much facilitated by the rocking movement due to circumnutation.

In the case of the radicles of several, probably of all seedling plants, sensitiveness to gravitation is confined to the tip, which transmits an influence to the adjoining upper part, causing it to bend towards the centre of the earth. That there is transmission of this kind was proved in an interesting manner when horizontally extended radicles of the bean were exposed to the attraction of gravity for 1 or 1½ h., and their tips were then amputated. Within this time no trace of curvature was exhibited, and the radicles were now placed pointing vertically downwards; but an influence had already been transmitted from the tip to the adjoining part, for it soon became bent to one side, in the same manner as would have occurred had the radicle remained horizontal and been still acted on by geotropism. Radicles thus treated continued to grow out horizontally for two or three days, until a new tip was re-formed; and this was then acted on by geotropism, and the radicle became curved perpendicularly downwards.

It has now been shown that the following important classes of movement all arise from modified circumnutation, which is omnipresent whilst growth lasts, and after growth has ceased, whenever pulvini are present. These classes of movement consist of those due to epinasty and hyponasty,—those proper to climbing plants, commonly called revolving nutation,—the nyctitropic or sleep movements of leaves and cotyledons,—and the two immense classes of movement excited by light and gravitation. When we speak of modified circumnutation we mean that light, or the alternations of light and darkness, gravitation, slight pressure or other irritants, and certain innate or constitutional states of the plant, do not directly cause the movement; they merely lead to a temporary increase or diminution of those spontaneous changes in the turgescence of the cells which are already in progress. In what manner, light, gravitation, etc., act on the cells is not known; and we will here only remark that, if any stimulus affected the cells in such a manner as to cause some slight tendency in the affected part to bend in a beneficial manner, this tendency might easily be increased through the preservation of the more sensitive individuals. But if such bending were injurious, the tendency would be eliminated unless it was overpoweringly strong; for we know how commonly all characters in all organisms vary. Nor can we see any reason to doubt, that after the complete elimination of a tendency to bend in some one direction under a certain stimulus, the power to bend in a directly opposite direction might gradually be acquired through natural selection.[2]

[2] See the remarks in Frank's 'Die wagerechte Richtung von Pflanzentheilen' (1870, pp. 90, 91, etc.), on natural selection in connection with geotropism, heliotropism, etc.

Although so many movements have arisen through modified circumnutation, there are others which appear to have had a quite independent origin; but they do not form such large and important classes. When a leaf of a Mimosa is touched it suddenly assumes the same position as when asleep, but Brucke has shown that this movement results from a different state of turgescence in the cells from that which occurs during sleep; and as sleep-movements are certainly due to modified circumnutation, those from a touch can hardly be thus due. The back of a leaf of Drosera rotundifolia was cemented to the summit of a stick driven into the ground, so that it could not move in the least, and a tentacle was observed during many hours under the microscope; but it exhibited no circumnutating movement, yet after being momentarily touched with a bit of raw meat, its basal part began to curve in 23 seconds. This curving movement therefore could not have resulted from modified circumnutation. But when a small object, such as a fragment of a bristle, was placed on one side of the tip of a radicle, which we know is continually circumnutating, the induced curvature was so similar to the movement caused by geotropism, that we can hardly doubt that it is due to modified circumnutation. A flower of a Mahonia was cemented to a stick, and the stamens exhibited no signs of circumnutation under the microscope, yet when they were lightly touched they suddenly moved towards the pistil. Lastly, the curling of the extremity of a tendril when touched seems to be independent of its revolving or circumnutating movement. This is best shown by the part which is the most sensitive to contact, circumnutating much less than the lower parts, or apparently not at all.[3]

[3] For the evidence on this head, see the 'Movements and Habits of Climbing Plants,' 1875, pp. 173, 174.

Although in these cases we have no reason to believe that the movement depends on modified circumnutation, as with the several classes of movement described in this volume, yet the difference between the two sets of cases may not be so great as it at first appears. In the one set, an irritant causes an increase or diminution in the turgescence of the cells, which are already in a state of change; whilst in the other set, the irritant first starts a similar change in their state of turgescence. Why a touch, slight pressure or any other irritant, such as electricity, heat, or the absorption of animal matter, should modify the turgescence of the affected cells in such a manner as to cause movement, we do not know. But a touch acts in this manner so often, and on such widely distinct plants, that the tendency seems to be a very general one; and if beneficial, it might be increased to any extent. In other cases, a touch produces a very different effect, as with Nitella, in which the protoplasm may be seen to recede from the walls of the cell; in Lactuca, in which a milky fluid exudes; and in the tendrils of certain Vitaceae, Cucurbitaceæ, and Bignoniaceae, in which slight pressure causes a cellular outgrowth.

Finally it is impossible not to be struck with the resemblance between the foregoing movements of plants and many of the actions performed unconsciously by the lower animals.[4] With plants an astonishingly small stimulus suffices; and even with allied plants one may be highly sensitive to the slightest continued pressure, and another highly sensitive to a slight momentary touch. The habit of moving at certain periods is inherited both by plants and animals; and several other points of similitude have been specified. But the most striking resemblance is the localisation of their sensitiveness, and the transmission of an influence from the excited part to another which consequently moves. Yet plants do not of course possess nerves or a central nervous system; and we may infer that with animals such structures serve only for the more perfect transmission of impressions, and for the more complete intercommunication of the several parts.

[4] Sachs remarks to nearly the same effect: "Dass sich die lebende Pflanzensubstanz derart innerlich differenzirt, dass einzelne Theile mit specifischen Energien ausgerüstet sind, ähnlich, wie die verschiedenen Sinnesnerven des Thiere" ('Arbeiten des Bot. Inst. in Würzburg,' Bd. ii. 1879, p. 282).

We believe that there is no structure in plants more wonderful, as far as its functions are concerned, than the tip of the radicle. If the tip be lightly pressed or burnt or cut, it transmits an influence to the upper adjoining part, causing it to bend away from the affected side; and, what is more surprising, the tip can distinguish between a slightly harder and softer object, by which it is simultaneously pressed on opposite sides. If, however, the radicle is pressed by a similar object a little above the tip, the pressed part does not transmit any influence to the more distant parts, but bends abruptly towards the object. If the tip perceives the air to be moister on one side than on the other, it likewise transmits an influence to the upper adjoining part, which bends towards the source of moisture. When the tip is excited by light (though in the case of radicles this was ascertained in only a single instance) the adjoining part bends from the light; but when excited by gravitation the same part bends towards the centre of gravity. In almost every case we can clearly perceive the final purpose or advantage of the several movements. Two, or perhaps more, of the exciting causes often act simultaneously on the tip, and one conquers the other, no doubt in accordance with its importance for the life of the plant. The course pursued by the radicle in penetrating the ground must be determined by the tip; hence it has acquired such diverse kinds of sensitiveness. It is hardly an exaggeration to say that the tip of the radicle thus endowed, and having the power of directing the movements of the adjoining parts, acts like the brain of one of the lower animals; the brain being seated within the anterior end of the body, receiving impressions from the sense-organs, and directing the several movements.

Milton Keynes UK
Ingram Content Group UK Ltd.
UKHW021711260524
443160UK00004B/231